Applied Probability and Statistics (Continued)

continued on back

Analysis and Control
of Dynamic
Economic Systems

A WILEY PUBLICATION IN APPLIED STATISTICS

Analysis and Control of Dynamic Economic Systems

GREGORY C. CHOW

Princeton University

John Wiley & Sons

New York · London · Sydney · Toronto

Library of Congress Cataloging in Publication Data
Chow, Gregory C   1929–
   Analysis and control of dynamic economic systems.

   (Wiley series in probability and mathematical statistics)
   "A Wiley-Interscience publication."
   Includes index.
   1. Econometrics.   2. Economic policy—Mathematical models.   I. Title.

HB141.C5 1975   338.9   74-22433
ISBN 0-471-15616-7

Printed in the United States of America

10 9 8 7 6 5 4 3 2 1

To Paula, John, James, and Mei-Mei

Preface

This book presents a set of related techniques for analyzing the properties of dynamic stochastic models in economics and for applying these models in the determination of quantitative economic policy. It also illustrates them with a number of examples and applications contained, in particular, in Chapters 5 and 9. Its contents fall naturally into two parts: Part I on the analysis of dynamic econometric systems and Part II on techniques of control in the use of such systems.

I have written this book primarily for graduate students and possibly for advanced undergraduates whose backgrounds include a course in econometric methods at the level of J. Johnston's *Econometric Methods* or Part II of R. J. Wonnacott and T. H. Wonnacott's *Econometrics*. Thus a broad enough knowledge of matrix algebra to master the treatment of econometrics at the level of these texts is required. The book can also serve as a reference work. To fill its multiple purpose more material has been included than can possibly be completed in a one-semester course by advanced undergraduate students. Each section that can be skipped without loss of continuity is starred, for only the nonstarred sections in Chapters 1 to 4 and 7 need be read to obtain a basic understanding of the two related topics of analysis and control. (Sections 8.1 and 8.2 can replace Sections 7.3, 7.4, and 7.6.)

By referring to the Contents the reader can observe the logical organization of the book. The first part contains six chapters. After an introduction (Chapter 1), it discusses methods of analyzing deterministic systems (Chapter 2) and stochastic systems (Chapters 3 and 4). Applications are treated in Chapter 5 and Chapter 6 deals with the analysis of nonlinear and nonstationary models. The second part on optimal control methods begins with two chapters, 7 and 8, on two basic techniques for obtaining optimal policies from *linear* models with *known* parameters. Chapter 9 discusses applications to economic policy problems. Chapters 10 and 11 are concerned with generalizations that apply to linear models with unknown parameters, and Chapter 12 presents methods of controlling nonlinear systems. In addition to the core chapters, 1 to 4 and 7 (or alternatively Sections 8.1 and 8.2), Chapters 5 and 9 on applications are highly recommended to all readers. The remaining chapters, 8 on dynamic programming, 6 and 12 on nonlinear systems, and 10 and 11 on control techniques for unknown linear systems, are optional. Chapters 10 and 11 are not required for the study of Chapter 12 but Chapter 10 is a prerequisite of Chapter 11. Chapters 6 and 12 are mathematically less advanced than Chapters 8 (excepting Sections 8.1 and 8.2), 10 and 11.

It is possible to treat the variety of topics covered in this book only by limiting the discussion to systems in discrete time. The mathematics for dealing with dynamic systems in continuous time is varied and more difficult. Methods for discrete time models are not only more useful because econometric models in the main take this form but are basic to a grasp of the elements of analysis and control without making unreasonable demands on the mathematical training of the reader.

The reader should understand that this book is concerned mainly with methods. He should not expect to master the art and science of applying them to practical economic problems by simply reading this book and not actually carrying out at least one substantial application, in the same way that studying an econometrics text is not a sufficient qualification for a practicing econometrician. Familiarity with the techniques discussed in this book, however, is a necessary step toward their application. It will also enable the reader with no intention of being a practitioner to appreciate the works of others that rely on these techniques and are becoming more and more important in the literature of economics.

I should like to express my sincere thanks to Andrew B. Abel, Ray C. Fair and Richard E. Quandt for having read the entire manuscript and for giving me their valuable comments. Abel has also prepared the index. Stephen M. Goldfeld, A. C. Harberger, and Edwin S. Mills read parts of the manuscript and contributed ideas to its development. Students who have used the drafts in various forms in class and raised questions also deserve my deep appreciation. Betty Kaminski typed most of the manuscript efficiently and in good spirit, her task having been shared by Diana Hauver and Grace Lilley.

Much of the research reported in this book was supported by grants from the National Science Foundation to the Econometric Research Program at Princeton University for which I am grateful. Part of my work is based on research completed and published while I was associated with the IBM Research Center at which I was provided with an extremely congenial environment. I am also indebted to the *American Economic Review*, *Annals of Economic and Social Measurement*, *International Economic Review*, *Journal of the American Statistical Association*, *Quarterly Journal of Economics*, and *Review of Economics and Statistics* for permission to use materials published in their respective journals.

GREGORY C. CHOW

Princeton, New Jersey
September 1974

Contents

CHAPTER

CHAPTER

PART 2.   CONTROL OF DYNAMIC ECONOMIC SYSTEMS

CHAPTER

Analysis and Control
of Dynamic
Economic Systems

Part 1
Analysis of Dynamic Economic Systems

CHAPTER 1

Problems of Stochastic Dynamic Economics

1.1 THE SUBJECT MATTER OF THIS BOOK IN ECONOMICS

Since the late 1940s much progress has been made in devising methods of estimating systems of interdependent relationships in macroeconomics and in applying them to construct econometric models of national economies. What are the main uses of these econometric models? To provide economic forecasts, to explain the dynamic behavior of the economy in question, and to help make better quantitative economic policies. This book is concerned with the second and third of these uses.

Besides being a natural development of econometrics, our subject matter is also an extension of macroeconomic theory and policy. Macroeconomic theory has become more quantitative in recent years, and sound macroeconomic policy will be based on the knowledge of quantitative economic relationships and on methods of utilizing that knowledge to achieve desired objectives. Furthermore, our subject is a natural development of economic analysis in general, from the simpler static, deterministic form to the more complicated dynamic, stochastic form that can be used to understand and control a changing economy. The evolution from static to dynamic and from deterministic to stochastic economics needs further elaboration.

A large portion of economic theory, especially microeconomic but to a lesser extent macroeconomic, is static. A static theory determines the values of certain endogenous variables, given the values of the exogenous variables and the parameters without specifying how the variables evolve through time. A theory is called a model if it is expressed in mathematical form; for example, a simple static theory or model of the macroeconomy consists of two equations:

$$Y = C + A,$$

$$C = \alpha_0 + \alpha_1 Y,$$

where Y is aggregate expenditures or Gross National Product, C is consumption expenditures, and A is autonomous expenditures; C and Y are endogenous variables and A is an exogenous variable; α_0 and α_1 in the consumption function are parameters whose values are taken as given. The model determines C and Y for any specified value of A.

Three related assumptions are implicit in the use of a static model like the one just mentioned, provided that unique equilibrium values of the endogenous variables can be determined by the equations in the model. First, if the exogenous variables are constant, the associated endogenous variables will take constant equilibrium values. Second, following from the first, if the exogenous variables were to take a new set of values, the associated endogenous variables would also take a new set of constant equilibrium values. The method of comparative statics is precisely to study the changes in the equilibrium values of the endogenous variables with respect to changes in the values of the exogenous variables. Third, a static model can be used for either a "short-run" or "long-run" analysis in the following sense. For a short-run analysis many other variables in the economy such as capital stock and population are assumed to remain constant. The effect of government expenditures (a part of autonomous expenditures) on GNP may then be studied by following the first two assumptions. For a long-run analysis some of the variables held constant in the short-run analysis are allowed to vary, and the model determines a larger number of endogenous variables, still obeying the first two assumptions. Short-run and long-run effects are assumed to differ.

Although static models may serve as a useful first approximation to explain or predict the effects of certain exogenous variables on the endogenous variables, holding constant the less relevant variables (depending on whether the short run or the long run is being considered) and ignoring timing relationships, their shortcomings in the study of economic fluctuations and growth should be obvious. Under the first assumption a static model cannot explain the phenomena of cyclical fluctuations in economic variables or their growth or decline. Furthermore, thinking in static terms, an economist might fall into the trap of believing that the economy is intrinsically stable in the sense of producing nearly constant or very smooth time paths for GNP, employment, the price level, and so on, only if the exogenous variables such as money supply are kept constant. This may be so, but a dynamic model (to be defined more precisely later in this chapter) is needed to ascertain its empirical validity. It is dangerous to identify the properties, especially the dynamic properties, of a static model used in abstract theorizing with the properties of the real economy.

The second assumption for static models is equally dangerous when applied to dynamic phenomena. We might be led, or misled, to believe that if government expenditures or money supply were to change from one level to another the response of GNP or the price level would be a smooth journey toward a new equilibrium level. Thinking in terms of static models might also lead to the conclusion that if money supply is controlled to grow at a constant percentage rate the resulting economic aggregates will also follow smooth growth paths. Anyone with experience in studying the dynamic properties of econometric models realizes that these presuppositions may not be valid. The resulting time path of GNP from a one-step increase in government expenditures is not likely to be smooth. Neither will it necessarily converge to a constant value as time goes on. From a dynamic econometric model one may also search in vain for the long-run multiplier of government expenditures on GNP (the rate of change of GNP with respect to government expenditures). The existence of a multiplier presumes that a new static equilibrium will be reached, a presumption that may not be valid for a dynamic economic model or for the real economy. Neither is the presumption that the paths of GNP and the price level will be smooth if money supply grows at a constant rate.

Now we come to the third assumption in using static models, that short-run and long-run effects may differ and that one can exist without the other. In a comparative static analysis the model may imply that changes in certain exogenous variables do not affect a certain subset of the endogenous variables in either the long or the short run; for example, according to one interpretation of the quantity theory of money, a change in money supply affects only the price level but not real output in the long run, although it may affect real output in the short run. If such a proposition is believed to hold for the real economic data, or for a dynamic econometric model that explains these data, certain difficulties will appear. Can one necessarily expect from the actual economy, or from the econometric model constructed, that if only money supply is increased from one quantity to another within a short period and is kept constant in the remaining periods, all other exogenous variables being kept constant, then real GNP will eventually return exactly to its initial level? If money supply affects economic life in the short run, that is, if real output will respond to money supply, how can the short-run effects be exactly cancelled in the long run? Increase in real output is likely to be associated with increase in capital goods. How can one assume that these capital goods will eventually be depreciated or disappear in some way so that the initial values of real GNP and all its components will prevail?

The shortcomings of static models and of the associated assumptions should be clear. So is the need for dynamic models that can explain the time paths of the endogenous variables. These time paths may show trends and fluctuations. There are no presumptions that they are smooth or will reach equilibrium values as time increases, that their responses to a one-step change in an exogenous variable are smooth and converging, and that their responses to smooth exogenous changes are also smooth. In this book we shall study first the dynamic properties of econometric models to find out whether they resemble the corresponding dynamic properties of the real economy. Constructing theories or models that can explain the phenomena observed is the main purpose of science. Economics is no exception. Second, we shall present quantitative methods for using econometric models for the purpose of improving the dynamic characteristics of the economy. Economic science is intimately related to economic policy. Policy analysis often follows economic analysis in both micro- and macroeconomics. We are mainly concerned with tools of macroeconomic policy, which can be applied to achieve smoother time paths and more desirable trends for the economic variables such as prices, employment, output, and balance of payments. Macroeconometric models are complex. Systematic methods are required for studying their dynamic properties and for improving their properties according to the stated goals of full employment, price stability, equilibrium in the balance of payments, and sufficient growth.

For introductory purposes let me be brief on the importance of introducing stochastic elements in dynamic economic models for the purposes of analysis and control. Anyone who has studied econometrics knows why stochastic models are needed. Economic life is more complicated than can be described completely by any deterministic economic model. Therefore stochastic residuals or disturbances are used partly to summarize the combined effects of the omitted variables in the equations. Furthermore, by using these random distrubances, a much richer theory of economic fluctuations will be obtained, as we demonstrate in Chapters 3, 4, and 5. Third, the use of a stochastic model enables one to bring the techniques of statistics to bear on problems of estimating the parameters in the model and of testing hypotheses concerning the model; or should one say that the techniques of statistical inference have been invented because stochastic models are needed in various branches of sciences? Fourth, to account for uncertainty, in the random disturbances and/or in the model parameters themselves, is important for the quantative study of economic policy simply because uncertainty actually exists.

In this book emphasis is placed on the development of methods for economic analysis and policy formulation by using econometric models. These methods are illustrated by applications, especially in Chapters 5 and 9, but elsewhere in the book as well. The stress on techniques remains, however. It is hoped that by using these techniques and by thinking in dynamic, stochastic terms the reader will eventually contribute to the evolution from static deterministic economics to dynamic stochastic economics that is still taking place today.

1.2 HOW TO USE THIS BOOK

Because this book is mainly concerned with methods of dynamic economic analysis and quantitative economic policy, its style is much like that of a text on econometric methods. There is an abundance of mathematical equations and derivations. Most chapters should be read as chapters in a text on econometrics or mathematical economics. It is necessary to work through many of the steps in the mathematical arguments and to fill in others in between. Reading may be a slow process, just as in an econometrics text. It is also necessary to do some of the problems at the end of each chapter to make sure that the material is well understood.

As pointed out in the preface, this book is written for a multiple audience. For the readers who are less oriented mathematically certain sections are marked by a star. There is no need to study the proofs or any mathematical arguments in these starred sections. Nevertheless, if the topic appeals to the reader, it is advisable to glance through enough of the English text to know the purpose of the section and the nature of any useful results.

It is often helpful to think in terms of concrete examples when confronted with a mathematical presentation. Economics is an applied science. Economic models should have some correspondence to reality, and optimization techniques in economic policy analysis should have some relevance to the solution of actual policy problems. It would therefore be useful for an economist to think about possible practical applications when studying mathematical models and techniques. Bear in mind the material in Section 1.1 and the remainder of this chapter which provides the motivation for studying these tools. To provide additional motivation Chapter 5 in Part I and Chapter 9 in Part II may be read simultaneously with the corresponding chapters on methods; they can be reread after the methods are well understood.

In the remainder of this chapter some basic concepts of dynamic economics are introduced and contrasted with related concepts in static economics. A number of problems in dynamic economics then serve as a preview of the book.

1.3 STATICS AND DYNAMICS

Much of existing economic theory is static. A static theory explains a set of endogenous variables by a system of simultaneous equations without specifying how the variables in different points of time are related. To take an elementary example, consider a simplified model of the determination of national income:

$$Y = C + I + G, \tag{1}$$

$$C = \alpha_0 + \alpha_1 Y, \tag{2}$$

where Y is total output or income, C is consumption expenditures, I is investment expenditures, and G is government purchase of goods and services. The first equation is an identity or definition. The second is a consumption function. These two equations form a *structure* to determine the two *endogenous* variables Y and C if I and G are treated as given outside the system, that is, as *exogenous* variables which the present theory does not explain. To obtain a solution for Y and C we may first express the *endogenous variables as functions of the exogenous variables*:

$$Y = \frac{\alpha_0}{1 - \alpha_1} + \frac{1}{1 - \alpha_1} I + \frac{1}{1 - \alpha_1} G, \tag{3}$$

$$C = \frac{\alpha_0}{1 - \alpha_1} + \frac{\alpha_1}{1 - \alpha_1} I + \frac{\alpha_1}{1 - \alpha_1} G. \tag{4}$$

Equations 3 and 4 are the *reduced-form equations*, derived from (1) and (2), the *structural equations*. Given the values of I and G and the parameters α_0 and α_1 of the structure, we can obtain the solution for Y and C. The theory, in the form of (1) and (2), thus explains national income and consumption. Time does not enter explicitly into the system. Even if a time subscript t is introduced for each variable, no relations among the variables in different points of time are specified.

A dynamic theory specifies the relations among the variables at different points of time and can thus yield the time paths of the endogenous variables as solutions to the system. In this book the dynamic equations are confined to *difference equations* in which the time subscripts can only be integers. A simple example results from modifying the consumption function of (2):

$$Y_t = C_t + I_t + G_t, \tag{5}$$

$$C_t = \beta_0 + \beta_1 Y_{t-1}. \tag{6}$$

The reduced form of this structure is

$$Y_t = \beta_0 + \beta_1 Y_{t-1} + I_t + G_t, \tag{7}$$

$$C_t = \beta_0 + \beta_1 Y_{t-1} + 0 + 0. \tag{8}$$

Note that when lagged endogenous variables such as Y_{t-1} are present in the structure they are also used as explanatory variables in the reduced form, together with the exogenous variables that may be current or lagged. The lagged endogenous variables, together with the exogenous variables, constitute the *predetermined variables* of the system, given which, the values of the current endogenous variables can be obtained from the reduced-form equations. For period 1 Y_1 and C_1 are determined by Y_0, together with I_1 and G_1. Given Y_1 so determined, Y_2 can be found if I_2 and G_2 are again specified. Hence, given the initial values of the lagged endogenous variables (Y_0 in this case), the lagged exogenous variables (if they appear in the system), and the values of I_t and G_t ($t = 1, 2, \ldots$), the time paths Y_t and C_t ($t = 1, 2, \ldots$) can be determined successively by the difference equations (7) and (8).

The solution of a dynamic theory, being a set of functions of time rather than a set of numbers, is more difficult to study or even to characterize than the solution of a static theory. The value of each endogenous variable changes through time. We may wish to know whether a certain variable increases or decreases with time, whether it oscillates, and whether it may eventually reach an *equilibrium* or steady-state value. The conditions under which the solution of a linear system of difference equations is explosive or damped, is oscillatory or not, and has an equilibrium as time increases are studied in Chapter 2. An equilibrium solution, if it exists, is the solution of the set of algebraic equations obtained by dropping all time subscripts in the difference equations. Thus in an equilibrium solution each (endogenous or exogenous) variable takes the same value at different points of time. When an equilibrium is reached, the values of Y_t and Y_{t-1} will be identical; for example, the equilibrium solution of the difference equation (7) is obtained by solving

$$Y_t = \beta_0 + \beta_1 Y_t + I_t + G_t, \tag{9}$$

where the t subscript might as well be omitted. Note that the algebraic equations so obtained can constitute a static theory. Equation 9 yields the same static theory as that of Equation 3. Thus a set of equations that characterizes the equilibrium or steady-state solution of a dynamic theory can be said to represent a static theory.

Perhaps in economic dynamics too much attention has been devoted to ascertaining whether an equilibrium will be reached. A theory is sometimes rejected because it does not yield a set of equilibrium values. From a static theory one does wish to obtain a set of equilibrium values for the endogenous variables. This is what a static theory is supposed to do. From a dynamic theory, however, we wish to ascertain how the endogenous variables evolve through time. For the problems treated by a dynamic theory, particularly those having to do with fluctuations in the economy or business cycles, a set of equilibrium values is not necessarily required. It is possible for the variables to be continuously fluctuating. A static theory is used in economics merely as an abstraction, often a useful one, from the changing conditions in order to understand in part how the forces work in the very short run. It is sometimes undesirable to carry over the aim of a static theory, that of obtaining a set of equilibrium values, to the realm of dynamic economics.

Before Equation 9 several questions were raised concerning the properties of the time paths of endogenous variables generated by a dynamic system. Several questions can, and should, also be raised regarding the relationships of these time paths. Will variables 1 and 2 move up and down together most of the time? Will they move in opposite directions, or do their movements bear any systematic relation to one another? If there is a systematic relation, will one tend to lead, or lag, the other? How do the amplitudes of their fluctuations differ? Around what trends will they fluctuate, if they do fluctuate? Thus a number of questions can be asked about the dynamic properties of the economic variables generated by a dynamic model. Methods will have to be found to characterize the time paths and their relationships.

1.4 COMPARATIVE STATICS AND COMPARATIVE DYNAMICS

A theory, static or dynamic, is used not only to provide a solution for the endogenous variables but also to determine how the solution changes when certain conditions change. These conditions include the exogenous variables, the parameters of the system, and, in the case of a dynamic theory, the initial values of the lagged endogenous and exogenous variables. The comparison of solutions for different given conditions by a static theory is comparative statics and by a dynamic theory, comparative dynamics.

At what rate will an endogenous variable change in response to the change in one exogenous variable, the values of the other exogenous variables and all the parameters being fixed? In the static model of Equations 3 and 4 national income changes by $1/(1-\alpha_1)$ units per unit change of investment/or government expenditures; consumption changes by $\alpha_1/(1-\alpha_1)$ units per unit change of either exogenous variable. These derivatives are the *multipliers* of a macroeconomic model. For a dynamic macroeconomic model, exemplified by (7) and (8), a derivative of this kind can be defined for each pair of endogenous and exogenous variables, both of which are functions of time. The derivative of Y_t with respect to I_t is called an *impact multiplier*, but there are derivatives of Y_t with respect to I_{t-k} for $k>0$. These derivatives can be found by applying Equation 7 repeatedly to eliminate the lagged endogenous variables $Y_{t-1}, Y_{t-2}, \ldots$, on the right-hand side, thus expressing Y_t as a function of I_{t-k}:

$$Y_t = \beta_0 + I_t + G_t + \beta_1(\beta_0 + I_{t-1} + G_{t-1} + \beta_1 Y_{t-2})$$

$$= (\beta_0 + \beta_1\beta_0) + I_t + G_t + \beta_1(I_{t-1} + G_{t-1}) + \beta_1^2 Y_{t-2} = \cdots$$

$$= (\beta_0 + \beta_1\beta_0 + \cdots + \beta_1^k\beta_0) + I_t + G_t + \beta_1(I_{t-1} + G_{t-1}) + \cdots$$

$$+ \beta_1^k(I_{t-k} + G_{t-k}) + \beta_1^{k+1} Y_{t-k-1}. \tag{10}$$

Hence the derivative of Y_t with respect to I_{t-k} is β_1^k. When I_s for a particular period s changes, the whole function Y_{s+k} of future time k ($k=0,1,2,\ldots$) will change; Y_{s+k} will change by β_1^k units per unit change in I_s. These derivatives are *delayed multipliers*.

We may also be interested in the combined effect on Y_t of changes in investment expenditures for several periods, say t, $t-1,\ldots,t-k$. This effect is measured by the sum of the derivatives of Y_t with respect to I_t, $I_{t-1},\ldots,I_{t-k}$. Using (10), we find that if each of I_t, $I_{t-1},\ldots$, and I_{t-k} is to change by one unit Y_t will change by

$$1 + \beta_1 + \beta_1^2 + \cdots + \beta_1^k = \frac{1 - \beta_1^{k+1}}{1 - \beta_1}. \tag{11}$$

This expression (11) is an example of an *intermediate-run multiplier*; it is the change in an endogenous variable per unit change of an exogenous variable for $k+1$ consecutive periods. When the hypothesized unit change in the exogenous variable is permanent, or for all periods considered, the resulting change in Y_t is given by the *long-run multiplier*. In the above example the long-run multiplier of investment on income is $1/(1-\beta_1)$, under the assumption that $0<\beta_1<1$. It is the limit of (11) as k approaches infinity. Alternatively, it could have been obtained by solving the algebraic equation (9) which the steady-state solution of (7) satisfies. This means that a long-run multiplier is the derivative of the steady-state solution for a dependent variable with respect to the permanent value of an exogenous variable.

So much for the effect of an exogenous variable on the solution. Consider also the effects of the parameters and the initial conditions. If a parameter or set of parameters were changed, or if the initial condition Y_0 were different, the whole time path of each dependent variable would be changed. Comparative dynamics is a study of the changes in the solution of a dynamic system when the given conditions change. If the dependent variable at each time is a linear function of the condition in question, as Equation 10 is a linear function of the exogenous variable I_{t-k} and the parameter β_0, we may conveniently speak of the multiplier. Note that the multiplier of the parameter β_0 in (10) is also $1 + \beta_1 + \cdots + \beta_1^k$. In general, however, the relation between the dependent variable and the condition, such as between Y_t and β_1 in (10), is nonlinear. We can still study the solutions under different given conditions, at least by plotting the solutions and examining their differences. It would be advantageous to employ certain summary measures, such as the mean rate of increase or variations around the mean, to characterize each solution.

1.5 STOCHASTIC AND COMPARATIVE STOCHASTIC DYNAMICS

As dynamics deals with time paths generated by dynamic systems, stochastic dynamics deals with stochastic time paths, or time series, generated by dynamic systems in which random elements enter. A simple and useful way to introduce random elements into a dynamic system is to add a random disturbance to an otherwise nonprobabilistic structural equation. To the right-hand side of Equation 6 may be added a random variable u_t which is assumed to have mean 0 and a variance v. The equation so modified is more realistic because empirical data on consumption and income never satisfy a linear function like (6) exactly. Besides Y_{t-1}, other variables also affect C_t. Even if we include more variables in the consumption function, a deviation or residual will still remain between data on C_t and the best fitted function of these variables partly because more variables are involved than can be suitably included in the model. The combined effect of the omitted variables on consumption can be captured by a random variable u_t. If the omitted variables enter linearly and there are many of them, we may appeal to a central limit theorem that the distribution of a linear combination of a large number of random variables is approximately normal under fairly general assumptions.

Therefore u_t is often assumed to be normally distributed. The resulting reduced-form equation for Y_t, with u_t added to the consumption equation, is

$$Y_t = \beta_0 + \beta_1 Y_{t-1} + I_t + G_t + u_t, \tag{12}$$

which is analogous to (7). Equation 12 is an example of a stochastic difference equation.

The solution of, or the *time series* generated by, a stochastic difference equation is a random function of time. Let u_t at different time periods t be uncorrelated random drawings from a distribution with mean zero and variance v. We can imagine generating an observation Y_1 by using (12), given Y_0, I_1, and G_1 and a number drawn at random from the specified distribution for u_1. The resulting Y_1 can be used to generate Y_2, given I_2, G_2, and a second random drawing for u_2. A series $Y_1, Y_2, Y_3, \ldots, Y_T$ can be so generated. Y_t is a random variable, at each time t. Because the random variable and its probability distribution depend on time, it is called a random function of time, or a *time series*. It is of interest to study the properties of the time series, or the set of time series, generated by a system of stochastic difference equations. Such a study can yield predictions from the model; it can be used to validate the model against comparable characteristics observed in the economy and can reveal the dynamic consequences of, and thus help to evaluate, government policies.

The time series Y_t $(t = 1, \ldots, T)$ can be described by the joint probability distribution of $(Y_1, \ldots, Y_T)$. Certain parameters of this joint distribution are of special interest. Consider the mean vector $(\overline{Y}_1, \ldots, \overline{Y}_T) \equiv E(Y_1, \ldots, Y_T)$, the symbol E denoting mathematical expectation. $EY_t \equiv \overline{Y}_t$ is a function of time. Although the stochastic difference equation (12) does not specify Y_t exactly, it does say what Y_t will be on the average. Taking expectations on both sides of (12), Eu_t being 0, we find

$$\overline{Y}_t = \beta_0 + \beta_1 \overline{Y}_{t-1} + I_t + G_t, \qquad (t = 1, \ldots, T). \tag{13}$$

Thus the mean function $\overline{Y}_t$ satisfies the same difference equation as (7) of the nonstochastic system.

Besides the mean function $\overline{Y}_t$ from the stochastic difference equation (12), other useful properties are the variances and covariances $E(Y_t - \overline{Y}_t)(Y_s - \overline{Y}_s)$ of the time series. In Chapter 3 dynamic properties of time series generated by systems of linear stochastic differences are defined, interpreted, and derived mathematically. They have to do with the mean, the variance, and the cyclical properties of each time series and with the relationships among them, which are what a stochastic dynamic theory is supposed to provide.

In comparative stochastic dynamics we study the dynamic properties and dynamic relationships of the time series generated by a system of stochastic difference equations under different assumptions about the time paths of the exogenous variables, the initial conditions, or the values of the parameters. After we learn how to characterize a stochastic time series and to derive the important characteristics under given assumptions we can compare these characteristics under different sets of assumptions.

1.6 FORCES AT WORK IN A MACROECONOMETRIC SYSTEM

Consider a system of reduced-form equations for a vector y_t of p dependent variables as a generalization of (12):

$$y_t = Ay_{t-1} + Cx_t + Bz_t + u_t, \tag{14}$$

where x_t is a $q \times 1$ vector of exogenous variables that are subject to control by the policy maker, such as government expenditures in (12), z_t is an $r \times 1$ vector of exogenous variables that are not subject to control such as investment expenditures in (12), u_t is a $p \times 1$ vector of random disturbances with mean 0 and covariance matrix V, and $A, C,$ and B are given matrices of parameters.

Given the initial vector y_0, the time series y_t can be solved by repeated substitutions of the lagged y's on the right-hand side of (14). Using A^2 for AA, etc., we obtain the result

$$Y_t = A^t y_0 + u_t + Au_{t-1} + A^2 u_{t-2} + \cdots + A^{t-1} u_1$$
$$+ Cx_t + ACx_{t-1} + A^2 Cx_{t-2} + \cdots + A^{t-1} Cx_1$$
$$+ Bz_t + ABz_{t-1} + A^2 Bz_{t-2} + \cdots + A^{t-1} Bz_1. \tag{15}$$

Equation 15 shows the vector time series y_t as the sum of four components. The first is $A^t y_0$ and is the solution of the system $y_t = Ay_{t-1}$. It shows what y_t would be if it were influenced only by its own lagged values. It results from a deterministic theory of economic fluctuations that rely only on the internal forces in the system. An example of this theory is the well-known model by Samuelson (1939) in which the interaction of the multiplier and the accelerator may produce oscillations in national income. The second component is $u_t + Au_{t-1} + \cdots + A^{t-1} u_1$ and is the result of cumulated random disturbances acting on the system. This component may also manifest certain characteristics of business cycles. It is a stochastic time series to be studied in Chapters 3 and 4. The third component, $Cx_t + \cdots + A^{t-1} Cx_1$, captures the result of government policies that affect the economy. The fourth component, $Bz_t + \cdots + A^{t-1}Bz_1$, is due to the influences of the unexplained and uncontrolled variables in the economy. The behavior of the variables z_t is taken as given. One variable in the vector z_t is the dummy variable 1, the coefficients of which are the intercepts of (14). If the first element of z_t is defined as 1, the first column b_1 of the matrix B will consist of the intercepts. The contribution of the vector b_1 to the solution (14) is $b_1 + Ab_1 + \cdots + A^{t-1}b_1$.

In summary, four types of force are at work in a dynamic economic system: the interactions of endogenous variables among themselves at different time periods because of delayed reactions, the random forces acting on the economy, the policy or control variables, and the unexplained and uncontrolled variables. It is an important purpose of government policy to choose the control variables x_t to improve the dynamic performance of the economy, with due allowances for the other forces at work.

1.7 OPTIMAL CONTROL OF STOCHASTIC SYSTEMS

Once the dynamic properties of a system like (14) can be derived, given the time paths of the unexplained exogenous variables, we can attempt to manipulate the control variables in order to achieve more desirable dynamic characteristics, such as specified growth for certain real variables, a stable price level, a low unemployment rate, and small fluctuations in the economic variables in general. The tools of stochastic dynamics are useful for the purpose of helping to analyze the outcomes of alternative government policies. They are far from sufficient, however. The behavior of a multivariate dynamic system is complicated. The control variables can be set through time in an infinite number of ways. But there are simply too many possibilities to try, and, given finite time and resources, we may not succeed in selecting a good policy without an efficient, systematic approach. Optimal control provides an efficient way to obtain good policies in conjunction with a dynamic stochastic model.

An essential step in optimization is to specify an objective function by which the outcome can be judged. One possible objective is to steer the economic variables y_t, or a subset thereof, toward a certain target vector a_t. One element of a_t may be a 5 percent annual growth path for real GNP. A second element may be a 2 percent annual growth path for the general price level. A third element may be a 4 percent unemployment rate, and so on. The welfare loss (the negative of the objective function) may be approximated by a weighted sum of the squared deviations of the elements of y_t from a_t, that is, $\sum_{t=1}^{T}(y_t - a_t)'K_t(y_t - a_t)$, where the diagonal matrices K_t specify the weights. Given such a welfare loss function and the stochastic dynamic model (14), it is desired to find a policy sequence x_t $(t = 1, \ldots, T)$ to minimize the expectation of the welfare loss from period 1 to period T. The solution, as shown in Chapters 7 and 8, takes the form of a feedback control equation $\hat{x}_t = G_t y_{t-1} + g_t$, which specifies the optimal policy $\hat{x}_t$ as a linear function of the recently observed data y_{t-1}, with feedback coefficients G_t and intercept g_t.

Having found an optimal policy, we can evaluate the expected welfare loss resulting from applying it, as compared with the loss from alternative policies, such as maintaining a constant rate of growth for the supply of money. The analysis does not end here. Although the above welfare loss function is a convenient tool in the search for a good policy, the policy maker may wish to examine other important dynamic characteristics of the solution. For this purpose the optimal control equation can be combined with the original system (14) to form a system under control:

$$y_t = (A + CG_t)y_{t-1} + (Cg_t + Bz_t) + u_t$$

$$= R_t y_{t-1} + b_t + u_t, \tag{16}$$

where $A + CG_t$ is denoted by the matrix R_t and $Cg_t + Bz_t$ by the vector b_t. The system (16) shows three sets of forces at work: the internally generated forces through the lagged effects of y_{t-1}, the random disturbances, and the exogenous factors b_t. We can then perform an analysis of the dynamic characteristics of this system by using the tools developed in Chapters 3, 4 and 6.

1.8 CHAPTER OUTLINE OF THIS BOOK

Solution to a system of nonstochastic linear difference equations is characterized and derived in Chapter 2 by elementary methods of matrix algebra. Chapter 3, which is devoted to the solution of a system of linear stochastic difference equations, is based on concepts of the mean functions and covariance functions in the time domain. Chapter 4 characterizes the solution in terms of the cyclical components of the time series in the frequency domain. Characteristics of individual time series and the relationships of time series are studied. Chapter 5 applies the methods in a study of the dynamic characteristics of a simple macroeconomic model constructed by the principles of the multiplier, the accelerator, and liquidity preference. Chapter 6 is concerned with nonstationary and nonlinear stochastic systems; it concludes the subject of dynamic analysis of stochastic systems in this book.

The subject of optimal control of linear systems with known parameters is taken up in Chapters 7 and 8. These chapters deal with deterministic systems and systems with additive stochastic disturbances and provide contrasts between the two. The methods used to obtain the optimal solution differ, however. Chapter 7 employs the method of Lagrange multipliers and Chapter 8, the method of dynamic programming. In Chapter 9 we study optimal control with the abovementioned simple macroeconometric model and discuss related problems of macroeconomic policy with the framework of optimal control.

The control of linear systems with unknown parameters is treated in Chapters 10 and 11. The method in Chapter 10, although accounting for the effect of uncertainty of the unknown parameters, ignores the possibility of reestimation of these parameters in the design of the current policy. The method in Chapter 11 takes reestimation or future learning into account and is a generalization of the methods in Chapters 10 and 8 in which even uncertainty is ignored. Chapter 12 deals with methods of controlling nonlinear systems.

PROBLEMS

1. Introduce total tax receipts T into the system in Equations 1 and 2 and assume that consumption C is a linear function of income after tax. Obtain the reduced form of this system, treating T as exogenous. Will income change if both government expenditures and government tax receipts are increased by one billion dollars? If so, by how much?

2. What is the reduced form of the system specified in Problem 1 if T is not exogenous but equals 20 percent of the national income? What is the income multiplier of government expenditures? How does it compare with the multiplier in Equation 3? Explain. How does it compare with the balanced-budget multiplier in Problem 1? Explain.

3. Introduce total tax receipts T into the system in Equations 5 and 6 and assume consumption that C is a linear function of lagged income after tax. Obtain the reduced form of this system, treating T as exogenous. Assume that in period 1 both government expenditures and tax receipts are increased by one billion dollars. By how much will income in period 1 be affected? Assuming that in each of periods 1 and 2 both G and T are increased by one billion, by how much will income in period 2 be affected? Assuming that the increases of one billion in G and T are permanent, by how much will income be affected in the long run?

4. Consider the model in Problem 3; T however, is no longer exogenous but equals 20 percent of Y in the same period. Calculate the effects of an increase of one billion in G_1, in both G_1 and $G_2,\ldots$, and in $G_1, G_2,\ldots G_k$, respectively, on $Y_1, Y_2,\ldots,Y_k$. What is the effect as k approaches infinity? Compare this effect with the multiplier in Equation 7. Compare it with the multiplier in Problem 2 above.

5. By simple graphic methods, or more sophisticated methods as you please, estimate the values of β_0 and β_1 in Equation 6 using the annual data of the United States from 1961 to 1970 (meaning that the data on C_t cover the years 1961 to 1970). With these estimates and the model of Equations 5 and 6 predict Y_t consecutively for $t = 1971, 1972$, and 1973, taking the values of the exogenous variables in these years as given. Compare your predictions with the actual values of Y_t.

6. In connection with Problem 5, estimate the standard deviation of the random residual around the consumption function (6). Draw five random numbers from a normal distribution with zero mean and the above standard deviation (or, alternatively, from a rectangular distribution with 0 mean and the above standard deviation). Using these random numbers and the estimates of β_0 and β_1 in conjunction with Equation 12, generate Y_t for $t = 1971$, 1972,...,1975. Compare the above time path with the time path of Y_t when the model is assumed to be nonstochastic, all random disturbances being 0.

7. Decompose the time path of Y_t generated from Problem 6 into the four components of Equation 15. Graph the time path and its four components. Comment on this decomposition.

8. Suppose that you are using the model in Equation 12 to control the economy by manipulating G_t in the years 1971 to 1975, and using the estimates of β_0 and β_1 of Problem 5 and the values of I_t as observed or projected. Specify a target path for Y_t. Would a 6 percent annual increase from Y_{1970} be a good target path? What is your optimal policy for G_{1971}? Can you write a feedback control equation for this optimal policy? What is your optimal policy for G_{1972}? Can you write a feedback control equation for this optimal policy? Would you, at the beginning of 1971, have been willing to commit yourself on a numerical value for the optimal setting of G_{1972}? Why or why not?

9. What crude ways can you suggest to deal with the errors in your estimates of β_0 and β_1 in Problem 8. You have assumed that these estimates are the true values of β_0 and β_1 in calculating its optimal policy. Realizing that they are subject to estimation errors, would you correct for your optimal policies? Why? This difficult problem is treated in Chapter 10, but nothing should prevent you from trying to provide a crude answer or at least to argue why the answer to Problem 8 should or should not be applicable here. *Hint.* Given the variances and covariances of your estimates of β_1 and β_1, would it be useful to solve Problem 8 many times with different sets of values for β_0 and β_1 drawn at random from the specified distribution?

CHAPTER 2

Analysis of Linear
Deterministic Systems

Dynamic economics is concerned with the evolution of economic variables through time, that is, with the determination of the time paths or time functions of economic variables. It is well known that static economics is concerned with the determination of economic variables without specifying when their values will materialize. The foundation of a large body of static economic theory is the assumption that individual economic units maximize something. The individual consumer is assumed to maximize utility, and a static demand theory of the consumer is derived. The individual firm is assumed to maximize profit, and a static theory of the behavior of the firm is derived. The basis of many of the dynamic economic theories today is less rigorous.

Quite often a dynamic theory is obtained by introducing certain time delays into an otherwise static theory. The pattern of time delays is not necessarily derived from an assumption that certain individuals maximize something over time, although the static theory forming the basis of dynamics may be derived from a maximizing assumption. One example is the set of two equations that explains the cobweb phenomena in agricultural economics. A static linear demand equation relates the quantity demanded negatively to the price of the commodity of the same period. A dynamic linear supply equation relates the quantity supplied positively to the price of the commodity in the *last period*. By solving these two equations, assuming that quantities demanded and supplied are equal, we get an equation relating price at period t to the price at $t-1$. In this dynamic model the static theory of demand and supply is modified by introducing a time delay in the supply equation. We can find many examples in macroeconomics in which time delays are introduced into the

19

relations between consumption and income, between investment expenditures and investment plans (the latter possibly based on certain theory of maximizing behavior), and between demand for money and income. Whether the time delays themselves should *always* be justified by a theory of maximizing behavior over time, is a methodological, and somewhat philosophical, question that is not discussed in this book.

From the practical point of view we have dynamic macroeconomic models that are built from a combination of theory (of whatever kind) and statistical measurement, and these econometric models deserve to be studied by the methods of dynamics described. Furthermore, we introduce methods of optimization over time for the purpose of improving the dynamic performance of an economy—methods that will be useful for those theorists who wish to derive dynamic economic theories from optimization over time and presumably under uncertainty. There is no need to entertain the question whether the basis of static economics under certainty, namely maximizing behavior, should always be extended to the study of dynamic behavior over time under uncertainty. Possibly, or possibly not, maximizing behavior is a better approximation to human situations in which time and uncertainty are less relevant.

In this chapter we shall study dynamic models in the form of systems of linear difference equations. Linearity is assumed because of simplicity and because the most important concepts of dynamics can be understood via a linear model. Methods of linear dynamics can also be extended and modified for nonlinear systems, as in Chapter 6. Difference equations, rather than differential equations, are studied mainly because most existing macroeconometric models are in this form, and it would be difficult to cover the broad subject matter of this book by using both discrete-time and continuous-time models. The additional investment in mathematical skills to deal with models in continuous time, especially for the stochastic case, may not be worthwhile for most applied economists. Readers interested in the techniques of analysis and control of continuous-time systems may refer to Astrom (1970).

We confine ourselves to deterministic systems in this chapter. Solutions to systems of difference equations are provided and characterized. The solution to a univariate, first-order system is introduced first and then generalized to the multivariate, high-order case by matrix algebra. Properties of the solution are characterized and examined. Methods and concepts developed in this chapter will be useful for the study of dynamic properties of systems of stochastic difference equations in the next two chapters. This chapter may require several class periods to cover if the readers have no prior knowledge of linear difference equations. Readers who feel the need for some economic applications of linear dynamic models may consult Baumol (1970) or Kenkel (1974) or read Sections 5.1, 5.2, and 5.3 at this point.

2.1 HOMOGENEOUS LINEAR DIFFERENCE EQUATION OF FIRST ORDER

To understand how an economic variable may evolve through time consider the simplest model of a first-order linear difference equation

$$y_t = ay_{t-1}. \tag{1}$$

It is *first order* because, in the determination of y_t, only y_{t-1} is used. If y_{t-m} is required, the equation is said to be of *order m*. In general, an equation is of order m if m is the largest difference in subscripts occurring in the equation. Equation 1 is also *homogeneous* because only lagged values of the endogenous variables are used, and no exogenous variables or constant or other given functions of time are present in the equation. The model (1) determines y_t simply by applying a factor a to its value of the last period. Given an initial value of the variable, say y_0 at time 0, the solution to (1) can be simply stated by repeated substitutions for the lagged variables on the right-hand side:

$$y_t = a^2 y_{t-2} = a^k y_{t-k} = a^t y_0. \tag{2}$$

The solution $y_0 a^t$, which is a function of time, can be easily characterized. First, whether it is *explosive* or *damped* depends on whether the absolute value of a is greater than or smaller than 1. It is a constant if a is exactly. 1. Second, whether it *oscillates* or not depends on whether a is negative or positive. If a is positive, the solution is a monotone function of time; if a is negative, the solution oscillates between successive time periods. Of course, the two characteristics can be combined. A solution may oscillate and be explosive if a is negative and has absolute value greater than 1; for example, if a equals -1.10. Thus the absolute value and the sign of a characterize the solution of (1).

2.2 HIGHER ORDER AND MULTIVARIATE HOMOGENEOUS LINEAR SYSTEMS

If a difference equation is of higher order, it is convenient for the purpose of analysis to rewrite it as a first-order system of many variables. Thus the equation

$$y_t = a_1 y_{t-1} + a_2 y_{t-2} \tag{3}$$

can be rewritten as

$$\begin{bmatrix} y_t \\ y_{t-1} \end{bmatrix} = \begin{bmatrix} a_1 & a_2 \\ 1 & 0 \end{bmatrix} \begin{bmatrix} y_{t-1} \\ y_{t-2} \end{bmatrix}. \tag{4}$$

This system explains a vector of two endogenous variables y_t and y_{t-1}. If the equation is of order m, a vector of $y_t, y_{t-1}, \ldots, y_{t-m+1}$ will be used in the system. The resulting system will be of first order like Equation 4 because we can denote the vector $(y_t, y_{t-1})'$ by z_t and the matrix on the right of (4) by A; then (4) can be written as $z_t = A z_{t-1}$ which is precisely in the same form as (1).

The same method of transforming a system into first order can be applied to a multivariate system of difference equations of higher order. Let y_t be a column *vector* of p endogenous variables that satisfy the mth-order difference equation

$$y_t = A_1 y_{t-1} + A_2 y_{t-2} + \cdots + A_m y_{t-m}, \tag{5}$$

where $A_1, \ldots, A_m$ are $p \times p$ matrices of real coefficients. Equation 5 can be rewritten as

$$\begin{bmatrix} y_t \\ y_{t-1} \\ \vdots \\ y_{t-m+1} \end{bmatrix} = \begin{bmatrix} A_1 & A_2 & \cdots & & A_m \\ I & 0 & \cdots & & 0 \\ & \cdots\cdots\cdots\cdots\cdots & & \\ 0 & \cdots & 0 & I & 0 \end{bmatrix} \begin{bmatrix} y_{t-1} \\ y_{t-2} \\ \vdots \\ y_{t-m} \end{bmatrix}. \tag{6}$$

The first p components of (6) recaptures Equation 5. The remaining components of (6) are identities that explain the newly introduced endogenous variables $y_{t-1}, \ldots, y_{t-m+1}$. In a more compact form (6) can be rewritten as

$$y_t = A y_{t-1}, \tag{7}$$

where y_t is redefined to represent the left-hand vector of the original equation (6) without the use of a new symbol and A stands for the right-hand matrix of (6). Difference equations in the form of (7) will be studied.

Analogous to (2), the solution to (7) can be obtained as

$$y_t = A^2 y_{t-2} = A^k y_{t-k} = A^t y_0, \tag{8}$$

where the product $A \cdot A$ is written as A^2, and so on. Given the initial value of the vector y_0, successive values of $y_t (t = 1, 2, \ldots)$ can be calculated by using $A^t y_0$. The solution in this form is not very informative, however. It would be desirable to characterize the time paths $A^t y_0$ to ascertain whether they will be explosive or damped and whether they will oscillate or fluctuate in some way.

2.3 CHARACTERIZATION OF THE SOLUTION TO A HOMOGENEOUS LINEAR SYSTEM

One useful approach to characterizing the solution (8) of a homogeneous linear system of difference equations is to decompose it into the solutions of many first-order univariate equations in the form of (1). This can be done by utilizing the *characteristic roots* and *characteristic vectors* (also called *eigenvalues* and *eigenvectors*) of the matrix A. A *characteristic root* of A is a scalar λ that satisfies

$$|A - \lambda I| = 0, \tag{9}$$

where the two vertical bars denote the determinant of the matrix inside and I is an identity matrix of order p. As an example, consider a 2×2 matrix $A = (a_{ij})$ in which a_{ij} are elements. Equation 9 becomes

$$|A - \lambda I| = \left| \begin{pmatrix} a_{11} & a_{12} \\ a_{21} & a_{22} \end{pmatrix} - \begin{pmatrix} \lambda & 0 \\ 0 & \lambda \end{pmatrix} \right| = \begin{vmatrix} a_{11} - \lambda & a_{12} \\ a_{21} & a_{22} - \lambda \end{vmatrix}$$

$$= \lambda^2 - (a_{11} + a_{22})\lambda + a_{11}a_{22} - a_{12}a_{21} = 0. \tag{10}$$

The two roots, say λ_1 and λ_2, of (10) are the characteristic roots or eigenvalues of the 2×2 matrix $A = (a_{ij})$. In general, Equation 9 is a polynomial equation of degree p in λ. It is called the *characteristic equation* of the $p \times p$ matrix A. Therefore the matrix A has p characteristic roots which can be found by solving a polynomial equation of degree p. Thus the computation of characteristic roots is equivalent to the solution of polynomial equations. The computational aspects of finding characteristic roots are not pursued in this book. Numerous computational algorithms for finding these roots and vectors and computer programs to implement them are readily available.

With each characteristic root λ_i there is associated a *right characteristic vector* b_i. A right characteristic vector b of a matrix A is a $p \times 1$ vector that satisfies the equation

$$Ab = \lambda b, \tag{11}$$

where λ is some scalar. In other words, the characteristic vector b has the property that, when premultiplied by the matrix A or when subject to the linear transformation A, it remains a scalar multiple of itself. Equation 11 is equivalent to

$$(A - \lambda I)b = 0 \tag{12}$$

and Equation 12 is satisfied, for b not equal to a zero vector, if and only if the matrix $(A - \lambda I)$ is singular, that is, if and only if the characteristic equation (9) is satisfied. The solution to (9) is the set of characteristic roots $\lambda_1, \ldots, \lambda_p$. For each λ_i, a characteristic vector b_i is defined by equation (12), that is,

$$(A - \lambda_i I)b_i = 0. \tag{13}$$

Note that the vector b_i is defined only up to a factor of proportionality; if b_i satisfies (13), so will $2b_i$, for instance. For the purpose of computing the vector b_i which corresponds to a real root λ_i we may arbitrarily set its first element b_{1i} equal to 1 and solve the resulting equations from (13) for the remaining unknowns $b_{2i}, \ldots, b_{pi}$. These equations will be a set of $(p-1)$ nonhomogeneous linear equations in these $(p-1)$ unknowns. Another convention, frequently applied in computer programs, normalizes the elements of b_i so that the sum of their squares is unity.

It will be useful to combine the p equations relating the characteristic roots λ_i and vectors $b_i (i = 1, \ldots, p)$ into one system of p vector equations; $Ab_1 = \lambda_1 b_1$, $Ab_2 = \lambda_2 b_2$, and so on, can be combined as

$$A(b_1, \ldots, b_p) = (\lambda_1 b_1, \ldots, \lambda_p b_p) = (b_1, \ldots, b_p)\begin{pmatrix} \lambda_1 & & 0 \\ & \ddots & \\ 0 & & \lambda_p \end{pmatrix}. \tag{14}$$

Denoting by B the matrix $(b_1, \ldots, b_p)$, whose columns are the characteristic vectors of A, and by D_λ the diagonal matrix which consists of the corresponding characteristic roots, we can write (14) as

$$AB = BD_\lambda. \tag{15}$$

To simplify the analysis, we assume that all the characteristic roots of A are distinct and ignore the case of multiple roots. (The seriousness of this assumption for practical purposes is discussed in Chapter 5.) Under this assumption B is known to be nonsingular. Postmultiplying both sides of (15) by B^{-1} gives

$$A = BD_\lambda B^{-1}. \tag{16}$$

Writing the matrix A in this form (16) is useful for the study of the solution $A^t y_0$ of the difference equations (7) because A^t can also be written in a convenient form. Note that $A^2 = (BD_\lambda B^{-1})(BD_\lambda B^{-1}) = BD_\lambda^2 B^{-1}$, and similarly that

$$A^t = BD_\lambda^t B^{-1}, \tag{17}$$

where D_λ^t is simply a diagonal matrix that has λ_i^t as its ith diagonal element. The solution (8), rewritten as

$$y_t = BD_\lambda^t B^{-1} y_0, \tag{18}$$

consists of functions of the characteristic roots of A. To write out the functions explicitly, let $B = (b_{ij})$ be a 3×3 matrix and let the $i-j$ element of B^{-1} be denoted by b^{ij}. Expanding (18) in scalar variables gives

$$\begin{bmatrix} y_{1t} \\ y_{2t} \\ y_{3t} \end{bmatrix} = \begin{bmatrix} b_{11}\lambda_1^t & b_{12}\lambda_2^t & b_{13}\lambda_3^t \\ b_{21}\lambda_1^t & b_{22}\lambda_2^t & b_{23}\lambda_3^t \\ b_{31}\lambda_1^t & b_{32}\lambda_2^t & b_{33}\lambda_3^t \end{bmatrix} \begin{bmatrix} b^{11}y_{10} + y^{12}y_{20} + b^{13}y_{30} \\ b^{21}y_{10} + b^{22}y_{20} + b^{23}y_{30} \\ b^{31}y_{10} + y^{32}y_{20} + b^{33}y_{30} \end{bmatrix}$$

$$= \begin{bmatrix} b_{11}z_{10}\lambda_1^t + b_{12}z_{20}\lambda_2^t + b_{13}z_{30}\lambda_3^t \\ b_{21}z_{10}\lambda_1^t + b_{22}z_{20}\lambda_2^t + b_{23}z_{30}\lambda_3^t \\ b_{31}z_{10}\lambda_1^t + b_{32}z_{20}\lambda_2^t + b_{33}z_{30}\lambda_3^t \end{bmatrix}, \tag{19}$$

where z_{i0} is defined as

$$z_{i0} = b^{i1}y_{10} + b^{i2}y_{20} + b^{i3}y_{30}, \tag{20}$$

that is, as a linear combination of y_{10}, y_{20}, and y_{30}, using b^{i1}, b^{i2}, and b^{i3} as weights. Written in scalar form, the solution y_{jt} is thus a linear combination of λ_1^t, λ_2^t, and λ_3^t.

2.4 CHARACTERIZATION OF THE SOLUTION
USING CANONICAL VARIABLES

Because, by (19), the solution of each y_{jt} is a linear combination of λ_i^t and the solution of a univariate first-order equation $z_{it} = \lambda_i z_{i,t-1}$ is also proportional to λ_i^t, we can think of y_{jt} as a linear combination of a set of new variables z_{it}. Equation 20 suggests that a set of *canonical variables* z_{it} can be defined as

$$z_{it} = b^{i1}y_{1t} + b^{i2}y_{2t} + \cdots + b^{ip}y_{pt}, \qquad (i = 1, \dots, p). \qquad (21)$$

The p canonical variables are linear combinations of the p original variables and vice persa. In vector form their relationship is given by

$$z_t = B^{-1}y_t \quad \text{or} \quad y_t = Bz_t. \qquad (22)$$

Because of this relationship, the solution of one set of variables can be obtained as a linear combination of the solution of the other. Given the solutions for $z_{1t}, \dots, z_{pt}$, which are easy to obtain and interpret, the solution for y_{jt} is simply their linear combination, using the jth row of B as weights. Canonical variables are useful not only for deriving a solution for y_{jt} but also for treating an explosive system of difference equations which otherwise can hardly be characterized. We turn to the latter subject in Sections 6.1, 6.2, and 6.3.

To find the solution for z_{it} we rewrite the system (7) in terms of the canonical variables, using (16) and the definition (22):

$$z_t = B^{-1}y_t = B^{-1}(BD_\lambda B^{-1})y_{t-1} = D_\lambda z_{t-1}. \qquad (23)$$

Each canonical variable z_{it} satisfies the difference equation

$$z_{it} = \lambda_i z_{i,t-1}. \qquad (24)$$

Its solution is $z_{it} = z_{i0}\lambda_i^t$. The solution for y_{jt} is therefore

$$y_{jt} = b_{j1}(z_{10}\lambda_1^t) + b_{j2}(z_{20}\lambda_2^t) + \dots + b_{jp}(z_{p0}\lambda_p^t). \qquad (25)$$

Equation 25 agrees with the result of (19). Because the initial condition is often given in terms of y_{j0}, we need to calculate z_{i0} from y_{j0} by using (22) or (20).

To summarize, by defining the canonical variables z_t as linear combinations B^{-1} of y_t and observing that the solution of z_{it} is simply $z_{i0}\lambda_i^t$ we find that the solution for y_{jt} is a linear combination of the above solutions, using the jth row of B as weights. Accordingly, the solution for y_{jt} is a linear combination of the roots of A, each raised to a power t.

2.5 THE CASE OF COMPLEX ROOTS

The behavior of the solution can be quite interesting if some roots are complex. Complex numbers arise from solutions to polynomial equations such as $x^2 + 4 = 0$ and $x^2 - x + 2 = 0$. They will be required to express the solution of Equation 10 if $(a_{11} + a_{22})^2 - 4(a_{11}a_{22} - a_{12}a_{21})$ is negative. A *complex* number λ takes the form

$$\lambda = a + bi, \tag{26}$$

where $i = \sqrt{-1}$ and a and b are real numbers. It has two parts, the real part a and the imaginary part bi. By using a two-dimensional diagram to represent the real part along the horizontal axis and the imaginary part along the vertical axis we can represent λ by the point (a, b), as in Figure 2.1. The *absolute value* (also termed the *modulus*) of λ is the length of the vector (a, b) or the distance of the point (a, b) from the origin.

$$|\lambda| = \sqrt{a^2 + b^2} \ . \tag{27}$$

Let θ denote the angle between the line from the origin to the point (a, b)

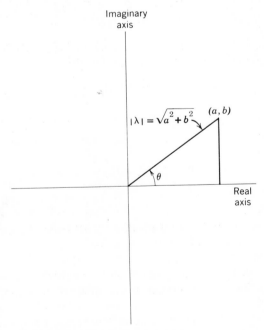

Figure 2.1 Diagrammatic representation of a complex number.

and the horizontal axis. By definition

$$a = |\lambda| \cos\theta; \qquad b = |\lambda| \sin\theta. \tag{28}$$

Using (28), we can rewrite the complex number (26) as

$$\lambda = |\lambda|(\cos\theta + i\sin\theta) = |\lambda|e^{i\theta}. \tag{29}$$

The second equality sign is due to the identity

$$e^{i\theta} = \cos\theta + i\sin\theta. \tag{30}$$

This identity can be made convincing, if not proved, by expanding the functions $e^{i\theta}$, $\cos\theta$, and $\sin\theta$ in Taylor series and matching the terms on both sides of (30). (See Problem 3.) Thus a complex number λ can be represented by the real and imaginary parts a and b or by the absolute value $|\lambda|$ and the angle θ, as in (29).

When complex roots of a matrix A of real numbers occur, they appear as pairs of *conjugate* numbers. The *conjugate* of the complex number λ, as defined by (26), is $a - bi$, or equivalently

$$|\lambda|(\cos\theta - i\sin\theta) = |\lambda|e^{-i\theta}.$$

It is denoted by $\bar\lambda$. The conjugate of $\bar\lambda$ is λ itself; λ and $\bar\lambda$ form a pair of conjugate complex numbers. Because the characteristic roots of A are the roots of a polynomial equation with real coefficients, they come in conjugate pairs when they are complex. This can be seen by factoring the polynomial equation (9) into

$$(\lambda - \lambda_1)(\lambda - \lambda_2)\cdots(\lambda - \lambda_p) = 0,$$

where λ is the unknown and $\lambda_1, \lambda_2, \ldots, \lambda_p$ are the roots. If λ_1 is complex, there must be another root, say λ_2, such that the product

$$(\lambda - \lambda_1)(\lambda - \lambda_2) = \lambda^2 - (\lambda_1 + \lambda_2)\lambda + \lambda_1\lambda_2$$

will have real coefficients $(\lambda_1 + \lambda_2)$ and $\lambda_1\lambda_2$, since the coefficients of the polynomial are real to begin with. These real coefficients imply that λ_1 and λ_2 are a pair of conjugate complex roots, as the reader may wish to verify in Problem 9 of this chapter.

In the solution to the system of difference equations, as given by (19), let λ_1 and λ_2 be a pair of conjugate complex roots, λ_3 being real. The contribution of the first two roots to the solution for y_{1t} is

$$b_{11}z_{10}\lambda_1^t + b_{12}z_{20}\lambda_2^t = c_{11}\lambda_1^t + c_{12}\lambda_2^t. \tag{31}$$

If this contribution is to be real, the coefficients c_{11} and c_{12} must be conjugate complex. The argument for this proposition is left to the reader as Problem 10 in this chapter. Let λ_1 and λ_2 be, respectively, $re^{i\theta}$ and $re^{-i\theta}$, r being the absolute value $|\lambda_1| = |\lambda_2|$. Let c_{11} and c_{12} be, respectively, $s_1 e^{i\psi_1}$ and $s_1 e^{-i\psi_1}$. The contribution of the first two roots to y_{1t} is then

$$c_{11}\lambda_1^t + c_{12}\lambda_2^t = s_1 r^t e^{i(\psi_1 + \theta t)} + s_1 r^t e^{-i(\psi_1 + \theta t)}$$

$$= 2 s_1 r^t \cos(\psi_1 + \theta t), \tag{32}$$

where the second equality sign is due to the identity

$$e^{ix} + e^{-ix} = \cos x + i \sin x + \cos x - i \sin x$$

$$= 2 \cos x. \tag{33}$$

It has been shown that when there is a pair of complex roots of the matrix A their contribution to the solution for each variable y_{it} can be written as a cosine function of time t, multiplied by a factor r^t. If the absolute value r of the roots is larger than 1, the cosine function will be magnified through time; if r is less than 1, the cosine function will be damped. If r is exactly 1, there will be no magnifying or damping effect. To obtain the entire solution for y_{it}, if there are more than two roots, this contribution will have to be added to the linear combination of the remaining roots raised to power t. The latter may involve a real root, such as $c_{13}\lambda_3^t$ for y_{1t} in (19). It may also involve another pair of complex roots. The solution would then be the sum of two magnified or damped cosine functions.

2.6 A NOTE ON COSINE FUNCTIONS

Because the solution to a system of linear difference equations may require a cosine function of time, it may be useful to review some properties of this function. The function $\cos \theta t$ takes the same value 1 when $\theta t = 0, 2\pi, 4\pi, 6\pi$, etc., that is, when $t = 0, 2\pi/\theta, 4\pi/\theta, 6\pi/\theta$, etc. In fact, it takes an identical value for $\theta t = k, 2\pi + k, 4\pi + k$, etc., for any constant k, that is, when $t = k/\theta, 2\pi/\theta + k/\theta, 4\pi/\theta + k/\theta$, etc. Because the function repeats itself every $2\pi/\theta$ time units, where θ is measured in radians, it is called a *periodic function*, and the *length of the cycle* or the *period* of $\cos \theta t$ is $2\pi/\theta$ time units. This means that for each time unit the function will repeat itself $\theta/2\pi$ times, $\theta/2\pi$ being a fraction in most cases; $\theta/2\pi$ is called the *frequency* of the function $\cos\theta t$ because it shows how frequently the function repeats itself per unit time. Frequency is the reciprocal of the length of the cycle.

The *amplitude* of the function $s\cos\theta t$ which is s shows the magnitude of the cyclical movements. The function may also be subject to a time shift, as given by $\cos(\theta t + \psi)$. Although $\cos\theta t$ starts with the value 1 at $t = 0$, the function $\cos(\theta t + \psi)$ takes the value 1 at $t = -\psi/\theta$. Thus $\cos(\theta t + \psi)$ leads $\cos\theta t$ by ψ/θ time units, or $\cos\theta t$ lags behind $\cos(\theta t + \psi)$ by ψ/θ time units; ψ/θ is the *phase shift* in number of time units for the function $\cos(\theta t + \psi)$ and indicates a lead compared with $\cos\theta t$ if ψ is positive, a lag if ψ is negative. Note that the absolute value of ψ should be smaller than π. By convention, we do not say that $\cos(t + 1.5\pi)$ leads $\cos t$ by 1.5π time units but rather that $\cos(t - .5\pi)$ lags behind $\cos t$ by $.5\pi$ time units.

To apply these concepts to the contribution (32) of a pair of complex roots $re^{i\theta}$ and $re^{-i\theta}$ to the solution of y_{1t} we note that this contribution is made up of a cosine function with cycle length equal to $2\pi/\theta$ time units or frequency equal to $\theta/2\pi$ per unit time. Its amplitude is a multiple of r^t that increases with t if the absolute value r of the roots is greater than 1 and decreases with t if the absolute value of r is smaller than 1. The phase ψ_1 of the function $\cos(\theta t + \psi_1)$ will depend on the initial conditions, discussed in the following section.

2.7 NUMERICAL EVALUATION OF THE SOLUTION

The solution of a homogeneous system of linear difference equations $y_t = Ay_{t-1}$ for the ith component of y_t has been found to be,

$$y_{it} = \sum_{k=1}^{p} b_{ik} z_{k0}\lambda_k^t = \sum_{k=1}^{p} c_{ik}\lambda_k^t, \tag{34}$$

where λ_k is the kth root of the $p \times p$ matrix A, b_{ik} is the ith element of the right characteristic vector b_k corresponding to the root λ_k, and z_{k0} is the initial value of the kth canonical variable which satisfies the univariate first-order difference equation $z_{kt} = \lambda_k z_{k,t-1}$. The solution y_{it} will be damped if all the roots $\lambda_k (k = 1, \ldots, p)$ are smaller than 1 in absolute value. It will be explosive if any root λ_k (with nonzero coefficient c_{ik}) is larger than 1 in absolute value. There will be oscillations if some roots are negative or complex. A negative root gives rise to a component with two periods between successive peaks. A pair of complex roots gives rise to a component with $2\pi/\theta$ periods between peaks, θ being the angle (in radians) of the complex roots $re^{i\theta}$ and $re^{-i\theta}$. If all the roots are real and positive, there can be some fluctuations in the linear combination (34) of the positive functions λ_k^t if the coefficients c_{ik} are partly positive and partly negative. But the fluctuations can hardly be prolonged because the behavior of the solution will eventually be dominated by the largest root as t increases. The behavior of the solution is thus mainly characterized by the characteristic roots of A as stated.

To compute the solution numerically three methods can be mentioned. The first is direct computation by using the equation $y_t = Ay_{t-1}$ successively, assuming that y_0 is known. This method requires repeated matrix multiplications. The second method employs the canonical variables discussed in Section 2.4. It is more complicated than the first, but it permits the decomposition of the solution into components of the form $c_{ik}\lambda_k^t$ ($k = 1,\ldots,p$). For practical purposes the user needs only an efficient computer program to find the characteristic roots λ_k and the corresponding right characteristic vectors b_k of the real matrix A. The program should also provide B^{-1}. He will then have all b_{ik} and $z_{k0} = \Sigma_j b^{kj} y_{j0}$ to compute the coefficients c_{ik} used in the linear combination for the solution.

In the third method the solution is obtained in the form of (34) by finding the roots λ_k without going through the calculation of the characteristic vectors. By using this form, together with the initial vales of y_{it} directly, we can evaluate the coefficients c_{ik}. Consider the case of a 2×2 matrix A for illustrative purpose. The solution for y_{1t} is $c_{11}\lambda_1^t + c_{12}\lambda_2^t$. Given two values of y_{1t}, say y_{10} and y_{11}, we will have two equations,

$$c_{11} + c_{12} = y_{10},$$

$$\lambda_1 c_{11} + \lambda_2 c_{12} = y_{11},$$

for the two unknown c_{11} and c_{12}. If λ_1 and λ_2 are conjugate complex, being, say, $a + bi$ and $a - bi$, we can solve for the unknowns by letting $c_{11} = c + di$ and $c_{12} = c - di$ and substituting into the above equations.

$$c + di + c - di = y_{10}$$

$$(a + bi)(c + di) + (a - bi)(c - di) = y_{11},$$

or

$$ac - bd + (ad + bc)i + ac - bd - (ad + bc)i = y_{11}.$$

The solution is $c = y_{10}/2$ and $d = (ay_{10} - y_{11})/2b$. Similarly, if there are p roots, p initial values of y_{it} can be used to determine the p coefficients $c_{i1},\ldots,c_{ip}$.

2.8 A NUMERICAL EXAMPLE INVOLVING COMPLEX ROOTS

This section provides a numerical example of a first-order system of two difference equations with a pair of complex roots:

$$\begin{bmatrix} y_{1t} \\ y_{2t} \end{bmatrix} = \begin{bmatrix} 1 & 1 \\ -1.62 & -.80 \end{bmatrix} \begin{bmatrix} y_{1,t-1} \\ y_{2,t-1} \end{bmatrix}. \tag{35}$$

The characteristic equation is

$$(1-\lambda)(-.80-\lambda) + 1.62 = \lambda^2 - .20\lambda + 0.82 = 0.$$

The characteristic roots are

$$\lambda_1 = .1 + .9i \cong .906e^{1.460i} \quad \text{and} \quad \lambda_2 = .1 - .9i \cong .906e^{-1.460i}.$$

To find the solution $y_t = A^t y_0 = B D_\lambda^t B^{-1} y_0$ we apply the second method mentioned in Section 2.7 by finding the right characteristic vectors of A.

Denote the elements of the characteristic vector b_1 corresponding to λ_1 by $\alpha_1 + \beta_1 i$ and $\alpha_2 + \beta_2 i$; b_1 cannot be a real vector. If it were real, Ab_1 would also be real and could not be equal to $\lambda_1 b_1$ which would be complex. Writing out the equation for b_1, we have

$$\begin{bmatrix} 1 & 1 \\ -1.62 & -.80 \end{bmatrix} \begin{bmatrix} \alpha_1 + \beta_1 i \\ \alpha_2 + \beta_2 i \end{bmatrix} = (.1 + .9i) \begin{bmatrix} \alpha_1 + \beta_1 i \\ \alpha_2 + \beta_2 i \end{bmatrix},$$

or

$$(\alpha_1 + \alpha_2) + (\beta_1 + \beta_2)i = (.1\alpha_1 - 0.9\beta_1) + (.9\alpha_1 + .1\beta_1)i$$

$$(-1.62\alpha_1 - .80\alpha_2) + (-1.62\beta_1 - .80\beta_2)i = (.1\alpha_2 - .9\beta_2) + (0.9\alpha_2 + .1\beta_2)i,$$

implying four homogeneous linear equations in α_1, α_2, β_1, and β_2:

$$\begin{aligned} 0.9\alpha_1 + \alpha_2 + 0.9\beta_1 &= 0 \\ 0.9\alpha_1 - 0.9\beta_1 - \beta_2 &= 0 \\ 1.62\alpha_1 + .9\alpha_2 - .9\beta_2 &= 0 \\ .9\alpha_2 + 1.62\beta_1 + .9\beta_2 &= 0. \end{aligned}$$

These equations are not linearly independent because they have come from $(A - \lambda_1 I)b_1 = 0$, and $(A - \lambda_1 I)$ is singular. Note that .9 times the sum of the first and second equations gives the third equation; .9 times the difference of the second and first equations gives the fourth equation. Because a characteristic vector is determined only up to a factor of proportionality, we let $\alpha_1 = 1$ and $\beta_2 = 0$ and solve the first two equations to get $\beta_1 = 1$ and $\alpha_2 = -1.8$.

Similarly, we can find a complex characteristic vector which corresponds to the root $\lambda_2 = .1 - .9i$. Its two elements are $1 - i$ and -1.8, respectively. This vector is the complex conjugate of the vector associated with the first root $\lambda_1 = .1 + 9i$, which means that each element is the complex conjugate of the corresponding element of the latter vector. In general, the *characteristic vectors corresponding to a pair of conjugate complex roots are conjugate complex*, which can be shown by doing Problem 14 of this chapter. Thus the matrix B which consists of the right characteristic vectors in our example is

$$B = \begin{bmatrix} 1+i & 1-i \\ -1.8 & -1.8 \end{bmatrix}.$$

To find the inverse of B we use the identity $B^{-1}B = I$. Denoting the first row of B^{-1} by $(\gamma_1 + \delta_1 i, \gamma_2 + \delta_2 i)$, we have

$$\gamma_1 - \delta_1 + (\gamma_1 + \delta_1)i - 1.8\gamma_2 - 1.8\delta_2 i = 1,$$

$$\gamma_1 + \delta_1 + (-\gamma_1 + \delta_1)i - 1.8\gamma_2 - 1.8\delta_2 i = 0,$$

which implies four linear equations in the unknowns γ_1, γ_2, δ_1, and δ_2:

$$\begin{aligned}
\gamma_1 & - 1.8\gamma_2 & - \delta_1 & & = 1, \\
\gamma_1 & & + \delta_1 & - 1.8\delta_2 & = 0, \\
\gamma_1 & - 1.8\gamma_2 & + \delta_1 & & = 0, \\
-\gamma_1 & & + \delta_1 & - 1.8\delta_2 & = 0.
\end{aligned}$$

The solution is $\gamma_1 = 0$, $\gamma_2 = -3.6^{-1}$, $\delta_1 = -.5$ and $\delta_2 = -3.6^{-1}$. Similarly, we can obtain the second row of B^{-1} to yield the result

$$B^{-1} = \begin{bmatrix} -.5i & -3.6^{-1} - 3.6^{-1}i \\ .5i & -3.6^{-1} + 3.6^{-1}i \end{bmatrix}.$$

Note that the two rows of B^{-1} are conjugate complex.

Let the initial condition be $y_{10}=2.0$ and $y_{20}=3.6$. The initial condition for the canonical variables will then be

$$
\begin{bmatrix} z_{10} \\ z_{20} \end{bmatrix} = \begin{bmatrix} -.5i & -3.6^{-1}-3.6^{-1}i \\ .5i & -3.6^{-1}+3.6^{-1}i \end{bmatrix} \begin{bmatrix} 2.0 \\ 3.6 \end{bmatrix} = \begin{bmatrix} -1-2i \\ -1+2i \end{bmatrix}.
$$

The solution for $y_t = BD_\lambda^t z_0$ is therefore

$$
\begin{bmatrix} y_{1t} \\ y_{2t} \end{bmatrix} = \begin{bmatrix} 1+i & 1-i \\ -1.8 & -1.8 \end{bmatrix} \begin{bmatrix} (-1-2i)\lambda_1^t \\ (-1+2i)\lambda_2^t \end{bmatrix}
$$

$$
= \begin{bmatrix} (1-3i)\lambda_1^t+(1+3i)\lambda_2^t \\ (1.8+3.6i)\lambda_1^t+(1.8-3.6i)\lambda_2^t \end{bmatrix}
$$

$$
= \begin{bmatrix} 3.162e^{-1.249i}(.906e^{1.460i})^t+3.162e^{1.249i}(.906e^{1.460i})^t \\ 4.025e^{1.107i}(.906e^{1.460i})^t+4.025e^{1.107i}(.906e^{-1.460i})^t \end{bmatrix}
$$

$$
= \begin{bmatrix} 6.324(.906)^t\cos(1.460t-1.249) \\ 8.050(.906)^t\cos(1.460t+1.107) \end{bmatrix}.
$$

Each time series is a damped cosine function with a cycle length equal to $2\pi/1.460$, or 4.30 time units, the damping factor being $(.906)^t$. The first series has smaller amplitudes than the second. It has a lag of $1.249/1.460$, or .855 time units, compared with the damped cosine function $(.906)^t \times \cos(1.460t)$. The second series has a lead of $1.107/1.460$, or .758 time units compared with the same. The first series thus lags behind the second by 1.613 time units.

In this chapter the solution to a homogeneous system of linear difference equations has been obtained and characterized in terms of the characteristic roots of the coefficient matrix A or of the solutions to the difference equations for the canonical variables. If the system is not homogeneous or if there are exogenous variables w_t affecting y_t by Bw_t, our solution $A'y_0$ will have to be added to the cumulative effect

$$Bw_t + ABw_{t-1} + \cdots + A^{t-1}Bw_1$$

of these exogenous variables. In the special case in which w_t consists only of the dummy variable equal to 1 B will be a column vector. In any case the study of the solution to the homogeneous system is useful because it forms a part of the general solution to a nonhomogeneous system to which the effect of exogenous forces can be added. Before we consider the exogenous variables and the way to manipulate some of them to achieve desired policy objectives it is important to consider the combined effect of the random disturbance

$$u_t + Au_{t-1} + \cdots + A^{t-1}u_1$$

on the solution. This is the subject of the next two chapters.

Readers interested in pursuing the subject of nonstochastic difference equations may consult Goldberg (1958), Samuelson (1948), part II and Mathematical Appendix B, and Allen (1959). More elementary treatment with economic applications can be found in Baumol (1970) and Kenkel (1974).

PROBLEMS

1. Find the characteristic roots and the right characteristic vectors of the matrix

$$A = \begin{bmatrix} 5 & -1 \\ 2 & 2 \end{bmatrix}.$$

2. Obtain the solution to the following system of difference equations in terms of the relevant characteristic roots:

$$y_{1t} = 5y_{1,t-1} - y_{2,t-1},$$

$$y_{2t} = 2y_{1,t-1} + 2y_{2,t-1}.$$

Assuming that $y_{10} = 0$ and $y_{20} = 1$, graph the solution for y_{1t}, $t = 1, 2, \ldots, 5$. Plot the solution for y_{2t}, $t = 1, 2, \ldots, 5$.

3. Prove the identity (30) by Taylor series expansions of its components.

4. Under what conditions will the second-order difference equation (3) be explosive or damped and oscillatory? State your conditions in terms of the coefficients a_1 and a_2.

5. Construct a numerical example of the second-order difference equation (3) whose solution is explosive and oscillatory. Construct an example whose solution is damped and oscillatory. Plot these solutions. Vary the initial conditions for these solutions and comment on the differences in the results.

6. Write out the characteristic equation of a 3×3 matrix $A = (a_{ij})$. Provide a numerical example and obtain the characteristic roots and associated right characteristic vectors.

7. Provide a multiplier-accelerator model that can be transformed into a second-order difference equation. Insert reasonable values for the parameters in the light of your knowledge of the economy of the United States. (If you are interested, you may actually estimate the values of the parameters by using some real data by whatever econometric method you deem appropriate.) Graph the solution by using initial conditions from actual data of the most recent periods. Discuss the nature of the solution. You may wish to consult Samuelson (1939) or Baumol (1970) for this problem.

8. Interpret the solution to Problem 2, 5, or 7 in terms of the canonical variables.

9. Let $\lambda_1 = a + bi$ and $\lambda_2 = c + di$, with $b \neq 0$. Show that λ_1 and λ_2 are conjugate complex if both $\lambda_1 + \lambda_2$ and $\lambda_1 \lambda_2$ are real.

10. Let λ_1 and λ_2 be, respectively, $a + bi$ and $a - bi$, with $b \neq 0$. If the linear combination $c_1 \lambda_1^t + c_2 \lambda_2^t$ is real for $t = 1, 2, \ldots$, show that c_1 and c_2 must be conjugate complex. *Hint.* Let $c_1 = f_1 + g_1 i$ and $c_2 = f_2 + g_2 i$. Show that $f_1 = f_2$ and $g_1 = -g_2$ if this linear combination is real for $t = 0, 1, 2, \ldots$.

11. Let national income y_t be composed only of consumption y_{1t} and investment y_{2t}. Let these components satisfy, respectively, the following consumption and investment functions:

$$y_{1t} = a_1(y_{1t} + y_{2t}) + b_1 y_{1,t-1},$$

$$y_{2t} = a_2(y_{1t} + y_{2t}) + b_2 y_{2,t-1},$$

where all the coefficients a_i and b_i are positive and the sum $a_1 + a_2$ of the marginal propensities is less than unity. Find the reduced form equations. Show that the roots of the system are real and positive and therefore that the system cannot have prolonged fluctuations. [A generalization of this proposition to the multivariate case can be found in Chow (1968).]

12. Let the investment function of the model of Problem 11 be changed to

$$y_{2t} = a_2(y_t - y_{t-1}) + b_2 y_{2,t-1},$$

the consumption function remaining the same as before. Construct a numerical example of this system which will give rise to a pair of complex roots. Graph your solution for five periods, using the most recent available data for consumption and investment (including government) expenditures as initial values.

13. By adding a new variable rewrite the nonhomogeneous system of difference equations $y_t = Ay_{t-1} + b$ as a homogeneous system. Obtain the solution of the nonhomogeneous system from the solution of the homogeneous system.

14. Show that the (right) characteristic vectors corresponding to a pair of conjugate complex roots $a + bi$ and $a - bi$ are conjugate complex. *Hint.* Follow the development in Section 2.8. Let the first characteristic vector be $\alpha + \beta i$, where α and β are real vectors. Write out the equation $A(\alpha + \beta i) = (a + bi)(\alpha + \beta i)$. Do the same for the second characteristic vector $\gamma + \delta i$.

15. Find the solution to the system

$$\begin{bmatrix} y_{1t} \\ y_{2t} \end{bmatrix} = \begin{bmatrix} 1 & -1.62 \\ 1 & -.80 \end{bmatrix} \begin{bmatrix} y_{1,t-1} \\ y_{2,t-1} \end{bmatrix}$$

by using the canonical variables, given the initial condition $y_{10} = 3.6$ and $y_{20} = 6.0$.

16. Find the solution to the system given by (35), using the third method described in Section 2.7, namely the initial conditions $y_{10} = 2.0$ and $y_{20} = 3.6$, directly without finding B or going through the canonical variables.

CHAPTER 3

Analysis of Linear Stochastic Systems: Time Domain

3.1 INTRODUCTION

In this chapter and the following we incorporate random disturbances as a part of the explanation of economic fluctuations and a part of dynamic economics. The importance of random disturbances in dynamic economics can hardly be exaggerated. All econometric models of the real world are built with explicit recognition of stochastic disturbances. It is therefore essential to incorporate this element into the theory of dynamic macroeconomics.

As early as the 1930s the Norwegian economist Ragnar Frisch emphasized the importance of random disturbances in the theory of business cycles. Partly for his contributions to stochastic dynamic economics, Frisch won the first Nobel Prize in Economic Science in 1969. The prize was shared by the Dutch economist Jan Tinbergen, who did pioneering work in econometrics and quantitative economic policy, the subject of Part 2 of this book. We quote from Frisch's classical paper [1933, pp. 197 and 202 − 203]:

> The examples we have discussed...show that when an [deterministic] economic system gives rise to oscillations, these will most frequently be damped. But in reality the cycles... are generally not damped. How can the maintenance of the swings be explained?... One way which I believe is particularly fruitful and promising is to study what would become of the solution of

a determinate dynamic system if it were exposed to a stream of erratic shocks....

Thus, by connecting the two ideas: (1) the continuous solution of a determinate dynamic system and (2) the discontinuous shocks intervening and supplying the energy that may maintain the swings—we get a theoretical setup which seems to furnish a rational interpretation of those movements which we have been accustomed to see in our statistical time data.

In the late 1950s, Irma Adelman and Frank Adelman (1959) used computer simulations to study the dynamic properties of the time series generated from an econometric model constructed by Klein and Goldberger (1955) and compared them with the properties of the economic time series actually observed in the United States and characterized by the work of the National Bureau of Economic Research. The method of computer simulations in economics is to solve the system of econometric equations for the values of the endogenous variables, given the values of the exogenous variables and random disturbances. If the random disturbances are excluded, it is called a nonstochastic simulation; otherwise, a stochastic simulation. The study by the Adelmans concluded that, without introducing random disturbances into the computer simulations, it was not possible to reproduce the dynamic characteristics of the economy by using the Klein-Goldberger model. With the random disturbances incorporated, the time series generated by the model looked remarkably similar to those actually observed from the viewpoint of measurements such as the mean time interval from peak to peak and trough to trough. In this chapter methods are provided to deduce some of these dynamic characteristics from a system of linear stochastic difference equations analytically rather than by computer simulations.

What are some of the dynamic characteristics of a stochastic time series that should be examined? A stochastic *time series* is a random function of time; that is, given time t, the time series is a random variable and the distribution of this random variable has time as a parameter. One important property is the mean, defined as a function of time, of each time series generated by a system of stochastic difference equations. It provides information on the trend of each series. The degree of variation around the mean should also be interesting. It may be measured by the standard deviation. Some measure of the length of the cycle may also be desirable. It could be the mean time interval from peak to peak or from trough to trough or the mean time interval when the time series crosses the trend, but there are other possible measures. We might wish to examine the correlations between successive time-series values. If the correlation between y_t

and y_{t-1} is high and positive, the time series may be considered slow moving or smooth; if the correlation is high and negative, the time series must oscillate. Besides their individual characteristics, we shall study the dynamic relations between time series. Does one time series grow faster than another? Does it fluctuate more than another? Is it made up of shorter cycles than another? Does it tend to lead or lag another or do the two tend to move up and down with approximately the same timing? These are some of the questions to be answered in this and the following chapter.

In this chapter we first study the simple case of a univariate first-order linear stochastic difference equation. Such an equation results from adding a random disturbance u_t to the right-hand side of the difference equation $y_t = ay_{t-1} + b$. Several useful concepts are defined. The discussion is generalized to a multivariate system of higher order linear difference equations. Canonical variables are also introduced to help characterize the solution, as in Chapter 2. Several dynamic characteristics concerning individual time series and the relations between them will be studied.

3.2 FIRST-ORDER LINEAR STOCHASTIC DIFFERENCE EQUATION

A simple example of a stochastic difference equation is

$$y_t = ay_{t-1} + b + u_t, \tag{1}$$

where y_t is a scalar and u_t is, for any integer t, a random variable with mean 0 and variance v; it is statistically independent of u_s for $t \neq s$. This model is not only simple but useful. In many applications of economic forecasting we may, as a crude approximation, predict the value of a variable by a linear function of its own value of the last period. The predictions from this model will almost always contain an error. The error, hopefully, is captured by the random variable u_t. Of course, we can make the model more complicated by introducing more lagged variables, y_{t-2}, y_{t-3}, and so on, and by specifying a more complicated random structure for u_t, but it is useful to consider the simplest case first. Besides being called a *first-order linear stochastic difference equation*, the model in (1) is also called a *first-order autoregression* or a *first-order autoregressive process*, for obvious reasons.

Going from a nonstochastic difference equation to a *stochastic* difference equation like (1), we would need different concepts for the solution. It is clear that the solution is no longer a deterministic function of time and the equation no longer specifies the time path of y_t exactly. It does, however, give the probability distribution of the solution y_t for each time t. A *time series* y_t, being a stochastic function of time, is specified if we know the joint probability distribution of any subset of the random variables, say $(y_1, \ldots, y_k)$, in the same way that a scalar random variable y is specified if we know its probability distribution. Rather than dealing with all the parameters of the joint distribution of any subset of y_t that are of interest, we choose to concentrate on the *means*, the *variances*, and the *covariances*. If the joint distribution of any subset of y_t is multivariate normal, it is well known that these parameters are sufficient to determine the distribution completely. It is easy to see from Equation 2 that if u_t are normal the joint distribution of $y_1, y_2, \ldots, y_k$ (for any k), given y_0, is multivariate normal as a consequence of the theorem that linear combinations of normal random variables are jointly normal. In this case the means, variances, and covariances contain all the information required. Even if u_t are not normal, these parameters still contain much useful information that we exploit in our study of the properties of the time series.

To derive the mean of y_t, given y_0, we may first express y_t as a function of y_0 and the past u's by successive substitutions for the lagged y's on the right-hand side of (1):

$$y_t = b + ab + a^2 b + \cdots + a^{t-1} b + a^t y_0 + u_t + a u_{t-1} + \cdots + a^{t-1} u_1$$

$$= b(1-a)^{-1}(1-a^t) + a^t y_0 + u_t + a u_{t-1} + \cdots + a^{t-1} u_1 \tag{2}$$

(provided $a \neq 1$; otherwise $bt + y_0$ replaces the sum of the first two items on the last line). Taking expectations on both sides of (2), we have

$$E y_t = \bar{y}_t = b(1-a)^{-1}(1-a^t) + a^t y_0, \tag{3}$$

because the expectation of each u_j is 0. Alternatively, we could have obtained the mean $E y_t \equiv \bar{y}_t$ by taking expectations on both sides of (1) to yield

$$\bar{y}_t = a \bar{y}_{t-1} + b \tag{4}$$

and then solving the nonstochastic difference equation (4). Note that $\bar{y}_0 \equiv E y_0 = y_0$, for y_0 is treated as a given constant. Equation 3 or 4 shows that the mean function of the time series satisfying the stochastic difference equation (1) is identical with the solution to the nonstochastic difference equation $y_t = a y_{t-1} + b$, ignoring the random disturbances.

To find the variance of y_t and covariances between y_t and y_{t-k} we can use (2) and subtract the mean $\bar{y}_t$ from both sides:

$$y_t - \bar{y}_t \equiv y_t^* = u_t + au_{t-1} + a^2 u_{t-2} + \cdots + a^{t-1}u_1, \tag{5}$$

where y_t^* denotes the deviation of y_t from its mean. Similarly, we can express y_{t-k}^* as a weighted sum of u_{t-k}, $u_{t-k-1}, \ldots$, and u_1. Taking the expectation of the product $y_t^* y_{t-k}^*$ and noting $E u_i u_j = 0$ for $i \neq j$, we get the covariance

$$\gamma(t,k) \equiv E y_t^* y_{t-k}^* = E(u_t + au_{t-1} + \cdots + a^{t-1}u_1)$$

$$\times (u_{t-k} + au_{t-k-1} + \cdots + a^{t-k-1}u_1)$$

$$= E(a^k u_{t-k}^2 + a^{k+2} u_{t-k-1}^2 + a^{k+4} u_{t-k-2}^2 + \cdots + a^{k+2(t-k-1)}u_1^2)$$

$$= va^k(1 + a^2 + a^4 + \cdots + a^{2(t-k-1)})$$

$$= v(1-a^2)^{-1}(1 - a^{2(t-k)})a^k. \tag{6}$$

The variance of y_t is a special case of (6) when $k = 0$. Expression 6 is called the *autocovariance function* $\gamma(t,k)$ of the time series y_t, for it shows the covariance between y_t and its own value lagged k periods.

3.3 COVARIANCE STATIONARY TIME SERIES AND ITS AUTOCOVARIANCE FUNCTION

If the coefficient a in (1) is smaller than 1 in absolute value, the mean function (3) and the autocovariance function (6) will, as t increases, approach, respectively, the limits

$$\lim_{t \to \infty} \bar{y}_t = \bar{y} = b(1-a)^{-1}, \tag{7}$$

$$\lim_{t \to \infty} \gamma(t,k) = \gamma_k = v(1-a^2)^{-1}a^k. \tag{8}$$

In (7) and (8) we have dropped t as a subscript or argument in the mean and autocovariance functions because they are constant through time when t is sufficiently large.

A time series is called *weakly stationary* if its mean and autocovariance function are independent of time. A time series is *stationary* if the joint distribution of any subset of observations of the series remains unchanged when the same constant is added to the time subscript of each observation. Stationarity implies weak stationarity. A time series is said to be at a *steady state* or in equilibrium if and only if it is stationary. This definition contrasts with the deterministic case in which *steady state* means unchanging values of the observations themselves. For a stochastic time series the steady state means unchanging probability distributions. A time series is *covariance stationary* if its autocovariance function is constant through time. The autocovariance, of course, is still a function of the time lag k between y_t and y_{t-k}. If the model (1) is modified by changing the intercept b into a function b_t of time, the mean function will be

$$\bar{y}_t = b_t + ab_{t-1} + \cdots + a^{t-1}b_1 + a^t y_0,$$

which may not approach a limit as t increases, but the time series will still have a constant covariance function around its mean.

Let us examine the autocovariance function γ_k given by (8) for the covariance stationary case. For $k=0$, $\gamma_0 = v(1-a^2)^{-1}$ is the variance of the time series. It is a measure of the dispersion of the time series around its mean and can be used to provide interval prediction for the time series. The variance γ_0 is larger, the larger the variance v of the random disturbance u_t; in fact, γ_0 is proportional to v. The factor of proportionality $(1-a^2)^{-1}$ depends on the coefficient a of the stochastic difference equation (1); it is larger, the larger the absolute value of a (which should not exceed 1 for the time series to remain covariance stationary). This relationship is reasonable because the larger the coefficient a, the more the past variations in the time series will be carried over to the present; if a is 0, the only variations in y_t are from u_t, for y_t will be identical with the random series u_t itself.

If a is positive and large, say it equals .9, the covariance between y_t and its lagged value y_{t-1} will be large, being .9 of the variance of y_t. The covariance between y_t and y_{t-k} will be $(.9)^k$ of the variance; it decreases with the lag k at a slow rate; that is, although the covariance between y_t and y_{t-k} weakens as the time interval between the two observations increases, it weakens only slowly and retains substantial degrees of association between the observations at different times.

It is sometimes useful to speak of the correlation or correlation coefficient, rather than the covariance, between y_t and y_{t-k} in order to free the measure of association from the choice of units of measurement. The correlation is defined as the ratio of the covariance to the product of the standard deviations of the two variables. In the case of equal standard deviations it is the ratio of the covariance to the variance of either variable. Thus the *autocorrelation function* of a time series is the ratio of its autovariance function to its variance. For the model in (1), in the covariance-stationary case, the autocorrelation function is

$$\rho_k = \frac{\gamma_k}{\gamma_0} = a^k. \tag{9}$$

If a equals .9, the correlation coefficient between y_t and y_{t-k} is $(.9)^k$; it remains substantial for observations several time units apart. This means that the time series does not change rapidly from period to period. It tends to behave like a time path with long or slow-moving cycles. The meaning of this statement is clarified in Chapter 4.

If the coefficient a in (1) is small, say equals .05, the time series behaves like a random series. If a is very small, y_t is almost equal to the random disturbance u_t. From the viewpoint of the autocorrelation function y_t and y_{t-1} have a correlation coefficient of only .05, and y_t and y_{t-2} have a correlation coefficient of only .0025.

If the coefficient a is negative but large in absolute value, say equals $-.9$, successive values of y_t will tend to be highly negatively correlated, the correlation coefficient between y_t and y_{t-1} being $-.9$. The time series is certainly not random through time but rather shows a cycle of about two time units in length. It tends to go up in one period and down in the next. The movement is just opposite to that of the case in which $a = .9$. The time series now fluctuates rapidly with high frequencies rather than moving slowly. It reveals short cycles rather than long cycles. This example also suggests the important idea that *fluctuations in the autocovariance function can give indications to fluctuations in the time series itself.*

3.4* EXPECTED TIMES BETWEEN MEAN CROSSINGS AND BETWEEN MAXIMA

Because the autocovariance function provides information on the cyclical characteristics of a time series, it can be used to calculate certain measures of the average length of the cyclical movements. These measures are important because they can be used to compare the dynamic characteristics of a model with those calculated directly from economic time-series data for the purpose of checking the validity of the model. There are several ways to measure the mean length of the cycles of a stochastic time series. Two of these measures are discussed in this section. The first is the mean time interval when the time series crosses its mean function. The second is the mean time interval between successive maxima. It is certainly possible for the time series to have several maxima before crossing the mean function so that the first measure may be expected to be greater than the second for the mean cycle length. Our discussion is confined to stationary time series. Only the results, not the proofs, of this section are required for the remainder of this book.

To obtain the expected time between successive down-crossings of the mean, that is, between crossings from an above-trend value to a below-trend value, we can find the probability at any discrete time point that the time series will change from an above-trend value to a below-trend value. If the probability is .2 per period for a time series to experience a downward movement crossing the trend, it will take an average of $1/.2$ or 5 periods to have a down-crossing. Thus the mean time for a down-crossing is the reciprocal of the probability of a down-crossing per period. To calculate the probability it is convenient to assume a 0 mean for the time series. Thus, instead of studying the time series of Equation 1, whose mean is $b(1-a)^{-1}$ in the stationary case, we shall study the deviations y_t^* from mean which satisfies the equation $y_t^* = ay_{t-1}^* + u_t$ and has a mean equal to 0. We have subtracted the mean from the time series and study the probability of down-crossing of 0 for the resulting series. The time series is assumed to be *normal*.

Let ρ_k be the autocorrelation function of a stationary time series which is normally distributed with 0 mean. In a normal time series with a constant mean covariance stationarity implies that the joint distribution of any subset of observations remains the same through time. The reason is that the mean and the autocovariance function specify the joint distribution completely. If the distribution of successive observations is invariant through time, we can consider any two successive observations, say, y_1 and y_2, and find the probability that y_1 is positive and y_2 is negative, using the bivariate normal distribution of y_1 and y_2.

$$P(y_1 > 0, y_2 < 0) = \left(2\pi\sqrt{1-\rho_1^2}\right)^{-1} \int_{-\infty}^{0} \int_{0}^{\infty} \exp\left\{-\left[2(1-\rho_1^2)\right]^{-1}\right.$$

$$\left. \times (y_1^2 - 2\rho_1 y_1 y_2 + y_2^2)\right\} dy_1\, dy_2. \tag{10}$$

In equation 10 the variance of the time series is assumed to be 1. By changing the variables we can easily check that the double integral in (10) equals the integral for the general case in which the variance γ_0 is not 1. To argue verbally, if the original time series has a variance γ_0 not equal to 1, we divide by the standard deviation or change the unit of measurement to achieve a variance of 1. This should not affect the probability of 0 down-crossing.

To integrate the right-hand side of (10) with respect to y_1 we complete the square for the exponent

$$y_1^2 - 2\rho_1 y_1 y_2 + y_2^2 = (y_1 - \rho_1 y_2)^2 + y_2^2 - \rho_1^2 y_2^2 \tag{11}$$

and rewrite the integral

$$\int_0^\infty \exp\left\{ -\left[2(1-\rho_1^2)\right]^{-1}(y_1 - \rho_1 y_2)^2 \right\} dy_1$$

$$= \sqrt{1-\rho_1^2} \int_{-\rho_1 y_2/\sqrt{1-\rho_1^2}}^\infty \exp(-\tfrac{1}{2}z^2)\,dz \tag{12}$$

by changing the variable y_1 to the standard normal deviate $z = (y_1 - \rho_1 y_2)/\sqrt{1-\rho_1^2}$, with $dy_1 = \sqrt{1-\rho_1^2}\, dz$. Substitution of (11) and (12) into (10) gives

$$P(y_1 > 0, y_2 < 0) = \frac{1}{2\pi} \int_{-\infty}^0 \int_{-\rho_1 y_2/\sqrt{1-\rho_1^2}}^\infty \exp\left[-\tfrac{1}{2}(y_2^2 + z^2) \right] dz\, dy_2. \tag{13}$$

Thus the desired result is the probability that the first random variable (whose value is denoted by z) will be larger than $-\rho_1/\sqrt{1-\rho_1^2}$ times the second random variable and that the second variable will be negative, given that the two variables are standard or unit normal and statistically independent.

If z is measured along the horizontal axis and y_2 along the vertical axis of a two-dimensional diagram, as in Figure 3.1, this event is represented by the set of points in the shaded area between the horizontal axis and the line $z = -\rho_1(1-\rho_1^2)^{-\frac{1}{2}} \cdot y_2$. The probability of this event is simply the ratio of the shaded area to the total area because, the two normal random variables being independent, the equal-probability contour lines are circles of different radii from the origin in Figure 3.1. In other words, the ratio of the angle θ (measured in radians) between the horizontal axis and the line $z = -\rho_1(1-\rho_1^2)^{-\frac{1}{2}} y_2$ to 2π will give the probability of down-crossing the mean. The angle θ is calculated by $\tan\theta = \rho_1^{-1}(1-\rho_1^2)^{\frac{1}{2}}$. By trigonometry it also satisfies $\cos\theta = \rho_1$. Therefore the probability of down-crossing the mean at any time unit is $\cos^{-1}\rho_1/2\pi$, and the expected time interval between *down-crossings of the mean for a stationary normal time series is* $2\pi/\cos^{-1}\rho_1$ *time units, where* ρ_1 *is the autocorrelation between* y_t *and* y_{t-1}. This expected time is obviously the same as the expected time for an up-crossing from a value below the mean to a value above because an up-crossing must be preceded by a down-crossing and vice versa.

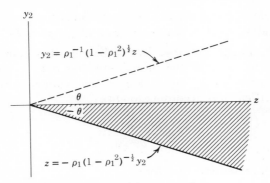

Figure 3.1 Probability of mean crossing.

Using the above formula, we find that a time series that obeys a first-order stochastic difference equation with coefficient .9 will take on the average, 13.93 time units between successive down-crossings. A coefficient of .8 would reduce the expected time between down-crossings to 9.76 time units.

A second measure of average cycle length is the expected time between relative maxima. A relative maximum occurs when the time series increases in the preceding period but decreases in the current period; that is, when $y_t > y_{t-1}$ and $y_t > y_{t+1}$. As before, we shall find the probability of this event, by assuming a multivariate normal distribution for the stationary time series with 0 mean, unit variance, and autocorrelation function ρ_k. If we define $z_1 = y_1 - y_0$ and $z_2 = y_1 - y_2$, the required probability is that of $z_1 > 0$ and $z_2 > 0$. First, the joint probability distribution of z_1 and z_2 must be derived. Being linear combinations of the normal random variables y_0, y_1, and y_2, the variables z_1 and z_2 are jointly normal. Their means are easily seen to be 0, as the means of the ys are all 0. The covariance matrix of z_1 and z_2 is

$$\text{cov}\begin{bmatrix} z_1 \\ z_2 \end{bmatrix} = \begin{bmatrix} 2-2\rho_1 & 1-2\rho_1+\rho_2 \\ 1-2\rho_1+\rho_2 & 2-2\rho_1 \end{bmatrix}. \tag{14}$$

The proof (14) is left as an exercise for Problem 6 in this chapter. We now evaluate the probability that the two random variables that satisfy a bivariate normal distribution with means 0 and covariance matrix given by (14) are both positive. We can integrate the specified bivariate density over positive values for the two variables.

To simplify this procedure we change the units for the two variables z_1 and z_2 by dividing by their standard deviation, that is, by using $x_1 = z_1/(2-2\rho_1)^{\frac{1}{2}}$ and $x_2 = z_2/(2-2\rho_1)^{\frac{1}{2}}$. The probability that both z_1 and z_2 will be positive is the same as the probability that both x_1 and x_2 will be positive. The means of x_1 and x_2 are 0; their variances are 1 and their correlation coefficient is, according to (14),

$$r = \frac{1-2\rho_1+\rho_2}{2-2\rho_1}. \tag{15}$$

The required probability is therefore

$$P(x_1 > 0, x_2 > 0) = (2\pi\sqrt{1-r^2})^{-1} \int_0^\infty \int_0^\infty \exp\left\{-[2(1-r^2)]^{-1}\right.$$
$$\left. \times (x_1^2 - 2rx_1x_2 + x_2^2)\right\} dx_1 dx_2, \tag{16}$$

which can be evaluated by the same method as in (10). By completing the square for the exponent as in (11) and by changing the variable x_1 to $z = (x_1 - rx_2)/\sqrt{1-r^2}$ as in (12) we can rewrite (16) as the following expression, analogous to (13):

$$P(x_1 > 0, x_2 > 0) = \frac{1}{2\pi} \int_0^\infty \int_{-rx_2/\sqrt{1-r^2}}^\infty \exp[-\tfrac{1}{2}(x_2^2 + z^2)] dz\, dx_2. \tag{17}$$

By the argument used for the evaluation of (13), the probability (17) is the ratio of the angle θ between the horizontal axis and the line $x_2 = -r^{-1} \times (1-r^2)^{\frac{1}{2}} z$ in the (z, x_2) diagram shown in Figure 3.2. The angle θ satisfies $\tan\theta = -r^{-1}(1-r^2)^{\frac{1}{2}}$ or, by trigonometry, $\cos\theta = -r$.

By using (15) for r we have obtained the probability $\cos^{-1}[(1-2\rho_1+\rho_2)/(2\rho_1-2)]/2\pi$ for a time series to be a maximum at any discrete time point. The *expected time between relative maxima is therefore* $2\pi/\cos^{-1}[(1-2\rho_1+\rho_2)/(2\rho_1-2)]$ *if the time series* y_t *is stationary normal and has autocorrelations* ρ_1 *and* ρ_2 *with* y_{t-1} *and* y_{t-2}, *respectively.* For a time series that satisfies (1) with $a = .9$ the expected time is 3.88, which is much shorter than the expected time between crossings of the mean.

There is a simple explanation why for many aggregate economic time series the average length of cycles, measured from peak to peak, is approximately four years. If the first difference of an annual time series is assumed to be serially independent and normally and identically distributed, the probability for the time series to be at a peak in any year, that

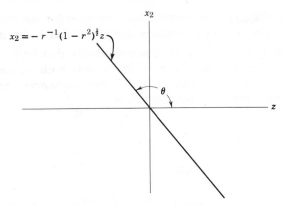

Figure 3.2   Probability of a relative maximum.

is, the probability of a positive first difference followed by a negative first difference, is one-half times one-half, or one quarter. Thus on the average a peak would be observed every four years. This result is a special case of the above formula, when $\rho_2 = \rho_1^2$ and ρ_1 is approximately equal to 1.

3.5 SYSTEMS OF LINEAR STOCHASTIC DIFFERENCE EQUATIONS

Having pointed out some important uses of the autocovariance function for a univariate time series, we shall now generalize the discussion to a situation that involves many time series generated by a system of linear stochastic difference equations:

$$y_t = A_1 y_{t-1} + A_2 y_{t-2} + \cdots + A_m y_{t-m} + b + u_t. \tag{18}$$

In (18) y_t is a multivariate or vector time series, A_i are matrices of real coefficients, b is a vector of intercepts, and u_t is a random vector with mean 0 and covariance matrix V which is statistically independent of u_s for $t \neq s$. We shall study the means of the time series as functions of time. For the study of linear models with autocorrelated disturbances u_t the reader is referred to Section 3.11.

As in the deterministic case in Chapter 2, it is convenient to rewrite (18) as a first-order system,

$$
\begin{bmatrix} y_t \\ y_{t-1} \\ \vdots \\ y_{t-m+1} \end{bmatrix} = \begin{bmatrix} A_1 & A_2 & \cdots & A_m \\ I & 0 & \cdots & 0 \\ \cdots\cdots\cdots\cdots\cdots\cdots \\ 0 & \cdots & 0 & I & 0 \end{bmatrix} \begin{bmatrix} y_{t-1} \\ y_{t-2} \\ \vdots \\ y_{t-m} \end{bmatrix} + \begin{bmatrix} b \\ 0 \\ \vdots \\ 0 \end{bmatrix} \begin{bmatrix} u_t \\ 0 \\ \vdots \\ 0 \end{bmatrix},
$$

$$(19)$$

and in a more compact form as

$$y_t = Ay_{t-1} + b + u_t. \tag{20}$$

The vector y_t in (20) stands for the vector on the left-hand side of (19), without the use of a new symbol. Similarly, b and u_t in (20) are also redefined. It is thus sufficient to study the first-order system (20).

The mean $Ey_t = \bar{y}_t$ of the time series that obeys (20) can be derived in two ways, as in the univariate time series in Equation 1. First, we eliminate the lagged y on the right-hand side of (20) successively to yield

$$y_t = b + Ab + A^2 b + \cdots + A^{t-1} b + A^t y_0 + u_t + Au_{t-1} + \cdots + A^{t-1} u_1. \tag{21}$$

By taking mathematical expectations on both sides of (21) we obtain the mean function

$$Ey_t = \bar{y}_t = b + Ab + A^2 b + \cdots + A^{t-1} b + A^t y_0. \tag{22}$$

Second, we take expectations of both sides of (20) and find that the mean function $\bar{y}_t$ satisfies the nonstochastic system of difference equation

$$\bar{y}_t = A\bar{y}_{t-1} + b. \tag{23}$$

The solution of (23) is (22), with $y_0 = Ey_0$.

The mean function (22) will reach a steady state or a vector equilibrium value as t increases if and only if all the characteristic roots of the matrix A are smaller than 1 *in absolute value.* To show this, we write the matrix A as $BD_\lambda B^{-1}$, following the discussion in Chapter 2. Here D_λ is a diagonal matrix consisting of the characteristic roots of A, and B is a matrix whose columns are the corresponding right characteristic vectors. Equation 22 can then be rewritten as

$$\bar{y}_t = b + BD_\lambda B^{-1} b + BD_\lambda^2 B^{-1} b + \cdots + BD_\lambda^{t-1} B^{-1} b + BD_\lambda^t B^{-1} y_0$$

$$= B(I + D_\lambda + D_\lambda^2 + \cdots + D_\lambda^{t-1}) B^{-1} b + BD_\lambda^t B^{-1} y_0. \tag{24}$$

Note that the sum of the matrices in parentheses is a diagonal matrix with diagonal elements

$$1 + \lambda_i + \lambda_i^2 + \cdots + \lambda_i^{t-1}.$$

As t approaches infinity, the limits of the diagonal elements are $(1 - \lambda_i)^{-1}$ if and only if the absolute values of all λ_i are smaller than 1. Under this assumption the matrix in parentheses in (24) will approach $(I - D_\lambda)^{-1}$ as the limit, and D_λ^t will approach a matrix of zeroes as t approaches infinity, making the last term of (24) vanish. The steady-state value of $\bar{y}_t$ is

$$\lim_{t \to \infty} \bar{y}_t = B(I - D_\lambda)^{-1}B^{-1}b. \tag{25}$$

An alternative, and simpler, way to write the steady state (25) is obtained by using the inverse of $B(I - D_\lambda)B^{-1} = I - A$, yielding

$$\lim_{t \to \infty} \bar{y}_t = (I - A)^{-1}b. \tag{26}$$

In connection with our proof of (25) and (26) it has been shown that if and only if all the characteristic roots of the matrix A are smaller than 1 in absolute value

$$\lim_{t \to \infty} (I + A + A^2 + \cdots + A^{t-1}) = B(I - D_\lambda)^{-1}B^{-1} = (I - A)^{-1} \tag{27}$$

and

$$\lim_{t \to \infty} A^t = 0. \tag{28}$$

The identity (27) can also be obtained in the following manner: let the finite sum of t terms $I + A + A^2 + \cdots + A^{t-1}$ be denoted by S_t. It is easy to see that $S_t - AS_t = I - A^t$, which implies that

$$S_t = I + A + A^2 + \cdots + A^{t-1} = (I - A)^{-1}(I - A^t), \tag{29}$$

itself a useful identity. By using (28) we find that the limit of (29) as t approaches infinity is $(I - A)^{-1}$.

3.6 AUTOCOVARIANCE MATRIX OF STOCHASTIC DIFFERENCE EQUATIONS

Having considered the means, we shall now study the variances and covariances of y_{it} and $y_{j,t-k}$. We have already dealt with the interpretation and uses of the autocovariance function of an individual time series. As for the covariances between y_{it} and $y_{j,t-k}$ for different values of k, they indicate the degrees of association between the two time series at different times and are called *cross-covariances*. If the covariance between y_{it} and y_{jt} is large, the two time series observed at the same time are highly associated. To eliminate the arbitrary effects of the units of measurement *cross-correlations* between y_{it} and $y_{j,t-k}$ are used and defined, as usual, as ratios of the cross-variances to the products of the respective standard deviations, that is, to $\sqrt{\operatorname{var} y_{it} \cdot \operatorname{var} y_{j,t-k}}$. The cross-covariance or the cross-correlation function also reveals the lead-lag relationships between the two time series. Its value, for example, may be highest for $k = 3$, which suggests that y_{it} and $y_{j,t-3}$ are most highly correlated or that y_{jt} perhaps leads y_{it} by about three time units.

To study the autocovariances and cross-covariances of a set of interrelated time series it is useful to define a matrix function whose $i - j$ element is $\operatorname{cov}(y_{it}, y_{j,t-k})$; that is,

$$\Gamma(t,k) = [\operatorname{cov}(y_{it}, y_{j,t-k})] = [\gamma_{ij}(t,k)]. \tag{30}$$

The diagonal elements of (30) are autocovariance functions of the individual time series. Equation 30 is a matrix generalization of the scalar function defined by (6) and is called the *autocovariance matrix* of the vector time series y_t.

For the time series that obeys system (20) the autocovariance matrix can be derived as follows. First, define $y_t^* = y_t - \bar{y}_t$ as the random vector of deviations of the time series from their means. By subtracting (23) from (20) we find that

$$y_t^* = A y_{t-1}^* + u_t, \tag{31}$$

which implies

$$y_t^* = u_t + A u_{t-1} + A^2 u_{t-2} + \cdots + A^{t-1} u_1, \tag{32}$$

where we assume that $y_0 = \bar{y}_0$ is a vector of constants, which implies $y_0^* = 0$. The autocovariance matrix $\Gamma(t,k)$ for $k \geqslant 0$ is

$$E(y_t^* y_{t-k}^{*\prime}) = E(u_t + A u_{t-1} + \cdots + A^{t-1} u_1)$$

$$\times (u_{t-k}' + u_{t-k-1}' A' + \cdots + u_1' A'^{t-k-1})$$

$$= E(A^k u_{t-k} u_{t-k}' + A^{k+1} u_{t-k-1} u_{t-k-1}' A' + \cdots + A^{t-1} u_1 u_1' A'^{t-k-1})$$

$$= A^k V + A^{k+1} V A' + \cdots + A^{t-1} V A'^{t-k-1} \qquad (k \geqslant 0). \tag{33}$$

In (33) the relations $E u_t u_s' = 0$ for $(t \neq s)$ and $E u_t u_t' = V$ have been used.

A necessary and sufficient condition for the autocovariance matrix (33) to approach a limit for any nonnegative integer k, as t approaches infinity, is that all the characteristic roots of A are smaller than one in absolute value. To prove this theorem

$$\Gamma(t,k) = E(y_t^* y_{t-k}^{*\prime}) = A^k[V + AVA' + A^2VA'^2 + \cdots + A^{t-k-1}VA'^{t-k-1}]$$

$$= A^k[V + BD_\lambda B^{-1}VB'^{-1}D_\lambda B' + \cdots + BD_\lambda^{t-k-1}B^{-1}VB'^{-1}D_\lambda^{t-k-1}B']$$

$$= A^k[B(B^{-1}VB'^{-1} + D_\lambda B^{-1}VB'^{-1}D_\lambda + \cdots$$

$$+ D_\lambda^{t-k-1}B^{-1}VB'^{-1}D_\lambda^{t-k-1})B']$$

$$= A^k[B(W + D_\lambda WD_\lambda + D_\lambda^2 WD_\lambda^2 + \cdots + D_\lambda^{t-k-1}WD_\lambda^{t-k-1})B'], \qquad (34)$$

where we have defined

$$W = B^{-1}VB'^{-1} \quad \text{or} \quad BWB' = V. \qquad (35)$$

Note that the $i-j$ element of the matrix in parentheses in the last line of (34) is

$$w_{ij} + w_{ij}\lambda_i\lambda_j + w_{ij}(\lambda_i\lambda_j)^2 + \cdots + w_{ij}(\lambda_i\lambda_j)^{t-k-1} = w_{ij}(1 - \lambda_i\lambda_j)^{-1}(1 - \lambda_i^{t-k}\lambda_j^{t-k}). \qquad (36)$$

The elements of this matrix will therefore approach limits equal to

$$w_{ij}(1 - \lambda_i\lambda_j)^{-1} \qquad (37)$$

if and only if all roots are smaller than 1 in absolute value. This proves the theorem.

While proving the important theorem of the last paragraph we have also provided an explicit expression for the autocovariance matrix $\Gamma(t,k)$ of a system of linear stochastic difference equations in the form of (20). To repeat the last line of (34), using (36) also, we write

$$\Gamma(t,k) = A^k\left\{B\left[w_{ij}(1 - \lambda_i\lambda_j)^{-1}(1 - \lambda_i^{t-k}\lambda_j^{t-k})\right]B'\right\} \qquad (k \geqslant 0), \quad (38)$$

where w_{ij} is defined by (35) and where the expression in brackets stands for the $i-j$ element of the matrix; B, as before, denotes a matrix consisting of columns of the right characteristic vectors of A. For $k=0$, in particular, we have a covariance matrix of the contemporaneous variables $y_{1t}, \ldots, y_{pt}$:

$$\Gamma(t,0) = B\left[w_{ij}(1 - \lambda_i\lambda_j)^{-1}(1 - \lambda_i^t\lambda_j^t)\right]B'. \qquad (39)$$

Equation 38 also shows that the autocovariance matrix satisfies the matrix difference equation in the variables t and k; that is,

$$\Gamma(t,k) = A\Gamma(t-1,k-1) = \cdots = A^k\Gamma(t-k,0), \quad (k \geqslant 0) \tag{40}$$

which can be used for the computation of $\Gamma(t,k)$ from $\Gamma(t-k,0)$, the latter given by (39).

If, and only if, all the roots of A are smaller than 1 in absolute value, the autocovariance matrix approaches a limit as t increases. The multivariate time series y_t is then *covariance-stationary*. The autocovariance function, at the steady state, can be written as

$$\lim_{t\to\infty} \Gamma(t,k) \equiv \Gamma_k = A^k\Gamma_0 = A^k \left\{ B\left[w_{ij}(1-\lambda_i\lambda_j)^{-1} \right]B' \right\} \quad (k \geqslant 0). \tag{41}$$

If we do not wish to use the characteristic roots and vectors of A to compute Γ_0, we can go back to the first line of (34) and use

$$\Gamma(t,0) = V + AVA' + \cdots + A^{t-1}VA'^{t-1}. \tag{42}$$

For the steady-state of a covariance-stationary vector time series, we use the infinite series

$$\lim_{t\to\infty} \Gamma(t,0) \equiv \Gamma_0 = V + AVA' + A^2VA'^2 + \cdots. \tag{43}$$

The additional terms on the right-hand side of (43) will get smaller and eventually become negligible if the roots of A are all smaller than 1 in absolute value. (43) can thus be used to compute Γ_0.

Explicit expressions have been given to evaluate the autocovariance matrix of a linear stochastic system. It suffices to evaluate the matrices for nonnegative integers k, using (40) and (42) for the general case and (41) and (43) for the stationary case. If k is negative, we can utilize the identity

$$\text{cov}(y_{it}, y_{j,t-k}) = \text{cov}(y_{j,t-k}, y_{i,t-k-(-k)}),$$

or in matrix terms

$$\Gamma(t,k) = \Gamma'(t-k, -k), \tag{44}$$

and for the steady state

$$\Gamma_k = \Gamma'_{-k}. \tag{45}$$

In particular, for the steady state

$$\gamma_{ii,k} \equiv \text{cov}(y_{it}, y_{i,t-k}) = \gamma_{ii,-k}. \tag{46}$$

Equation 46 simply means that for any individual time series at the steady state the covariance between y_{it} and $y_{i,t-k}$ is the same as the covariance between y_{it} and $y_{i,t+k}$ because the observations in either pair are k time units apart.

3.7 THE AUTOCOVARIANCE MATRIX VIA CANONICAL VARIABLES

The derivation of the autocovariance matrix of a system of linear stochastic difference equations is intimately tied up with characteristic roots and vectors and therefore not surprisingly with the canonical variables of the system. Recall the construction of canonical variables in Chapter 2 as linear combinations of the original variables y_t; that is,

$$z_t = B^{-1}y_t \quad \text{or} \quad y_t = Bz_t. \tag{47}$$

The construction can be applied whether y_t is stochastic or not.

For stochastic y_t, taking expectations on both sides of either equation in (47), we get

$$Ez_t \equiv \bar{z}_t = B^{-1}\bar{y}_t \quad \text{or} \quad \bar{y}_t = B\bar{z}_t; \tag{48}$$

that is, the mean vector of z_t consists of the same linear combinations of the elements of the mean vector of y_t as z_t itself of y_t. Also, the deviations of z_t from mean are the same linear combinations of the deviations of y_t from mean as

$$z_t^* \equiv z_t - \bar{z}_t = B^{-1}(y_t - \bar{y}_t) = B^{-1}y_t^* \quad \text{or} \quad y_t^* = Bz_t^*. \tag{49}$$

Therefore the autocovariance matrices of z_t and y_t are related by

$$\Gamma_z(t,k) = Ez_t^* z_{t-k}^{*\prime} = B^{-1}(Ey_t^* y_{t-k}^{*\prime})B^{-1\prime} = B^{-1}\Gamma_y(t,k)B^{-1\prime};$$
$$\Gamma_y(t,k) = B\Gamma_z(t,k)B'. \tag{50}$$

We could have derived the autocovariance matrix of y_t from that of z_t by using (50). The latter matrix is easy to obtain. By premultiplying (31) by B^{-1} and using (49), we have

$$z_t^* = D_\lambda z_{t-1}^* + v_t, \tag{51}$$

where v_t is defined as

$$v_t = B^{-1}u_t \tag{52}$$

and has a covariance matrix

$$W \equiv Ev_t v_t' = B^{-1}(Eu_t u_t')B^{-1\prime} = B^{-1}VB^{-1\prime} = (w_{ij}). \tag{53}$$

Each component of (51) satisfies a first-order stochastic difference equation

$$z_{it}^* = \lambda_i z_{i,t-1}^* + v_{it} = v_{it} + \lambda_i v_{i,t-1} + \lambda_i^2 v_{i,t-2} + \cdots + \lambda_i^{t-1} v_{i1}. \tag{54}$$

By taking expectations of the products $z_{it}^* z_{j,t-k}^*$ the auto- and cross-covariances of the z_{it} and z_{jt} are

$$Ez_{it}^* z_{j,t-k}^* = w_{ij} \lambda_i^k (1 + \lambda_i \lambda_j + \lambda_i^2 \lambda_j^2 + \cdots + \lambda_i^{t-1} \lambda_j^{t-1})$$

$$= w_{ij} \lambda_i^k (1 - \lambda_i \lambda_j)^{-1} (1 - \lambda_i^t \lambda_j^t). \tag{55}$$

Therefore by using (55) and (50) we obtain the autocovariance matrix of the original variable y_t as

$$\Gamma_y(t,k) = B\left[w_{ij} \lambda_i^k (1 - \lambda_i \lambda_j)^{-1} (1 - \lambda_i^t \lambda_j^t) \right] B'$$

$$= BD_\lambda^k \left[w_{ij} (1 - \lambda_i \lambda_j)^{-1} (1 - \lambda_i^t \lambda_j^t) \right] B'$$

$$= A^k B\left[w_{ij} (1 - \lambda_i \lambda_j)^{-1} (1 - \lambda_i^t \lambda_j^t) \right] B', \tag{56}$$

which is identical with the result in (38) obtained without explicit mention of the canonical variables.

3.8 A RELATION BETWEEN STOCHASTIC AND NONSTOCHASTIC TIME SERIES

In Chapter 2 it was found that the solution of a system of nonstochastic linear difference equations is a linear combination of the characteristic roots λ_i, each raised to a power t. This is so because each of the solutions $z_{jt} = z_{j0}\lambda_j^t$ of the canonical variables is a multiple of λ_j^t and the solution of the original variable y_{it} is a linear combination of the above solutions, that is, $\sum_j b_{ij} z_{j0}\lambda_j^t$.

For a stochastic time series y_{it} we can no longer think in terms of a deterministic time path for the variable itself. It would be useful to think of its autocovariance function, which is a function of the time lag k. If y_{it} is a weighted sum $\sum_j b_{ij} z_{jt}$ of the canonical variables, the covariance between y_{it} and $y_{i,t-k}$ will be a weighted sum of the auto- and cross-covariances of the canonical variables, all with lag k:

$$Ey_{it}^* y_{i,t-k}^* = \sum_{j,m} b_{ij} b_{im} Ez_{jt}^* z_{m,t-k}^*. \qquad (57)$$

Equation 57 is simply the ith diagonal element of the second line of (50), but we know, from (55), that the auto- and cross-covariances between the jth and other canonical variables lagged k time units are multiples of λ_j^k. Therefore by (57) and (55) the autocovariance function of y_{it} is a linear combination of λ_j^k:

$$
\begin{aligned}
Ey_{it}^* y_{i,t-k}^* &= \sum_{j,m} b_{ij} b_{im} w_{jm} (1 - \lambda_j \lambda_m)^{-1} (1 - \lambda_j^t \lambda_m^t) \lambda_j^k \\
&= \sum_{j,m} b_{ij} b_{im} (Ez_{jt}^* z_{mt}^*) \cdot \lambda_j^k.
\end{aligned}
\qquad (58)
$$

In the covariance stationary case the autocovariance function is reduced to

$$
\begin{aligned}
Ey_{it}^* y_{i,t-k}^* &= \sum_{j,m} b_{ij} b_{im} w_{jm} (1 - \lambda_j \lambda_m)^{-1} \cdot \lambda_j^k \\
&= \sum_{j,m} b_{ij} b_{im} (Ez_{jt}^* z_{mt}^*) \cdot \lambda_j^k.
\end{aligned}
\qquad (59)
$$

Therefore, whereas the ith deterministic time path from a system of non-stochastic difference equations is a linear combination of λ_j^t ($j = 1,...,p$), the autocovariance function of the ith time series from a system of stochastic difference equations is a linear combination of λ_j^k ($j = 1,...,p$).

3.9 PERIODICITY IN THE AUTOCOVARIANCE FUNCTION

The result just obtained with the autocovariance function has an interesting implication. If a pair of complex roots λ_1 and λ_2 exist, being, respectively, $re^{i\theta}$ and $re^{-i\theta}$, they would in the deterministic case contribute a component to the solution of the ith time path, which is a multiple of a cosine function with cycle length equal to $2\pi/\theta$, as we pointed out in Chapter 2. In the stochastic case the pair of complex roots would contribute a component to the autocovariance function of the ith time series, which is also a multiple of a cosine function with cycle length equal to

$2\pi/\theta$. By (58) or (59) the component is

$$b_{i1}\sum_m b_{im}(Ez^*_{1t}z^*_{mt})\lambda_1^k + b_{i2}\sum_m b_{im}(Ez^*_{2t}z^*_{mt})\lambda_2^k = 2q(t)r^k\cos[\theta k + \Phi(t)], \quad (60)$$

where the two coefficients of λ_1^k and λ_2^k are written as $q(t)e^{i\Phi(t)}$ and $q(t)e^{-i\Phi(t)}$.

If the absolute value r of the roots is greater than 1, the contribution to the time path in the deterministic case will be magnified by r^t through time; if the absolute value is smaller than 1, the combination will be damped, as pointed out in Section 2.5. In the stochastic case, if r is greater than 1, the contribution (60) to the autocovariance function of y_{it} will be a function of t because the terms $Ez^*_{1t}z^*_{mt}$ and $Ez^*_{2t}z^*_{mt}$ in (58) and (60) are in general functions of t. If r is smaller than 1 and all other roots are also smaller than 1 in absolute value, the contribution (60) to the auto-covariance function is constant through time because the terms $1-\lambda_1^t\lambda_m^t$ and $1\lambda_2^t\lambda_m^t$ in (58) will approach unity. From (60) this contribution will be a damped cosine function $2qr^k\cos(\theta k + \Phi)$ of the lag k. Thus the angle θ of a pair of complex roots tells something about the length of the cycles in both the deterministic and the stochastic cases. If the autocovariance function shows high degrees of association between observations of a time series $2\pi/\theta$ time units apart, we may be tempted to say that the time series itself has important cycles of lengths approximately equal to $2\pi/\theta$. The statement implies another possible measure to the length of cycles for a stochastic time series via the angle of the complex roots. The elaboration of this statement, and a general discussion of time series in terms of their cyclical components will be found in the following chapter.

3.10 A NUMERICAL EXAMPLE

This section provides a numerical example of an autocovariance matrix for a first-order system of two stochastic difference equations:

$$\begin{bmatrix} y_{1t} \\ y_{2t} \end{bmatrix} = \begin{bmatrix} 1 & 1 \\ -1.62 & -.80 \end{bmatrix}\begin{bmatrix} y_{1,t-1} \\ y_{2,t-1} \end{bmatrix} + \begin{bmatrix} u_{1t} \\ u_{2t} \end{bmatrix}, \quad (61)$$

where the covariance matrix V of u_{1t} and u_{2t} is assumed to be the identity matrix. This example is obtained by adding the random disturbances to the nonstochastic equations in (35) in Section 2.8. We shall evaluate the autocovariance matrix Γ_k by $BD_\lambda^k B^{-1}\Gamma_0$, as Section 2.8 already contains the roots λ_1 and λ_2 and the matrices B and B^{-1}.

To obtain Γ_0 we can use (43):

$$
\begin{bmatrix} 1 & 0 \\ 0 & 1 \end{bmatrix} + \begin{bmatrix} 1 & 1 \\ -1.62 & -.80 \end{bmatrix}\begin{bmatrix} 1 & -1.62 \\ 1 & -.80 \end{bmatrix}
$$

$$
+ \begin{bmatrix} 1 & 1 \\ -1.62 & -.80 \end{bmatrix}^2 \begin{bmatrix} 1 & -1.62 \\ 1 & -.80 \end{bmatrix}^2 + \cdots
$$

$$
= \begin{bmatrix} 1 & 0 \\ 0 & 1 \end{bmatrix} + \begin{bmatrix} 2 & -2.42 \\ -2.42 & 3.2644 \end{bmatrix} + \begin{bmatrix} .4244 & .00488 \\ .00488 & 1.065376 \end{bmatrix}
$$

$$
+ \begin{bmatrix} 1.499,536 & -1.551,638 \\ -1.551,638 & 1.808,285 \end{bmatrix} + \cdots .
$$

This series will converge, but not too rapidly because the absolute value of the roots is .906. We can apply (56) alternatively by first finding the covariance matrix W of the residuals in the autoregressions for the canonical variables:

$$
W = B^{-1}VB^{-1\prime} = \begin{bmatrix} -.5i & -3.6^{-1} - 3.6^{-1}i \\ .5i & -3.6^{-1} + 3.6^{-1}i \end{bmatrix}
$$

$$
\times \begin{bmatrix} -.5i & .5i \\ -3.6^{-1} - 3.6^{-1}i & -3.6^{-1} + 3.6^{-1}i \end{bmatrix}
$$

$$
= \begin{bmatrix} -.25 + .1543i & .4043 \\ .4043 & -.25 - .1543i \end{bmatrix},
$$

where the matrix B^{-1} is taken from Section 2.8. Because the residuals are complex, their variances can be complex. At the steady state the covariance matrix of the canonical variables is

$$\left[w_{ij}(1-\lambda_i\lambda_j)^{-1} \right] = \begin{bmatrix} (-.25+.1543i)(1.80-.18i)^{-1} & .4043(.18)^{-1} \\ .4043(.18)^{-1} & (-.25-.1543i)(1.80+.18i)^{-1} \end{bmatrix}$$

$$= \begin{bmatrix} .2938e^{-.5530i}(1.809e^{-.09967i})^{-1} & 2.246 \\ 2.246 & .2938e^{.5530i}(1.809e^{.09967i})^{-1} \end{bmatrix}$$

$$= \begin{bmatrix} .1624e^{-.4533i} & 2.246 \\ 2.246 & .1624e^{.4533i} \end{bmatrix}.$$

The autocovariance matrix of y_{1t} and y_{2t} is $BD_\lambda^k\left[w_{ij}(1-\lambda_i\lambda_j)^{-1} \right]B'$ by the second line of (56):

$$\begin{bmatrix} 1+i & 1-i \\ -1.8 & -1.8 \end{bmatrix} \begin{bmatrix} (.1460-.07112i)\lambda_1^k & 2.246\lambda_1^k \\ 2.246\lambda_2^k & (.1460+.07112i)\lambda_2^k \end{bmatrix} \begin{bmatrix} 1+i & -1.8 \\ 1-i & -1.8 \end{bmatrix}$$

$$= \begin{bmatrix} (4.634+.292i)\lambda_1^k & (-4.434-4.178i)\lambda_1^k \\ +(4.634-.292i)\lambda_2^k & +(-4.434+4.178i)\lambda_2^k \\ \\ (-4.434+3.908i)\lambda_1^k & (7.750-.230i)\lambda_1^k \\ +(-4.434-3.908i)\lambda_2^k & +(7.750+.230i)\lambda_2^k \end{bmatrix}.$$

The autocovariance function $\gamma_{11,k}$ of the first time series is

$$Ey_{1t}^*y_{1,t-k}^* = 4.643e^{.0629i}(.9055e^{1.460i})^k + 4.643e^{-.0629i}(.9055e^{-1.460i})^k$$

$$= 9.286(.9055)^k\cos(1.460k+.0629) \quad (k\geqslant 0).$$

The autocovariance function $\gamma_{22,k}$ of the second time series is

$$Ey_{2t}^*y_{2,t-k}^* = 7.754e^{-.0297i}(.9055e^{1.460i})^k + 7.754e^{.0297i}(.9055e^{-1.460i})^k$$

$$= 15.508(.9055)^k\cos(1.460k-.0297) \quad (k\geqslant 0).$$

By evaluating these functions at $k=0$ we obtain the variances of the two time series, 9.268 and 15.501, respectively. Both autocovariance functions are damped cosine functions with frequency $1.460/2\pi$ or cycle length of 4.3 time units. The periodicity is the same as that of the damped cosine functions showing the time paths of the deterministic series in Section 2.8.

The cross-covariance function $\gamma_{12,k}$ is

$$Ey_{1t}^{*}y_{2,t-k}^{*}=6.092e^{.756i}(.9055e^{1.460i})^{k}+6.092e^{-.756i}(.9055e^{-.1460i})^{k}$$

$$=12.184(.9055)^{k}\cos(1.460k+.756)\qquad(k\geqslant0).$$

The cross-covariance function $\gamma_{21,k}$ is

$$Ey_{2t}^{*}y_{1,t-k}^{*}=5.910e^{-.722i}(.9055e^{1.460i})^{k}+5.910e^{.722i}(.9055e^{-1.460i})^{k}$$

$$=11.820(.9055)^{k}\cos(1.460k-.722)\qquad(k\geqslant0).$$

By evaluating either of these functions at $k=0$ we obtain the contemporaneous covariance 8.87, between the two time series. The correlation coefficient between these two series is 8.87 $(9.268\times15.501)^{\frac{1}{2}}$, or .740. This fairly high correlation is not surprising in view of the way the two series are interrelated in the model (61).

3.11 TREATMENT OF AUTOCORRELATED RESIDUALS

This section provides a method of converting a system of linear stochastic difference equations with autocorrelated residuals u_t to a system with serially uncorrelated residuals. This method of conversion can be applied not only for the study of the dynamic properties in this and later chapters but also for the solution to optimal control problems from Chapter 7 on. It is therefore convenient to consider the linear model

$$y_t=Ay_{t-1}+Cx_t+b+u_t,\qquad(62)$$

which is made more general than the model (20) by including a vector of exogenous variables x_t, some of which may be subject to the control of a policy maker. Assume that a higher order system to begin with has been transformed to first-order and that lagged exogenous variables are also eliminated by identities (see Section 7.2). The following method has been used in Pagan (1973).

Consider first the case of a residual vector u_t which satisfies an autoregressive system

$$u_t = \Phi_1 u_{t-1} + \cdots + \Phi_q u_{t-q} + e_t = (\Phi_1 \cdots \Phi_q) \begin{bmatrix} u_{t-1} \\ \vdots \\ u_{t-q} \end{bmatrix} + e_t, \qquad (63)$$

where e_t is serially uncorrelated. The system (63) can be transformed into first-order by writing

$$\begin{bmatrix} u_t \\ u_{t-1} \\ \vdots \\ u_{t-q+1} \end{bmatrix} = \begin{bmatrix} \Phi_1 & \Phi_2 & \cdots & & \Phi_q \\ 1 & 0 & \cdots & & 0 \\ \cdots & \cdots & \cdots & \cdots & \cdots \\ 0 & 0 & \cdots & 1 & 0 \end{bmatrix} \begin{bmatrix} u_{t-1} \\ u_{t-2} \\ \vdots \\ u_{t-q} \end{bmatrix} + \begin{bmatrix} e_t \\ 0 \\ \vdots \\ 0 \end{bmatrix} \qquad (64)$$

and by using (64) to define the first-order system

$$v_t = F v_{t-1} + e_t^*. \qquad (65)$$

The definition of v_t also permits us to write

$$u_t = (\Phi_1 \cdots \Phi_q) v_{t-1} + e_t = \Phi v_{t-1} + e_t. \qquad (66)$$

By substituting the right-hand side of (66) for u_t in the original system (62), we have

$$y_t = A y_{t-1} + \Phi v_{t-1} + C x_t + b + e_t, \qquad (67)$$

and by combining this result with (65) we form a first-order system with serially uncorrelated residuals:

$$\begin{bmatrix} y_t \\ v_t \end{bmatrix} = \begin{bmatrix} A & \Phi \\ 0 & F \end{bmatrix} \begin{bmatrix} y_{t-1} \\ v_{t-1} \end{bmatrix} + \begin{bmatrix} C \\ 0 \end{bmatrix} x_t + \begin{bmatrix} b \\ 0 \end{bmatrix} + \begin{bmatrix} e_t \\ e_t^* \end{bmatrix}. \qquad (68)$$

If the residual vector u_t satisfies a moving average process

$$u_t = e_t + \Phi_1 e_{t-1} + \cdots + \Phi_q e_{t-q} = e_t + \Phi v_{t-1}^*, \qquad (69)$$

where the e_t are serially uncorrelated as before, we observe that v_t^* is specified by

$$
v_t^* = \begin{bmatrix} e_t \\ e_{t-1} \\ \vdots \\ e_{t-q+1} \end{bmatrix} = \begin{bmatrix} 0 & 0 & \cdots & & 0 \\ 1 & 0 & \cdots & & 0 \\ \multicolumn{5}{c}{\dotfill} \\ 0 & 0 & \cdots & 1 & 0 \end{bmatrix} \begin{bmatrix} e_{t-1} \\ e_{t-2} \\ \vdots \\ e_{t-q} \end{bmatrix} + \begin{bmatrix} e_t \\ 0 \\ \vdots \\ 0 \end{bmatrix} \tag{70}
$$

or by the first-order equation

$$
v_t^* = F^* v_{t-1}^* + e_t^*. \tag{71}
$$

Substituting the right-hand side of (69) for u_t in (62) and combining the result with (71) will form a first-order system in y_t and v_t^* with serially uncorrelated residuals.

Numerous references to the subject of time series relevant to various parts of this chapter could be cited. Among the many books on time series analysis in general perhaps mention should be made of Anderson (1971) and Box and Jenkins (1970). Both books cover much more ground than this chapter and contain in particular material on the statistical estimation of parameters in time series models, a subject that has been ignored here, and other kinds of time series models than systems of stochastic difference equations. One important early contribution to time series analysis is Quenouille (1947), but it was not written as a textbook.

On the subject of expected times between down-crossings of the mean and between relative maxima the reader may refer to Kendall (1945). These measurements were used by Johnston (1955) to study the average duration of business cycles in the United States. Howrey (1968) provided an economic application of several measurements of average cycle length, and Chow (1968) made an attempt to integrate stochastic elements into business cycle theory, using the tools of this chapter.

PROBLEMS

1. Plot the mean function of the time series $y_t = .9y_{t-1} + 1 + u_t$, given that the variance of u_t is $\frac{1}{2}$ and given $y_0 = 10$. What is the stationary value for the mean? When will this stationary value be a good approximation to the mean.

2. Plot the autocovariance function $\gamma(t, k)$, $0 \leqslant k \leqslant 3$, $t = 1, 2, 3, \ldots$, for the time series specified by Problem 1. What is the autocovariance function when the steady state is reached? How long does it take to reach the steady state, approximately? What is the autocorrelation function at the steady state?

3. Plot the stationary autocorrelation function for the time series $y_t = ay_{t-1} + u_t$, given that the variance of u_t is 1, for $a = .8$, $a = .2$, $a = 0$, $a = -.2$, and $a = -.8$. Comment on the differences.

4. What are the expected times between down-crossings of the mean for the time series specified in Problem 3, assuming u_t to be normal?

5. What are the expected times between maxima for the time series specified in Problem 3, assuming u_t to be normal?

6. Let y_0, y_1, and y_2 be jointly normal with means 0 and variances 1. The correlation between y_0 and y_1 and between y_1 and y_2 are both ρ_1; the correlation between y_0 and y_2 is ρ_2. Define $z_1 = y_1 - y_0$ and $z_2 = y_1 - y_2$. Show that the covariance matrix of z_1 and z_2 is that given by Equation 14.

7. Let y_t satisfy the second-order stochastic difference equation

$$y_t = .9y_{t-1} - .8y_{t-2} + u_t,$$

where u_t has mean 0, variance 1, and is uncorrelated with u_s for $t \neq s$. Find the autocorrelation function of y_t in the steady state, using (41) and (43). Plot this function for $k = 0$, $\pm 1, \pm 2, \ldots$, and other selected values.

8. Find the canonical variables of the equation in Problem 7.

9. Express the autocorrelation function of the time series in Problem 7 in terms of the characteristic roots and as a modified cosine function of the lag k.

10. What is the expected time interval between down-crossings of the mean for the time series specified in Problem 7?

11. What is the expected time interval between relative maxima for the time series specified in Problem 7?

12. What is your estimate of the approximate length of cycles for the time series specified in Problem 7, using the angle of the pair of complex roots?

13. Compare the answers to Problems 10, 11, and 12, or any pair of them. Comment on the differences.

14. Compare the behavior of the time series

$$y_t = .99y_{t-1} - .99y_{t-2} + u_t$$

with the time series specified in Problem 7 in terms of any one of the following:
   a.   the autocorrelation function;
   b.   the associated canonical variables and the difference equations for them;
   c.   the length of cycles in terms of the angle of the roots;
   d.   the expected time between down-crossings of mean;
   e.   the expected time between maxima.

15. Construct a time series which, when compared with that of Problem 7, will have longer cycles. Check your answer.

16. Find the autocovariance matrix of the system

$$\begin{bmatrix} y_{1t} \\ y_{2t} \end{bmatrix} = \begin{bmatrix} 1 & -1.62 \\ 1 & -.80 \end{bmatrix} \begin{bmatrix} y_{1,t-1} \\ y_{2,t-1} \end{bmatrix} + \begin{bmatrix} u_{1t} \\ u_{2t} \end{bmatrix},$$

where the covariance matrix V of u_{1t} and u_{2t} is assumed to be the identity matrix.

CHAPTER 4

Analysis of Linear Stochastic Systems: Frequency Domain

In Chapter 3 it was pointed out that periodic or approximately periodic movements in the autocovariance function may reveal periodic movements in the time series itself. In this chapter the discussion of the dynamic properties of stochastic time series stresses their periodic or cyclical movements, both individually and in relation to one another. Because the notion of the autocorrelation function is already familiar, it is used first to derive a spectral density function for the time series. This function shows the importance of different periodic movements of which the time series is composed. A time series is viewed as a weighted sum of many cosine or sine functions of time with different periodicities or frequencies. The spectral density, being a function of frequency, measures the importance of the cosine function of that frequency as a component of the time series. The analysis of a time series through its periodic components of different frequencies is said to be an analysis in the *frequency domain*.

Having obtained a spectral density function by way of the auto-covariance function of a time series, we then perform a similar operation on the cross-covariance function to obtain a cross-spectral density function to show the relation between the cyclical movements of two time series. A time series is defined as a weighted sum of many periodic or frequency components. Spectral and cross-spectral densities are defined directly in terms of the time series and the relations between two time series without using the autocovariance and cross-covariance functions. The two definitions are shown to be equivalent. The spectral and cross-spectral densities

65

of time series generated by a system of linear stochastic difference equations are also derived. Throughout the discussion, the time series is assumed to be covariance stationary, implying, in the case of linear stochastic difference equations, that all roots are smaller than 1 in absolute value. Traditionally, spectral and cross-spectral densities are defined only for covariance stationary time series. We postpone the definition and derivation of these functions for linear systems with roots greater than 1 in absolute value until Chapter 6. It will then be possible to study stochastic systems of cyclical growth by the useful concepts of spectral and cross-spectral densities.

4.1 SPECTRAL DENSITY FUNCTION VIA THE AUTOCOVARIANCE FUNCTION

As suggested in Chapter 3, periodic or approximately periodic movements in the autocovariance function reveal periodic movements of similar frequencies in the time series itself. A striking illustration is the time series $y_t = -.9y_{t-1} + u_t$, where u_t is random, with mean 0 and variance 1, and uncorrelated with u_s for $t \neq s$. We can imagine that this series moves up and down from one time unit to the next, exhibiting an approximately periodic movement of about two time units in length. It is only approximately periodic because there are random disturbances contributing to the series and because only .9 of the preceding observation repeats itself, with opposite sign, in each time unit. Accordingly, it is not precisely a periodic function of exactly two time units in cycle length but only approximately so.

What about its autocovariance function γ_k? By the method in Chapter 3 $\gamma_k = (-.9)^k \gamma_0 (k = 0, 1, 2, \ldots)$. Thus γ_k, as a function of the time lag k, satisfies the difference equation $\gamma_k = -.9\gamma_{k-1}$. This difference equation is identical to the difference equation $y_t = -.9y_{t-1}$ for the original variable y_t in time t omitting the random disturbance u_t; γ_k must behave like y_t of the nonstochastic difference equation $y_t = -.9y_{t-1}$. Note that the initial conditions γ_0 and y_0 differ: γ_0, the variance in the stochastic time series at the steady state is $(1 - .9^2)^{-1}$ for the above example; y_0 is the value of the deterministic series at time 0, whatever it may be. In any case, periodicities in the autocovariance function do reveal similar periodicities in the time series itself.

This point can be generalized to a time series generated by a system of linear stochastic difference equations such as the first component y_{1t} of the vector time series $y_t = Ay_{t-1} + u_t$. The deterministic system omitting the random disturbances satisfies $y_t = Ay_{t-1}$. The autocovariance matrix of the stochastic system satisfies the matrix difference equation $\Gamma_k = A\Gamma_{k-1}$, implying, for the first column and with $\gamma_{i1,k}$ denoting $\text{cov}(y_{it}, y_{1,t-k})$,

$$
\begin{bmatrix}
\gamma_{11,k} \\
\gamma_{21,k} \\
\vdots \\
\gamma_{p1,k}
\end{bmatrix}
=
\begin{bmatrix}
a_{11} & a_{12} & \cdots & a_{1p} \\
a_{21} & a_{22} & \cdots & a_{2p} \\
\multicolumn{4}{c}{\cdots\cdots\cdots\cdots\cdots} \\
a_{p1} & a_{p2} & \cdots & a_{pp}
\end{bmatrix}
\begin{bmatrix}
\gamma_{11,k-1} \\
\gamma_{21,k-1} \\
\vdots \\
\gamma_{p1,k-1}
\end{bmatrix}.
$$

Thus the autocovariance function for the first variable $\gamma_{11,k}$, along with other functions $\gamma_{21,k}, \ldots, \gamma_{p1,k}$, satisfies the same system of difference equations in the time lag k as the variable y_{1t} of the deterministic system $y_t = Ay_{t-1}$, along with other variables $y_{2t}, \ldots, y_{pt}$. The initial conditions of the two systems are different. The first is a set of covariances $\gamma_{11,0}$, $\gamma_{21,0} \ldots, \gamma_{p1,0}$. The second is a set of initial observations of the multivariate time series itself, $y_{10}, \ldots, y_{p0}$, but the two sets of variables should have similar cyclical characteristics.

To extract the periodic movements in the autocovariance function $\gamma_{11,k}$ we can weight it by the periodic function $\cos \omega k$ and form the weighted sum

$$
f_{11}(\omega) = \frac{1}{\pi} \sum_{k=-\infty}^{\infty} \gamma_{11,k} \cos \omega k. \tag{1}
$$

If the periodic movements of $\gamma_{11,k}$ and of $\cos \omega k$ coincide, this weighted sum will be large. In particular, imagine that $\gamma_{11,k}$ is itself a cosine function of the same frequency $\omega/2\pi$ as the weighting function. When $\gamma_{11,k}$ equals 1 (for $k=0$, $2\pi/\omega$, $4\pi/\omega, \ldots$), the weighting function will also be 1. When $\gamma_{11,k}$ equals -1 (for $k = \pi/\omega$, $3\pi/\omega$, $5\pi/\omega, \ldots$), the weighting function will be -1, giving a product of 1. On the other hand, if the periodic movements of $\gamma_{11,k}$ and of $\cos \omega k$ do not coincide, the weighted sum (1) will be small, for when $\gamma_{11,k}$ is large $\cos \omega k$ may be small or even negative and when $\gamma_{11,k}$ is small or negative $\cos \omega k$ may be large, so that the products of the two functions for different values of k will cancel out to yield a small sum. In other words, if $\gamma_{11,k}$ includes an important periodic movement of frequency $\omega/2\pi$, the weighted sum (1) will be large; otherwise the sum will be small. To the extent that periodic movements in the autocovariance function reflect similar periodic movements in the time series itself, we can say that if (1) is large the time series itself will contain important periodic movements of frequency $\omega/2\pi$ or of cycle length $2\pi/\omega$.

The function (1) is called the *power spectrum* of the time series y_{1t}. It is a weighted sum, over k, of the autocovariance function $\gamma_{11,k}$ with $\cos \omega k$ as weights. The power spectrum is a function of ω. A large value of the function at $\omega = \omega_1$ indicates that the time series contains important cycles of frequency $\omega_1 / 2\pi$ or of cycle length equal to $2\pi / \omega_1$ time units. For a discrete time series the argument ω can range from 0 to π, as the corresponding length of the cycle $2\pi / \omega$ ranges from an infinitely large value to 2. Observations in discrete time, with t an integer, cannot reveal cycles shorter than two time units or frequencies larger than $\frac{1}{2}$. If we divide the power spectrum by the variance of the time series, we will obtain the *spectral density function* of the time series. In other words, the spectral density function of y_{1t} is the weighted sum of the autocorrelation function $\rho_{11,k}$, for $k = -\infty$ to $+\infty$, using $\cos \omega k$ as weights. In the literature, however, the term spectral density function is often used more generally to denote both the power spectrum (1) and the normalized power spectrum that results from dividing (1) by the variance $\gamma_{11,0}$ of the time series. This more general use of the term spectral density is adopted in this book. If the spectral density $f_{11}(\omega)$ is normalized, the integral $\int_0^\pi f_{11}(\omega) d\omega$ will be equal to 1. That is why π^{-1} appears in the definition (1). For a proof of this normalization constant see (28) below, with ρ_k replacing γ_k and $k = 0$.

As an example, let us find the spectral density function of the time series $y_t = a y_{t-1} + u_t$. For this time series the autocorrelation function is $\rho_k = a^k$ for $k = 0, 1, 2, \ldots$, and $\rho_k = \rho_{-k}$; that is, $\rho_k = a^{|k|}$. We evaluate the weighted sum of $\rho_k = a^{|k|}$ by rewriting the spectral density function, using complex weights:

$$f(\omega) = \frac{1}{\pi} \sum_{k=-\infty}^{\infty} \rho_k \cos \omega k = \frac{1}{\pi} \left(\sum_{k=-\infty}^{\infty} \rho_k e^{-i\omega k} + i \sum_{k=-\infty}^{\infty} \rho_k \sin \omega k \right)$$

$$= \frac{1}{\pi} \sum_{k=-\infty}^{\infty} \rho_k e^{-i\omega k}, \tag{2}$$

where we have used the identities

$$e^{-i\omega k} = \cos \omega k - i \sin \omega k$$

and

$$\sum_{k=-\infty}^{\infty} \rho_k \sin \omega k = 0,$$

because $\rho_{-k} = \rho_k$ and $\sin \omega(-k) = -\sin \omega k$. The last expression in (2) serves as an alternative definition of spectral density. For $\rho_k = a^{|k|}$ it can be evaluated thus:

$$f(\omega) = \frac{1}{\pi}\left(\sum_{k=0}^{\infty} \rho_k e^{-i\omega k} + \sum_{k=0}^{\infty} \rho_{-k} e^{i\omega k} - \rho_0 \right)$$

$$= \frac{1}{\pi}\left(\sum_{k=0}^{\infty} a^k e^{-i\omega k} + \sum_{k=0}^{\infty} a^k e^{i\omega k} - 1 \right)$$

$$= \frac{1}{\pi}\left(\frac{1}{1 - ae^{-i\omega}} + \frac{1}{1 - ae^{i\omega}} - 1 \right)$$

$$= \frac{1}{\pi} \cdot \frac{2 - ae^{i\omega} - ae^{-i\omega} - (1 - ae^{-i\omega})(1 - ae^{i\omega})}{(1 - ae^{-i\omega})(1 - ae^{i\omega})}$$

$$= \frac{1}{\pi} \cdot \frac{1 - a^2}{1 + a^2 - a(e^{-i\omega} + e^{i\omega})} = \frac{1}{\pi} \cdot \frac{1 - a^2}{(1 + a^2 - 2a\cos\omega)}. \tag{3}$$

This is the spectral density function for the autoregressive time series $y_t = ay_{t-1} + u_t$.

The shapes of the spectral density functions for different values of the coefficient a can be asertained from the last expression in (3). Please see Problem 1 in this chapter. For $0 < \omega < \pi$, $\cos\omega$ is a decreasing function of ω. If a is *positive*, the denominator $(1 + a^2 - 2a\cos\omega)$ increases with ω and therefore the spectral density function $f(\omega)$ *decreases with* ω. This means that the components of the time series with the smaller frequencies $\omega/2\pi$ or the longer cycles dominate those with greater frequencies or shorter cycles. A time series satisfying $y_t = .9y_{t-1} + u_t$ is a slow moving time series in which 90 percent of its value in the last period carries over to the present. The series does not change much from one period to the next. Therefore long cycles dominate short cycles. If a is *negative*, the denominator $(1 + a^2 - 2a\cos\omega)$ decreases with ω and the spectral density function $f(\omega)$ *increases with* ω. In this case short cycles dominate long cycles. The time series shows fluctuations with high frequencies. The cycles with the highest frequency $\frac{1}{2}$ or with the shortest length of two time units are the most pronounced. In the extreme case in which $a = .99$, say, the function $f(\omega)$ has densities heavily concentrated for ω near π or for cycles with lengths close to two time units. When a equals 0, y_t is just the *random series* u_t and $f(\omega) = 1/\pi$ *is a constant*. It shows that cycles of different lengths are equally important. Because $f(\omega)$ is normalized, the area under the function from $\omega = 0$ to $\omega = \pi$ is exactly 1.

4.2 SPECTRAL DENSITY FUNCTIONS OF A BIVARIATE SYSTEM

The spectral density function of a time series that satisfies a first-order autoregression $y_t = ay_{t-1} + u_t$ does not have a maximum at ω between 0 and π. In other words, either very long or very short cycles dominate, depending on whether a is positive or negative. The spectral density function of a time series generated by $y_t = Ay_{t-1} + u_t$, A being a 2×2 matrix, can have a relative maximum at a value of ω between 0 and π. To pursue this point and to set the stage for the discussion of spectral density functions of a multivariate stochastic system we consider next the spectral density functions of time series that satisfy a first-order bivariate system of stochastic difference equations.

The autocorrelation function ρ_k of a time series generated by a bivariate system of first-order stochastic difference equations takes the form $d_1\lambda_1^k + d_2\lambda_2^k$ ($k \geqslant 0$) and $\rho_{-k} = \rho_k$, as we saw in Chapter 3. Here λ_1 and λ_2 are the characteristic roots of the system; d_1 and d_2 are suitable weights, which are conjugate complex if λ_1 and λ_2 are conjugate complex. To derive the spectral density function from this autocorrelation function we can use the result from (3) that for any λ with absolute value smaller than 1

$$\sum_{k=-\infty}^{\infty} \lambda^{|k|}e^{-i\omega k} = \frac{1-\lambda^2}{(1-\lambda e^{i\omega})(1-\lambda e^{-i\omega})} = \frac{1-\lambda^2}{1+\lambda^2 - 2\lambda\cos\omega}, \tag{4}$$

and obtain

$$f(\omega) = \frac{1}{\pi}\sum_{k=-\infty}^{\infty}(d_1\lambda_1^{|k|} + d_2\lambda_2^{|k|})e^{-i\omega k}$$

$$= \frac{1}{\pi}\left[\frac{d_1(1-\lambda_1^2)}{1+\lambda_1^2 - 2\lambda_1\cos\omega} + \frac{d_2(1-\lambda_2^2)}{1+\lambda_2^2 - 2\lambda_2\cos\omega}\right]. \tag{5}$$

This is the spectral density function of a time series that satisfies a bivariate system of stochastic difference equations.

To find out whether the time series contains any periodic components of particular importance we can try to locate a relative maximum for the spectral density function (5) for $0 < \omega < \pi$. If a relative maximum exists at ω_1, periodic movements at frequencies near $\omega_1/2\pi$ are more important than at other frequencies. Of course, how pronounced these periodic movements are depends on the sharpness of the peak. To search for a relative maximum we differentiate (5) with respect to ω and obtain

$$\frac{df(\omega)}{d\omega} = \frac{1}{\pi}\left[\frac{-d_1(1-\lambda_1^2)\cdot 2\lambda_1\sin\omega}{(1+\lambda_1^2-2\lambda_1\cos\omega)^2} + \frac{-d_2(1-\lambda_2^2)\cdot 2\lambda_2\sin\omega}{(1+\lambda_2^2-2\lambda_2\cos\omega)^2}\right] = 0.$$

Because $\sin\omega > 0$ for $0 < \omega < \pi$, the foregoing necessary condition can be restated as

$$d_1\lambda_1(1-\lambda_1^2)(1+\lambda_2^2-2\lambda_2\cos\omega)^2 + d_2\lambda_2(1-\lambda_2^2)(1+\lambda_1^2-2\lambda_1\cos\omega)^2 = 0. \quad (6)$$

Equation 6 is a quadratic equation in $\cos\omega$. In order for a relative maximum to exist a solution of (6) must take a value between 1 and -1 for $\cos\omega$.

We illustrate the spectral density (5) with the numerical example in Section 3.10. The system of stochastic difference equations is

$$\begin{bmatrix} y_{1t} \\ y_{2t} \end{bmatrix} = \begin{bmatrix} 1 & 1 \\ -1.62 & -.80 \end{bmatrix}\begin{bmatrix} y_{1,t-1} \\ y_{2,t-1} \end{bmatrix} + \begin{bmatrix} u_{1t} \\ u_{2t} \end{bmatrix}, \quad (7)$$

where the covariance matrix V of u_{1t} and u_{2t} is assumed to be the identity matrix. According to Section 3.10, the autocorrelation functions of the two time series are, respectively,

$$\rho_{11,k} = d_{11}\lambda_1^k + d_{12}\lambda_2^k$$

$$= .5010e^{.0629i}(.9055e^{1.460i})^k + .5010e^{-.0629i}(.9055e^{-1.460i})^k$$

$$= 1.0020(.9055)^k\cos(1.460k + .0629) \quad (k \geqslant 0),$$

and

$$\rho_{22,k} = d_{21}\lambda_1^k + d_{22}\lambda_2^k$$

$$= .5002e^{-.0297i}(.9055e^{1.460i})^k + .5002e^{.0297i}(.9055e^{-1.460i})^k$$

$$= 1.0004(.9055)^k\cos(1.460k - .0297) \quad (k \geqslant 0).$$

As we have pointed out, both autocorrelation functions are damped cosine functions with periodicity $2\pi/1.460$ or 4.30 time units. If either function is weighted by $\cos 1.460k$ and the sum over k from $-\infty$ to $+\infty$ is formed, we would expect the result to be larger than the sum obtained by a weighting function $\cos \omega k$ whose argument ω is quite different from 1.460. In other words, we can expect the spectral density to have a relative maximum near $\omega = 1.460$. The relative maximum will not occur at exactly 1.460 because the autocorrelation function is not strictly a periodic function, having been blurred by the factor $(.9055)^k$.

The spectral density function for the first time series is

$$f_{11}(\omega) = \frac{1}{\pi}\left[\frac{.5010e^{.0629i}(1.8090e^{-.0997i})}{.2691e^{.7328i} - 1.811e^{1.460i}\cos\omega} \right.$$

$$\left. + \frac{.5010e^{-.0629i}(1.8090e^{.0997i})}{.2691e^{-.7328i} - 1.811e^{-1.460i}\cos\omega} \right]$$

For any ω between 0 and π the two components of the sum in square brackets are complex conjugates. Therefore the spectral density is real. To find a relative maximum we use (6):

$$\lambda_1 d_{11}(1-\lambda_1^2)\left[(1+\lambda_2^2)^2 - 4(1+\lambda_2^2)\lambda_2\cos\omega + 4\lambda_2^2\cos^2\omega \right]$$

$$+\text{complex conjugate} = 0.$$

Substituting the values for λ_1, d_{11}, and λ_2 and dividing the equation by the absolute value or modulus of $\lambda_1 d_{11}(1-\lambda_1^2)$, we have

$$e^{1.4232i}(.07241e^{-1.4656i} - .9746e^{-2.1928i}\cos\omega + 3.2797e^{-2.920i}\cos^2\omega)$$

$$+\text{complex conjugate} = 0.$$

The addition of each complex term to its conjugate by (33) in Chapter 2 yields

$$.07234 - .69995\cos\omega + .24246\cos^2\omega = 0.$$

The solutions to this quadratic equation are 2.780 and .1074. The first is greater than 1, hence unacceptable. The second implies $\omega = 1.463$. This figure is close to the value 1.460 of ω in the autocorrelation function. It implies a cycle length of $2\pi/1.463$ or 4.29 time units, compared with 4.30 units for the latter value. This example shows how a periodic weighting function $e^{-i\omega k}$ or $\cos\omega k$ picks up the periodic movements in the auto-correlation function.

4.3 CROSS-SPECTRAL DENSITY FUNCTION VIA
THE CROSS-CORRELATION FUNCTION

In the same way that a spectral density function is obtained as a weighted sum of the autocorrelation function by using periodic weights the cross-spectral density function is a periodically weighted sum of the cross-correlation function. There is one difference, however, in using the periodic weights. For the autocorrelation function $\cos\omega k$ can serve as the weighting function. There is no possibility that these two functions will be out of phase because both functions reach a maximum at $k=0$ and both take equal values at k and $-k$. One only has to match the two functions by their periodicities or frequencies and does not have to be concerned with their phase differences. The cross-correlation function $\rho_{12,k}$ may not have a maximum at $k=0$ and, in general, $\rho_{12,k}$ does not equal $\rho_{12,-k}$. The function $\rho_{12,k}$ may have a maximum at $k=2$, for instance, which suggests that $y_{2,t-2}$ is more highly correlated with y_{1t} than any $y_{2,t-k}$ for $k\neq2$. This may indicate that the second time series leads the first. In any case, the weighting function applied to the cross-correlation function should be $\cos(\omega k-\psi)$ for some ψ yet to be determined rather than just $\cos\omega k$.

To measure the importance of the periodic component with frequency $\omega/2\pi$ in the cross-correlation function $\rho_{12,k}$ we use the weighting function $\cos(\omega k-\psi)$, where ψ is chosen to make the weighted sum

$$\frac{1}{\pi}\sum_{k=-\infty}^{\infty}\rho_{12,k}\cos(\omega k-\psi)=\frac{1}{\pi}\sum_{k=-\infty}^{\infty}\rho_{12,k}(\cos\omega k\cdot\cos\psi+\sin\omega k\cdot\sin\psi) \quad (8)$$

as large as possible. By maximizing (8) with respect to ψ we find the appropriate phase shift for the periodic weighting function with frequency $\omega/2\pi$. Setting the derivative of (8) with respect to ψ equal to 0, we get

$$\frac{1}{\pi}\left(-\sin\psi\sum_{k=-\infty}^{\infty}\rho_{12,k}\cos\omega k+\cos\psi\sum_{k=-\infty}^{\infty}\rho_{12,k}\sin\omega k\right)=0,$$

which implies

$$\tan\psi_{12}(\omega)=\frac{q_{12}(\omega)}{c_{12}(\omega)}, \quad (9)$$

where

$$c_{12}(\omega) = \frac{1}{\pi} \sum_{k=-\infty}^{\infty} \rho_{12,k} \cos \omega k, \qquad (10)$$

and

$$q_{12}(\omega) = \frac{1}{\pi} \sum_{k=-\infty}^{\infty} \rho_{12,k} \sin \omega k. \qquad (11)$$

The solution of (9) for $\psi_{12}(\omega)$ is that phase shift in the weighting function $\cos(\omega k - \psi)$ which will make the weighted sum (8) as large as possible. It indicates that the periodic component of frequency $\omega/2\pi$ in the auto-correlation function $\rho_{12,k}$ has a phase shift of $\psi_{12}(\omega)$. This means that, as far as the periodic components of frequency $\omega/2\pi$ are concerned, the first time series is most highly correlated with the second time series lagged $k = [\psi_{12}(\omega)]/\omega$ periods. In other words, the cyclical component of frequency $\omega/2\pi$ in the first time series lags behind the corresponding cyclical component in the second time series by $[\psi_{12}(\omega)]/\omega$ periods or by $\psi_{12}(\omega)/2\pi$ cycles; $\psi_{12}(\omega)$, defined by (9), is called the *phase-difference cross-spectral density* of the two series. It shows that the first series lags behind the second series by $\psi_{12}(\omega)/\omega$ periods in relation to the cyclical components of frequency $\omega/2\pi$. It is computed from the functions $c_{12}(\omega)$ and $q_{12}(\omega)$, defined respectively by (10) and (11). The former is called the *in-phase cross-spectral density* and the latter the *out-of-phase cross-spectral density*. If the out-of-phase cross-spectral density $q_{12}(\omega)$ is 0, the phase-difference cross-spectral density will also be 0. In this case there is no phase shift required in the weighting function $\cos \omega k$.

Having obtained the appropriate phase shift $\psi_{12}(\omega)$ for the weighting function $\cos(\omega k - \psi)$, we can use it to compute the weighted sum (8). The result will measure the correlation between the periodic components of frequency $\omega/2\pi$ in the two time series, once their phase-difference is ironed out. Substituting (9) for ψ in (8) and using the definitions (10) and (11), we obtain

$$\frac{1}{\pi} \left[\cos \psi_{12}(\omega) \cdot \sum_{k=-\infty}^{\infty} \rho_{12,k} \cos \omega k + \sin \psi_{12}(\omega) \cdot \sum_{k=-\infty}^{\infty} \rho_{12,k} \sin \omega k \right]$$

$$= \frac{c_{12}(\omega)}{\sqrt{c_{12}^2(\omega) + q_{12}^2(\omega)}} \cdot c_{12}(\omega) + \frac{q_{12}(\omega)}{\sqrt{c_{12}^2(\omega) + q_{12}^2(\omega)}} \cdot q_{12}(\omega)$$

$$= \sqrt{c_{12}^2(\omega) + q_{12}^2(\omega)} \ . \qquad (12)$$

Equation 12 is called the *cross-amplitude spectral density*. It shows the degree of association between the periodic component of frequency $\omega/2\pi$ in the first series and the corresponding component in the second series after appropriate adjustment has been made of the phase difference.

The above development of the phase-difference cross-spectral density and the cross-amplitude spectral density helps to explain and motivate the definition of the *cross-spectral* density as

$$f_{12}(\omega) = \frac{1}{\pi} \sum_{k=-\infty}^{\infty} \rho_{12,k} e^{-i\omega k}, \tag{13}$$

where the complex weighting function $e^{-i\omega k}$ is used. The resulting function $f_{12}(\omega)$ is in general complex. It can be written as

$$f_{12}(\omega) = \frac{1}{\pi} \left[\sum_{k=-\infty}^{\infty} \rho_{12,k} \cos \omega k - i \sum_{k=-\infty}^{\infty} \rho_{12,k} \sin \omega k \right]$$

$$= c_{12}(\omega) - i q_{12}(\omega)$$

$$= \sqrt{c_{12}^2(\omega) + q_{12}^2(\omega)} \cdot e^{-i\psi_{12}(\omega)}. \tag{14}$$

The *absolute value of* $f_{12}(\omega)$ *is* $\sqrt{c_{12}^2(\omega) + q_{12}^2(\omega)}$ or the *cross amplitude spectral density*. *The angle* $\psi_{12}(\omega)$ *of* $f_{12}(\omega)$ *is* $\tan^{-1}[q_{12}(\omega)/c_{12}(\omega)]$ or the *phase-difference cross-spectral density*.

For a pair of time series satisfying a first-order system of stochastic difference equations the cross-spectral density function $f_{12}(\omega)$ can be derived from the cross-correlation function

$$\rho_{12,k} = c_{11}\lambda_1^k + c_{12}\lambda_2^k \qquad (k \geq 0),$$

$$\rho_{12,-k} = \rho_{21,k} = c_{21}\lambda_1^k + c_{22}\lambda_2^k \qquad (k \geq 0).$$

Because $\rho_{12,-k} \neq \rho_{12,k}$, the derivation of the cross-spectral density function does not simplify to the same extent as that of (5) for the spectral density function. The result can be written as

$$f_{12}(\omega) = \frac{1}{\pi} \left(\sum_{k=0}^{\infty} \rho_{12,k} e^{-i\omega k} + \sum_{k=0}^{\infty} \rho_{12,-k} e^{i\omega k} - \rho_{12,0} \right)$$

$$= \frac{1}{\pi} \left[\frac{c_{11}}{1-\lambda_1 e^{-i\omega}} + \frac{c_{12}}{1-\lambda_2 e^{-i\omega}} + \frac{c_{21}}{1-\lambda_1 e^{i\omega}} + \frac{c_{22}}{1-\lambda_2 e^{i\omega}} - (c_{11}+c_{12}) \right].$$

$$\tag{15}$$

Note that $c_{11} + c_{12} = c_{21} + c_{22}$ because $\rho_{12,0} = \rho_{21,0}$. Equation 15 is in general complex. It can be used to obtain the cross-amplitude spectral density and the phase-difference spectral density as recommended in Problem 5 in this chapter. We shall study the applications of the cross-spectral density functions in Sections 4.8 and 5.8.

4.4 DECOMPOSITION OF TIME SERIES DATA INTO PERIODIC COMPONENTS

Our discussion has hinted to the possibility of decomposing a time series into periodic components of cosine and sine functions. This decomposition provides a useful way to characterize a time series. We first perform the decomposition for a *set of data* y_{1t} $(t = 1,...,N)$, which are simply N *arbitrarily given numbers*. This decomposition will be suggestive to, and enhance the understanding of, a similar decomposition for a time series that is, as we have defined it, *a random function of time*. The first decomposition amounts to manipulating sample data. The second is concerned with analyzing population values and should be distinguished from the first; it is the subject of Section 4.5.

Let N arbitrary numbers $y_{11}, y_{12}, ..., y_{1N}$ be given. Imagine fitting this set of data by a weighted sum of sine and cosine functions. These functions should have cycle lengths equal to N, $\frac{1}{2}N$, $\frac{1}{3}N,...,2$, so that the N points will cover one cycle, two cycles, and so on, of the functions. The functions are $\cos \omega_j t$ and $\sin \omega_j t$, where $\omega_j = (2\pi/N) \cdot j$ $(j = 1,...,N/2)$. For convenience let N be even or $N/2 = n$. For $j = n$ we have $\omega_n = \pi$ and therefore $\sin \omega_n t = 0$. The decomposition is

$$y_{1t} = a_{10} + \sum_{j=1}^{n-1} (a_{1j} \cos \omega_j t + b_{1j} \sin \omega_j t) + a_{1n} \cos \omega_n t. \tag{16}$$

The N unknown coefficients a_{1j} and b_{1j} can be determined by performing a least-squares regression of y_{1t} on the explanatory variables $\cos \omega_j t$ $(j = 1,...,n)$ and $\sin \omega_j t$ $(j = 1,...,n-1)$.

To obtain the least-squares estimates the sums of squares and cross-products of the explanatory variables (measured from their means) are used. Note first that the sum of each variable over t is 0 because the positive and negative values of each function cancel out for every cycle and there are an integral number of cycles. Thus

$$\sum_{t=1}^{N} \cos \omega_j t = \sum_{t=1}^{N} \sin \omega_j t = 0 \qquad \left(\omega_j = \frac{2\pi}{N} \cdot j; j = 1,...,\frac{N}{2} \right). \tag{17}$$

The explanatory variables have 0 mean and we can deal directly with the sums of squares and cross-products:

$$\sum_{t=1}^{N} \cos^2 \omega_j t = \sum_{t=1}^{N} \sin^2 \omega_j t = \frac{N}{2} \qquad (j=1,\ldots,n-1),$$

$$\sum_{t=1}^{N} \cos^2 \omega_n t = N,$$

(18)

$$\sum_{t=1}^{N} (\cos \omega_j t)(\sin \omega_k t) = 0 \qquad (j,k=1,\ldots,n-1),$$

$$\sum_{t=1}^{N} (\cos \omega_j t)(\cos \omega_k t) = \sum_{t=1}^{N} (\sin \omega_j t)(\sin \omega_k t) = 0 \qquad (j \neq k).$$

(19)

For the proofs of (18) and (19) consult Problems 6 and 7 at the end of this chapter. Equations 18 and 19 are important identities. They show that each periodic component $\cos \omega_j t$ or $\sin \omega_j t$ $(j=1,\ldots,n-1)$ has the same variance and that each component is uncorrelated with any other component.

By the method of least squares the normal equations are

$$
\begin{bmatrix}
\frac{N}{2} & 0 & \cdots & & 0 \\
0 & \frac{N}{2} & \cdots & & 0 \\
\cdots & \cdots & \cdots & \cdots & \cdots \\
0 & & \cdots & \frac{N}{2} & 0 \\
0 & & \cdots & 0 & N
\end{bmatrix}
\begin{bmatrix}
a_{11} \\
b_{11} \\
\vdots \\
b_{1,n-1} \\
a_{1n}
\end{bmatrix}
=
\begin{bmatrix}
\sum_t y_{1t} \cos \omega_1 t \\
\sum_t y_{1t} \sin \omega_1 t \\
\vdots \\
\sum_t y_{1t} \sin \omega_{n-1} t \\
\sum_t y_{1t} \cos \pi t
\end{bmatrix}
$$

(20)

Solving these equations and using a well-known formula for the intercept a_{10}, we obtain

$$a_{1j} = \frac{2}{N} \sum_{t=1}^{N} y_{1t} \cos \omega_j t, \qquad b_{1j} = \frac{2}{N} \sum_{t=1}^{N} y_{1t} \sin \omega_j t \qquad (j=1,\ldots,n-1)$$

(21)

$$a_{10} = \frac{1}{N} \sum_{t=1}^{N} y_{1t} = \bar{y}_1; \qquad a_{1n} = \frac{1}{N} \sum_{t=1}^{N} (-1)^t y_{1t}.$$

Because the number of coefficients in (16) equals the number of data points, the equation will fit the observations y_{1t} $(t = 1, \ldots, N)$ exactly without leaving any residuals. A given set of data has thus been decomposed into a weighted sum of cosine and sine functions that are mutually uncorrelated.

To measure the importance of each periodic component in y_{1t} it seems natural to use the contribution of that component to the sample variance of y_{1t}. Squaring $y_{1t} - a_{10}$ from (16), summing over t and using (19) and (18), we have

$$\sum_{t=1}^{N} (y_{1t} - \bar{y}_1)^2 = \sum_{j=1}^{n-1} \left(a_{1j}^2 \sum_{t=1}^{N} \cos^2 \omega_j t + b_{1j}^2 \sum_{t=1}^{N} \sin^2 \omega_j t \right) + a_{1n}^2 \sum_{t=1}^{N} \cos^2 \omega_n t$$

$$= \frac{N}{2} \sum_{j=1}^{n-1} \left(a_{1j}^2 + b_{1j}^2 \right) + N a_{1n}^2. \tag{22}$$

The contribution of the periodic component of frequency $\omega_j/2\pi$ to the sample variance of y_{1t} is therefore $\frac{1}{2}(a_{1j}^2 + b_{1j}^2)$.

4.5 DECOMPOSITION OF THEORETICAL TIME SERIES INTO PERIODIC COMPONENTS

The idea of the last section can be extended and modified for the decomposition of a time series that is defined as a random function of time. No probability considerations were involved in the decomposition of (16). Now let y_{1t} *be a random function of* t defined by

$$y_{1t} = \sum_{j} \left[\alpha_1(\omega_j) \cos \omega_j t + \beta_1(\omega_j) \sin \omega_j t \right], \tag{23}$$

where $\alpha_1(\omega_j)$ and $\beta_1(\omega_j)$ are random variables with 0 mean and equal variance $E[\alpha_1^2(\omega_j)] = E[\beta_1^2(\omega_j)]$ for each j and all of these random variables are mutually uncorrelated. By introducing randomness in the coefficients $\alpha_1(\omega_j)$ and $\beta_1(\omega_j)$ we attempt to model a stochastic time series. Here ω_j may range over many values between 0 and π, so that there may be many and even an infinite number of periodic components. What is the contribution of the periodic component of frequency $\omega_j/2\pi$ to the total variance of y_{1t}? To answer this question we decompose the variance of y_{1t}:

$$Ey_{1t}^2 = E\left\{ \sum_j \left[\alpha_1(\omega_j)\cos\omega_j t + \beta_1(\omega_j)\sin\omega_j t \right] \right\}^2$$

$$= \sum_j \left\{ \left[E\alpha_1^2(\omega_j) \right]\cos^2\omega_j t + \left[E\beta_1^2(\omega_j) \right]\sin^2\omega_j t \right\}$$

$$= \sum_j \left[E\alpha_1^2(\omega_j) \right] = \sum_j \text{var}[\alpha_1(\omega_j)], \qquad (24)$$

where use has been made of $E[\alpha_1^2(\omega_j)] = E[\beta_1^2(\omega_j)]$ and the 0 correlations of the random coefficients. Thus the total variance of the time series is the sum of the variances of the periodic components. The contribution of the jth component is the common variance of the coefficient of $\cos\omega_j t$ or of $\sin\omega_j t$. Please compare this contribution with the contribution given in the last section by the jth component to the sample variance of an arbitrary set of data. In the latter case we have estimated coefficients a_{1j} and b_{1j}. The common variance of these coefficients can be estimated by $\frac{1}{2}(a_{1j}^2 + b_{1j}^2)$, which is analogous to $E\alpha_1^2(\omega_j)$ of (24).

For the time series (23) let us *redefine the power spectrum $f_{11}(\omega_j)$ as the common variance $E\alpha_1^2(\omega_j)$ of the random coefficient $\alpha_1(\omega_j)$ or $\beta_1(\omega_j)$ or equivalently as the contribution of the random periodic component of frequency $\omega_j/2\pi$ to the total variance of the time series.* In order to make this definition applicable to more general situations we have to allow the variable ω to be continuous. Thus ω_j in (23), even if countably infinite, will have to be replaced by a continuous variable ω. The sum of (23) can indeed be replaced by an appropriate integral such that all covariance stationary time series can be so represented. This can be done without abandoning the essential structure of the αs and the βs as specified above. We can then speak of a power spectrum or a spectral density function $f_{11}(\omega)$ for continuous ω between 0 and π. The area $\int_a^b f_{11}(\omega)d\omega$ under the spectral density function between two points a and b measures the contribution of the periodic components with ω between a and b to the total variance of the time series. If the spectral density is normalized, the contribution to total variance is given as a fraction and the total area under the function $\int_0^\pi f_{11}(\omega)d\omega$ is equal to 1.

This definition of the spectral density function enables us to understand the concept in terms of regression theory. As pointed out in Section 4.4, we can imagine decomposing a time series into many periodic components as performing a regression of the time series on many explanatory variables which are cosine and sine functions of time. The variances of the regression coefficients are the spectral densities, but we have to think of an infinite number of explanatory variables and of regression coefficients. These coefficients are also random coefficients.

4.6* EQUIVALENCE OF TWO DEFINITIONS OF SPECTRAL DENSITY

In Sections 4.1 and 4.5 two definitions have been given to the term spectral density. The first is through the autocovariance function. The second is in terms of the variances of the periodic components of the time series itself. Both are useful ways to view the concept of spectral density and should be equivalent.

Before we show the equivalence of the two definitions, one further mathematical relationship has to be developed from the first definition. In Equation 1 the spectral density function $f_{11}(\omega)$ is derived from the autocovariance function $\gamma_{11,k}$. It is possible to invert the operation, that is, to obtain the autocovariance function from the spectral density function. The inverse operation will be derived in a more general setting by the use of complex weights $e^{-i\omega k}$ and $e^{i\omega k}$ and by defining $g(\omega) = g(-\omega) = \frac{1}{2} f(\omega)$ $(0 \le \omega \le \pi)$ to extend the domain of the spectral density function to negative values of ω. Thus we propose to invert the operation

$$g(\omega) = \frac{1}{2\pi} \sum_{k=-\infty}^{\infty} \gamma_k e^{-i\omega k} \qquad (-\pi \le \omega \le \pi). \tag{25}$$

The inversion can be achieved by taking the integral

$$\int_{-\pi}^{\pi} g(\omega) e^{i\omega k} d\omega = \int_{-\pi}^{\pi} \left(\frac{1}{2\pi} \sum_{m=-\infty}^{\infty} \gamma_m e^{-i\omega m} \right) e^{i\omega k} d\omega$$

$$= \frac{1}{2\pi} \sum_{m=-\infty}^{\infty} \gamma_m \int_{-\pi}^{\pi} e^{i(k-m)\omega} d\omega$$

$$= \frac{1}{2\pi} \gamma_k \int_{-\pi}^{\pi} e^{i(k-k)\omega} d\omega = \gamma_k, \tag{26}$$

where it was observed that the integral of

$$e^{i(k-m)\omega} = \cos(k-m)\omega + i\sin(k-m)\omega$$

over ω from $-\pi$ to π is 0 for $m \neq k$ because the positive and the negative parts from these cosine and sine functions exactly cancel out. The function $g(\omega)$ defined by (25) is the *Fourier transform of* γ_k. The *inverse Fourier transform* of $g(\omega)$ is given by (26) and equals γ_k. Thus the spectral density function and the autocovariance function are a pair of Fourier transforms to each other. One can be defined in terms of the other if we adopt our earlier definition of the spectral density function of Section 4.1. It is also known that the pair of Fourier transforms is unique. Given either one of the transforms, the other is uniquely determined.

The development of (25) and (26) implies that the *cosine transform of the autocovariance function*

$$f(\omega) = \frac{1}{\pi} \sum_{k=-\infty}^{\infty} \gamma_k \cos \omega k = 2g(\omega) \qquad (0 \leqslant \omega \leqslant \pi), \qquad (27)$$

can be inverted by the *cosine transform of the spectral density*

$$\int_0^{\pi} f(\omega) \cos \omega k \, d\omega = 2 \int_0^{\pi} g(\omega) \cos \omega k \, d\omega$$

$$= \int_{-\pi}^{\pi} g(\omega) \cos \omega k \, d\omega = \int_{-\pi}^{\pi} g(\omega) e^{i\omega k} \, d\omega = \gamma_k, \qquad (28)$$

where the integral of $g(\omega)\sin \omega k$ vanishes because $g(\omega)\sin \omega k = -g(-\omega)\sin(-\omega k)$. Hence the spectral density function and the autocovariance function are related by a pair of cosine transforms. One can be obtained from the other.

To check the equivalence of the two definitions of the spectral density function for a model like (23) we evaluate the autocovariance function of the time series defined by (23). If the result is a cosine transform (28) of the spectral density function, as specified by the second definition, the two definitions are equivalent because of the uniqueness of the cosine transforms. The autocovariance function of y_{1t} by (23) is, with $\alpha_1(\omega_j)$ and $\beta_1(\omega_j)$

abbreviated by α_j and β_j,

$$\gamma_{11,k} = Ey_{1t}y_{1t-k}$$

$$= E\left[\sum_j (\alpha_j \cos\omega_j t + \beta_j \sin\omega_j t) \right]$$

$$\times \left\{ \sum_j [\alpha_j \cos(\omega_j t - \omega_j k) + \beta_j \sin(\omega_j t - \omega_j k)] \right\}$$

$$= \sum_j E(\alpha_j \cos\omega_j t + \beta_j \sin\omega_j t)[\alpha_j \cos(\omega_j t - \omega_j k) + \beta_j \sin(\omega_j t - \omega_j k)]$$

$$= \sum_j E(\alpha_j \cos\omega_j t + \beta_j \sin\omega_j t)(\alpha_j \cos\omega_j t \cos\omega_j k + \beta_j \sin\omega_j t \cos\omega_j k$$

$$+ \alpha_j \sin\omega_j t \sin\omega_j k - \beta_j \cos\omega_j t \sin\omega_j k)$$

$$= \sum_j \left(E\alpha_j^2 \cos^2\omega_j t + E\beta_j^2 \sin^2\omega_j t \right)\cos\omega_j k$$

$$= \sum_j [E\alpha_1^2(\omega_j)]\cos\omega_j k. \tag{29}$$

In (29) we used the assumption that α_i, α_j, β_i and β_j are all uncorrelated and observed, after the fourth equality sign, that certain terms with opposite signs cancel out. Presumably, if proper care is taken to define (23) as an integral and to make ω continuous, the result in (29) will be an integral like the first part of (28), with $E\alpha_1^2(\omega)$ identified with the spectral density $f_{11}(\omega)$. Thus the two definitions of the spectral density function coincide. Although this discussion is far from being a rigorous mathematical argument, it is useful for an intuitive understanding of the subject.

4.7* A SECOND DEFINITION OF CROSS-SPECTRAL DENSITY

Besides the time series (23), let there be another time series defined by

$$y_{2t} = \sum_j [\alpha_2(\omega_j) \cos\omega_j t + \beta_2(\omega_j) \sin\omega_j t]. \tag{30}$$

The coefficients $\alpha_2(\omega_j)$ and $\beta_2(\omega_j)$ are also assumed to be random with 0 mean, uncorrelated, and with the same variance for each j. In addition, it is specified that

$$E[\alpha_1(\omega_j)\alpha_2(\omega_j)] = E[\beta_1(\omega_j)\beta_2(\omega_j)];$$

$$E[\beta_1(\omega_j)\alpha_2(\omega_j)] = -E[\alpha_1(\omega_j)\beta_2(\omega_j)]. \tag{31}$$

These are covariances because the means are all 0.

Consider the covariance between the component of frequency $\omega_j/2\pi$ in y_{1t} and the corresponding component in $y_{2,t-k}$:

$$\text{cov}\left[\alpha_1(\omega_j)\cos\omega_j t + \beta_1(\omega_j)\sin\omega_j t, \alpha_2(\omega_j)\cos(\omega_j t - \omega_j k) + \beta_2(\omega_j)\sin(\omega_j t - \omega_j k)\right]$$

$$= E\left\{\left[\alpha_1(\omega_j)\cos\omega_j t + \beta_1(\omega_j)\sin\omega_j t\right] \times \left[\alpha_2\cos\omega_j t \cdot \cos\omega_j k\right.\right.$$

$$\left.\left. + \alpha_2\sin\omega_j t \cdot \sin\omega_j k + \beta_2\sin\omega_j t \cdot \cos\omega_j k - \beta_2\cos\omega_j t\sin\omega_j k\right]\right\}$$

$$= \cos\omega_j k E\left[(\alpha_1\cos\omega_j t + \beta_1\sin\omega_j t)(\alpha_2\cos\omega_j t + \beta_2\sin\omega_j t)\right]$$

$$+ \sin\omega_j k E\left[(\alpha_1\cos\omega_j t + \beta_1\sin\omega_j t)(\alpha_2\sin\omega_j t - \beta_2\cos\omega_j t)\right] \quad (32)$$

Define the *in-phase cross-spectral density* as

$$c_{12}(\omega_j) = \text{cov}\left[\alpha_1(\omega_j)\cos\omega_j t + \beta_1(\omega_j)\sin\omega_j t, \alpha_2(\omega_j)\cos\omega_j t + \beta_2(\omega_j)\sin\omega_j t\right]$$

$$= E\left[\alpha_1(\omega_j)\alpha_2(\omega_j)\right] = E\left[\beta_1(\omega_j)\beta_2(\omega_j)\right], \quad (33)$$

where the relations in (31) have been applied. Define the *out-of-phase cross-spectral density* as

$$q_{12}(\omega_j) = \text{cov}\left[\alpha_1(\omega_j)\cos\omega_j t + \beta_1(\omega_j)\sin\omega_j t, \alpha_2(\omega_j)\sin\omega_j t - \beta_2(\omega_j)\cos\omega_j t\right]$$

$$= E\left[\beta_1(\omega_j)\alpha_2(\omega_j)\right] = -E\left[\alpha_1(\omega_j)\beta_2(\omega_j)\right]. \quad (34)$$

The covariance (32) of the two components then becomes, for any ω,

$$\cos\psi c_{12}(\omega) + \sin\psi q_{12}(\omega), \quad (35)$$

where we have defined $\psi = \omega k$.

To find the phase shift ψ that will maximize the covariance (35) we set the derivative equal to 0 to obtain

$$\tan\psi_{12}(\omega) = \frac{q_{12}(\omega)}{c_{12}(\omega)}. \quad (36)$$

The phase shift $\psi_{12}(\omega)$ so defined shows that the ω-component of the second time series leads the ω-component of the first time series by $k = \psi_{12}(\omega)/\omega$ time units; $\psi_{12}(\omega)$ is the *phase-difference cross-spectral density*. Once this optimum phase shift is found, the covariance between the ω-component of the first series and the appropriately timed ω-component of the second series is, by (35) and (36),

$$\sqrt{c_{12}^2(\omega) + q_{12}^2(\omega)}. \quad (37)$$

This is the *cross-amplitude spectral density*.

By using (36) and (37) we can define the *cross-spectral density* as the complex function

$$f_{12}(\omega) = \sqrt{c_{12}^2(\omega) + q_{12}^2(\omega)} \; e^{-i\psi_{12}(\omega)} = c_{12}(\omega) - iq_{12}(\omega). \qquad (38)$$

In sum, the cross-spectral density shows in two parts the magnitude of the covariance between corresponding periodic components in the two time series and their relative lead or lag.

To check whether this definition of cross-spectral density is consistent with the one given in Section 4.3 via the cross-covariance function an analysis similar to that in Section 4.6 can be performed. By the same argument used in (25) and (26) the cross-spectral density function and the cross-covariance functions are a pair of Fourier transforms:

$$g_{12}(\omega) = \frac{1}{2\pi} \sum_{k=-\infty}^{\infty} \gamma_{12,k} e^{-i\omega k}, \qquad (-\pi \leqslant \omega \leqslant \pi); \qquad (39)$$

where $g_{12}(\omega) = \frac{1}{2}f_{12}(\omega)$ is seen to be the complex conjugate of $g_{12}(-\omega)$ for $0 \leqslant \omega \leqslant \pi$;

$$\gamma_{12,k} = \int_{-\pi}^{\pi} g_{12}(\omega)e^{i\omega k} \, d\omega$$

$$= \int_0^{\pi} g_{12}(\omega)e^{i\omega k} \, d\omega + \int_0^{\pi} g_{12}(-\omega)e^{-i\omega k} \, d\omega. \qquad (40)$$

Equation 40 implies that $\gamma_{12,k}$ is twice the real part of either of the last two integrals. For the two definitions to agree the covariance between y_{1t} and $y_{2,t-k}$, as given by (23) and (30), should be equal to the $\gamma_{12,k}$ of (40). To evaluate the $\gamma_{12,k}$ of (40) we take the real part of

$$\int_0^{\pi} f_{12}(\omega)e^{i\omega k} \, d\omega$$

$$= \int_0^{\pi} [c_{12}(\omega) - iq_{12}(\omega)](\cos \omega k + i \sin \omega k) d\omega$$

$$= \int_0^{\pi} c_{12}(\omega)\cos(\omega k) d\omega + \int_0^{\pi} q_{12}(\omega)\sin(\omega k) d\omega + \text{imaginary part.} \qquad (41)$$

If the definition proposed in this section is to agree with the former definition $c_{12}(\omega)$ and $q_{12}(\omega)$ in the above integrals should be, respectively, $E[\alpha_1(\omega)\alpha_2(\omega)]$ and $E[\beta_1(\omega)\alpha_2(\omega)]$, as given by (33) and (34). The reader may wish to check this by doing Problem 10.

4.8 GAIN AND COHERENCE

The variances and covariances of periodic components of time series have been defined as spectral and cross-spectral densities. It is natural to utilize these concepts to study the simple regression problem for the periodic components. Consider the regression of the ω-component of the first series on the ω-component of the second. The regression coefficient is the ratio of the covariance to the variance of the second series; that is,

$$.G_{12}(\omega) = \frac{|f_{12}(\omega)|}{f_{22}(\omega)} \, , \tag{42}$$

where $|f_{12}(\omega)|$ stands for the absolute value or the modulus of the cross-spectral density function. The regression coefficient $G_{12}(\omega)$ is called the *gain*. The squared correlation coefficient between the two components is

$$R_{12}^2(\omega) = \frac{|f_{12}(\omega)|^2}{f_{11}(\omega) \cdot f_{22}(\omega)} \, , \tag{43}$$

called the *coherence*.

The gain and coherence have the same interpretations as the regression coefficient and correlation squared, respectively, as they are applied to the relationship between the corresponding periodic components of two time series. They are functions of ω. It is possible for some components of one time series to be highly dependent on the corresponding components of a second series, but not for others; for example, the slow-moving components of consumption expenditures may have large coefficients in the regressions on the corresponding components of income, but the fast-moving components of consumption expenditures may not. We can extend the regression and correlation analyses via the periodic components to more than two variables. Multiple regression and partial correlations can be defined in a fairly straightforward manner, but we shall not pursue them in this book.

4.9 SPECTRAL DENSITY MATRIX OF STOCHASTIC DIFFERENCE EQUATIONS

Having defined and interpreted spectral and cross-spectral densities and derived them for univariate and bivariate systems of stochastic difference equations of the first-order, we now derive them for a multivariate system.

Define the *spectral density matrix* for a vector time series y_t as

$$F(\omega) = [f_{ij}(\omega)].$$

(44)

The diagonal elements of $F(\omega)$ are the spectral densities and the off-diagonal elements are the cross-spectral densities. The spectral-density matrix is related to the autocovariance matrix $\Gamma_k = \Gamma'_{-k}$ by

$$F(\omega) = \frac{1}{\pi} \sum_{k=-\infty}^{\infty} \Gamma_k e^{-i\omega k}$$

$$= \frac{1}{\pi} \left(\sum_{k=0}^{\infty} \Gamma_k e^{-i\omega k} + \sum_{k=0}^{\infty} \Gamma'_k e^{i\omega k} - \Gamma_0 \right).$$

(45)

In each term of a sum the weight $e^{-i\omega k}$ is to multiply every element of the autocovariance matrix Γ_k or Γ'_k. A sum of these matrices is obtained by adding corresponding elements as usual.

By using the results on Γ_k from Chapter 3 we can derive the spectral density matrix for the multivariate stochastic system

$$y_t = Ay_{t-1} + b + u_t,$$

(46)

where u_t has a covariance matrix V and is serially uncorrelated. Recall the matrix B that consists of the right characteristic vectors of A and the diagonal matrix D_λ that consists of the corresponding characteristic roots, with $A = BD_\lambda B^{-1}$. The autocovariance matrix has been found to be $\Gamma_k = BD_\lambda^k B^{-1}\Gamma_0$ for $k \geq 0$, where $\Gamma_0 = B\{w_{ij}(1 - \lambda_i\lambda_j)^{-1}\}B'$, $(w_{ij}) = W = B^{-1}VB^{-1'}$ being the covariance matrix of the residuals of the autoregressions for the canonical variables. The canonical variables were defined by $z_t = B^{-1}y_t$. Their autocovariance matrix is

$$\Gamma_k^z = D_\lambda^k \Gamma_0^z = D_\lambda^k \left[w_{ij}(1 - \lambda_i\lambda_j)^{-1} \right], \qquad (k \geq 0),$$

(47)

and the autocovariance matrix of y_t is

$$\Gamma_k = BD_\lambda^k \Gamma_0^z B' = BD_\lambda^k B^{-1}\Gamma_0, \qquad (k \geq 0).$$

(48)

By applying the first part of (48) in (45) and interchanging summation and matrix multiplications in each of the two sums we have

$$F(\omega) = \frac{1}{\pi}\left[B\left(\sum_{k=0}^{\infty} D_{\lambda}^{k}\Gamma_{0}^{z}e^{-i\omega k}\right)B' + B\left(\sum_{k=0}^{\infty}\Gamma_{0}^{z}D_{\lambda}^{k}e^{i\omega k}\right)B' - \Gamma_{0}\right]$$

$$= \frac{1}{\pi}B\left(\sum_{k=0}^{\infty}D_{\lambda}^{k}\Gamma_{0}^{z}e^{-i\omega k} + \sum_{k=0}^{\infty}\Gamma_{0}^{z}D_{\lambda}^{k}e^{i\omega k} - \Gamma_{0}^{z}\right)B'. \qquad (49)$$

The $i - j$ element of the matrix in parentheses in the second line of (49) is

$$\sum_{k=0}^{\infty}\lambda_{i}^{k}\gamma_{ij,0}^{z}e^{-i\omega k} + \sum_{k=0}^{\infty}\gamma_{ij,0}^{z}\lambda_{j}^{k}e^{i\omega k} - \gamma_{ij,0}^{z}$$

$$= \gamma_{ij,0}^{z}\left(\frac{1}{1-\lambda_{i}e^{-i\omega}} + \frac{1}{1-\lambda_{j}e^{i\omega}} - 1\right)$$

$$= \frac{w_{ij}}{1-\lambda_{i}\lambda_{j}}\left[\frac{1-\lambda_{i}\lambda_{j}}{\left(1-\lambda_{i}e^{-i\omega}\right)\left(1-\lambda_{j}e^{i\omega}\right)}\right]$$

$$= \frac{w_{ij}}{\left(1-\lambda_{i}e^{-i\omega}\right)\left(1-\lambda_{j}e^{i\omega}\right)}. \qquad (50)$$

The spectral density matrix of y_t is therefore

$$F(\omega) = \frac{1}{\pi}B\left[\frac{w_{ij}}{\left(1-\lambda_{i}e^{-i\omega}\right)\left(1-\lambda_{j}e^{i\omega}\right)}\right]B'. \qquad (51)$$

4.10* SPECTRAL DENSITY MATRIX IN TERMS OF CANONICAL VARIABLES

The derivation of the spectral density matrix in (51) is based on and intimately related to the idea of canonical variables. To make the idea explicit recall that the autocovariance matrix Γ_k^z of the canonical variables is related to the autocovariance matrix Γ_k of y_t by

$$\Gamma_k = B\Gamma_k^z B'. \qquad (52)$$

Because the spectral density matrix is defined in (45) as a weighted sum of Γ_k and because summation and matrix multiplications can be interchanged, the spectral density matrices of the two sets of variables are related by

$$F(\omega) = BF^z(\omega)B'. \tag{53}$$

Once the spectral density matrix $F^z(\omega)$ of the canonical variables is found the required result for $F(\omega)$ can be obtained by matrix multiplications as specified by (53).

To derive the cross-spectral density for two canonical variables z_{it} and z_{jt} we first obtain their cross-covariance function:

$$\gamma_{ij,k}^z = Ez_{it}z_{j,t-k}.$$

$$= E(v_{it} + \lambda_i v_{i,t-1} + \lambda_i^2 v_{i,t-2} + \cdots)(v_{j,t-k} + \lambda_j v_{j,t-k-1} + \cdots)$$

$$= (Ev_{it}v_{jt})(\lambda_i^k + \lambda_i^{k+1}\lambda_j + \lambda_i^{k+2}\lambda_j^2 + \cdots)$$

$$= \frac{w_{ij}}{1 - \lambda_i\lambda_j} \cdot \lambda_i^k$$

$$= \gamma_{ij,0}^z \lambda_i^k, \qquad (k \geqslant 0);$$

$$\gamma_{ij,-k}^z = \gamma_{ji,k}^z = \gamma_{ij,0}^z \lambda_j^k, \qquad (k \geqslant 0). \tag{54}$$

The Fourier transform of (54) can be performed as it was in (50) to give the cross-spectral density

$$f_{ij}^z(\omega) = \frac{1}{\pi} \sum_{k=-\infty}^{\infty} \gamma_{ij,k}^z e^{-i\omega k} = \frac{1}{\pi} \cdot \frac{w_{ij}}{(1 - \lambda_i e^{-i\omega})(1 - \lambda_j e^{i\omega})} \tag{55}$$

for the ith and jth canonical variables. Application of (55) to (53) yields the desired result (51).

Equation 55 gives the spectral or cross-spectral density functions $f_{ij}^z(\omega)$ of the canonical variables. If the spectral density function of the ith canonical variable is *normalized* and denoted by $g_i(\omega)$ [not the $g(\omega)$ of (25)], it is

$$g_i(\omega) = \frac{1}{\gamma_{ii,0}^z} f_{ii}^z(\omega) = \frac{1}{\pi} \frac{1 - \lambda_i^2}{(1 - \lambda_i e^{-i\omega})(1 - \lambda_i e^{i\omega})}$$

$$= \frac{1}{\pi} \frac{1 - \lambda_i^2}{(1 + \lambda_i^2 - 2\lambda_i \cos\omega)}. \tag{56}$$

This is the same result that we have obtained for the first-order autoregression in Section 4.1. If a root λ_i is real, the spectral density of the corresponding canonical variable is real. If λ_i is positive, the spectral density is a decreasing function of ω; if λ_i is negative, it is an increasing function that shows the relative importance of short cycles. It would be interesting to inquire into the relation between the spectral density of an original variable y_{it} and the spectral densities of the canonical variables.

By (51) the spectral density of y_{it} is a quadratic form in the ith row $b_{i.}$ of B:

$$f_{ii}(\omega) = b_{i.}F^z(\omega)b_{i.}' = \tfrac{1}{2}b_{i.}[F^z(\omega) + F^z(\omega)']b_{i.}', \qquad (57)$$

where the $j - m$ element of the matrix in square brackets is, by (50),

$$f_{jm}^z(\omega) + f_{mj}^z(\omega)$$

$$= \frac{1}{\pi}\gamma_{jm,0}^z\left(\frac{1}{1-\lambda_je^{-i\omega}} + \frac{1}{1-\lambda_me^{i\omega}} + \frac{1}{1-\lambda_me^{-i\omega}} + \frac{1}{1-\lambda_je^{i\omega}} - 2\right)$$

$$= \gamma_{jm,0}^z\big[\, g_j(\omega) + g_m(\omega)\big]. \qquad (58)$$

Denoting by D_g the diagonal matrix with the normalized spectral densities $g_j(\omega)$ of the canonical variables on the diagonal, we can combine (57) and (58) to write

$$f_{ii}(\omega) = \tfrac{1}{2}b_{i.}D_g\Gamma_0^zb_{i.}' + \tfrac{1}{2}b_{i.}\Gamma_0^zD_gb_{i.}'.$$

$$= b_{i.}D_g\Gamma_0^zb_{i.}' = \sum_j\sum_m b_{ij}b_{im}\gamma_{jm,0}^z g_j(\omega). \qquad (59)$$

The result is that the spectral density function of y_{it} is a linear combination of the normalized spectral densities $g_j(\omega)$ of the canonical variables. In fact, this is the same linear combination as the autocovariance function $\gamma_{ii,k}$ of y_{it} is of the autocorrelation functions λ_j^k of the canonical variables. Recall the latter relationship given by (59) in Chapter 3:

$$\gamma_{ii,k} = \sum_j\sum_m b_{ij}b_{im}\gamma_{jm,0}^z\lambda_j^k. \qquad (60)$$

By taking the Fourier transform on both sides of (60) we could have obtained (59) in this chapter directly. Thus we see that the spectral density function of a time series generated by a system of linear stochastic difference equations is a linear combination of the spectral densities of first-order univariate autoregressions with the form (56).

If all the roots in a linear deterministic system are real and positive, the time paths cannot have prolonged oscillations. An interesting question is whether the time series generated by a linear stochastic system can have pronounced cycles in the form of a peak in the spectral density function if all roots are real and positive. The answer is yes because a linear combination of the spectral densities $g_j(\omega)$ formed by real and positive roots λ_j can have a relative maximum for $0 < \omega < \pi$. Consider the following example of a bivariate first-order system.

$$\lambda_1 = .1 \quad \lambda_2 = .9$$

$$W = \begin{bmatrix} 1 & .8 \\ .8 & 1 \end{bmatrix} \tag{61}$$

$$(b_{11} \ b_{12}) = (1 \quad -.01)$$

The first time series y_{1t} will have a spectral density equal to π^{-1} times

$$[1 \quad -.01] \begin{bmatrix} g_1(\omega) & 0 \\ 0 & g_2(\omega) \end{bmatrix} \begin{bmatrix} \dfrac{1}{.99} & \dfrac{.8}{.91} \\ \dfrac{.8}{.91} & \dfrac{1}{.19} \end{bmatrix} \begin{bmatrix} 1 \\ -.01 \end{bmatrix}$$

$$= \frac{.9913}{1.01 - .2\cos\omega} - \frac{.001570}{1.81 - 1.8\cos\omega}. \tag{62}$$

Selected values of the function (62) are

$\omega/2\pi$	0	1/32	1/16	2/16	3/16	4/16	5/16	6/16	7/16	8/16
$\pi f_{11}(\omega)$	1.067	1.183	1.191	1.138	1.061	.981	.912	.860	.829	.819

The peak of the spectral density at $\omega/2\pi$ approximately equal to $\frac{1}{20}$ can be discerned. This example shows how much the incorporation of random disturbances can alter the conclusion from a deterministic theory.

4.11* A NOTE ON SPECTRAL ANALYSIS

The concepts of spectral and cross-spectral densities are useful for studying cyclical properties of econometric models. In addition, and more prevalently, they are useful for extracting cyclical properties of observed economic time series without the intervention of an econometric model.

Thus, assuming that certain time series are covariance stationary, or approximately so, probably after adjustments for trends by fitting trend functions or by taking first differences and the like, we may wish to estimate the spectral and cross-spectral densities directly from the data. This process of estimation is called *spectral analysis* of time series data. No econometric models in the form of systems of interdependent dynamic equations are needed. The only hypothesis employed is concerned with the form of the trend, if trend is taken out, with the appropriate transformation of variables such as taking logarithm and with the smoothness of the spectral density function. Little theory is required. Spectral analysis can also be applied, not to economic data directly but to data generated from stochastic simulations of an econometric model estimated from economic data. This will enable us to study certain dynamic properties of the econometric model. Such an undertaking differs from spectral analysis of raw data and also from the analytical derivation of spectral properties from a model as we have done in this chapter.

Given time series data for a variable y_{1t}, $(t = 1, \ldots, N)$, it seems reasonable to utilize the periodic regression coefficients a_{1j} and b_{1j} of (21) in Section 4.4 to estimate the spectral density at ω_j. The spectral density at ω_j is the common variance of the coefficients a_{1j} and b_{1j}. There are $N/2$ of these variances that can be calculated from a sample of size N. Each variance has to cover a range of ω values equal in width to $2\pi/N$ if ω is a continuous variable from 0 to π. Therefore a possible estimate of the spectral density at ω_j is the variance $\frac{1}{2}(a_{1j}^2 + b_{1j}^2)$ divided by the width $2\pi/N$:

$$
\begin{aligned}
I(\omega_j) &= \tfrac{1}{2}\left(a_{1j}^2 + b_{1j}^2\right)\frac{N}{2\pi} \\[2mm]
&= \frac{1}{\pi N}\left[\left(\sum_{t=1}^{N} y_{1t}\cos\omega_j t\right)^2 + \left(\sum_{t=1}^{N} y_{1t}\sin\omega_j t\right)^2\right] \\[2mm]
&= \frac{1}{\pi N}\left\{\sum_{t=1}^{N} y_{1t}y_{1t}(\cos\omega_j t\cos\omega_j t + \sin\omega_j t\sin\omega_j t)\right. \\[2mm]
&\quad + \sum_{t=1}^{N-1} y_{1t}y_{1,t+1}[\cos\omega_j t\cos(\omega_j t+\omega_j) + \sin\omega_j t\sin(\omega_j t+\omega_j)] \\[2mm]
&\quad \left. + \sum_{t=1}^{N-2} y_{1t}y_{1,t+2}[\cos\omega_j t\cos(\omega_j t+2\omega_j) + \sin\omega_j t\sin(\omega_j t+2\omega_j)] + \cdots\right\} \\[2mm]
&= \frac{1}{\pi}\sum_{k=-N+1}^{N-1} c_{11,k}\cdot\cos\omega_j k,
\end{aligned}
\tag{63}
$$

where $c_{11,k}$ is the sample autocovariance

$$c_{11,k} = c_{11,-k} = \frac{1}{N} \sum_{t=1}^{N-k} y_{1t} y_{1,t+k}. \tag{64}$$

Thus the use of the arithmetic mean of the squares of the regression coefficients a_{1j} and b_{1j} amounts to the same thing as the application of a cosine transform to the sample autocovariance $c_{11,k}$.

What is the sampling property of the estimate of spectral density given by (63)? If y_{1t} is normally and independently distributed, each regression coefficient a_{1j} and b_{1j} computed by the method of least squares, using (20), will also be normally and independently distributed. Each will have a variance equal to $2/N$ times the variance of y_{1t}, according to least-squares theory. The variables $a_{1j}\sqrt{N/2}$ and $b_{1j}\sqrt{N/2}$ are independently normal and have variance equal to the variance of y_{1t}. The estimate $I(\omega_j)$ in (63) can be written as

$$\left[\left(a_{ij}\sqrt{\frac{N}{2}}\right)^2 + \left(b_{2j}\sqrt{\frac{N}{2}}\right)^2\right](2\pi)^{-1}.$$

Except for the factor $(2\pi)^{-1}$, it is the sum of the squares of two independent normal random variables with a common variance. Therefore, except for a factor, it is distributed as a chi-square variable with two degrees of freedom. The main point is that, no matter how large the sample size N, the estimate $I(\omega_j)$ for each ω_j will be distributed as a chi-square random variable with two degrees of freedom and thus will not converge to a constant. This means that a consistent estimate of the spectral density cannot be obtained by using $I(\omega_j)$.

In order to obtain a consistent estimate of the spectral density, it is required to modify (63) by applying a set of weights w_k to the sample autocovariance functions $c_{11,k}$ before performing the cosine transform:

$$\hat{f}_{11}(\omega) = \frac{1}{\pi} \sum_{k=-\infty}^{\infty} w_k c_{11,k} \cdot \cos \omega k. \tag{65}$$

The weighting function w_k is called the *lag window*. Specialists in spectral analysis have proposed various windows and investigated their properties. A simple one is the Bartlett window. For some m smaller than N this window is

$$
w_k = \begin{cases} 1 - \dfrac{|k|}{m}, & 0 \leqslant |k| \leqslant m, \\ 0, & \text{otherwise.} \end{cases}
$$

It ignores those sample autocovariances $c_{11,k}$ with lag k larger than m that are computed from fewer and fewer observations as k increases in absolute value. For this particular window the weights w_k are linearly declining with the absolute value of k. Others have proposed windows that are different decreasing functions of $|k|$, but there is no need for us to pursue the subject here. These windows can also be applied to the sample cross-covariance functions before the appropriate cosine and sine transforms are formed for the consistent estimation of cross-spectral densities. This process is termed *cross-spectral analysis*. Computer programs are available to perform spectral analysis and cross-spectral analysis of time series data.

In this chapter our main concern has been to introduce the important tools of spectral and cross-spectral densities for the analysis of stochastic dynamic systems and to derive these functions for linear systems of stochastic difference equations. We turn to some applications of these tools in economics in Chapter 5.

Readers interested in studying time series analysis from the frequency point of view may consult Anderson (1971), Box and Jenkins (1970), Cox and Miller (1965), Dhrymes (1970), Hannan (1960), Kendall and Stuart (1966), and Whittle (1963). Blackman and Tukey (1958) and Parzen (1961) contain early contributions to spectral analysis. On applications of spectral analysis in economics important references include Fishman (1969), Granger and Hatanaka (1964), Granger and Morgenstern (1970), and Nerlove (1964), as well as many other studies of special economic problems. Part of the material in this chapter is based on Chow (1968).

PROBLEMS

1. Plot the spectral densities of the time series $y_t = a y_{t-1} + u_t$ for $a = .8$, $.2$, $.0, -.2$ and $-.8$.

2. Find and plot the spectral densities of the time series

$$
y_t = .9 y_{t-1} - .8 y_{t-2} + u_t.
$$

3. Consider y_t and y_{t-1} as two separate time series that satisfy a first-order system of stochastic difference equations obtained by transforming the equation $y_t = .9 y_{t-1} - .8 y_{t-2} + u_t$. Plot the cross-spectral density function for these two time series. Comment on your answer.

4. Find and plot the spectral density function of y_{2t} as defined by (7).

5. Calculate two values at $\omega = 0$ and $\omega = \pi/2$ for the phase-difference spectral density

function and the cross-amplitude spectral density function of the two time series defined by (7), using (15) and the result from Section 3.10. Interpret your answer.

6. Prove Equations 18 in this chapter. *Hint:* You may write $\cos \omega_j t = \frac{1}{2}(e^{i\omega_j t} + e^{-i\omega_j t})$; $\sin \omega_j t = -(i/2)(e^{i\omega_j t} - e^{-i\omega_j t})$. Use (17). Once it is shown that either sum is equal to $N/2$, the other sum can be evaluated by using $\sin^2 \omega_j t + \cos^2 \omega_j t = 1$.

7. Prove (19) in this chapter. *Hint:* See the hint in Problem 6.

8. Obtain a quarterly series of real GNP in recent decades consisting of an even number of N (≥ 40) observations. Use a computer to decompose the series into cosine and sine functions as in (16). Plot the contribution of each periodic component to the sample variance of the series.

9. Fit the model $y_t = ay_{t-1} + u_t$ to the data obtained in Problem 8. Plot the power spectrum of the time series generated by the fitted model. Compare it with the graph of Problem 8, if you have done that problem.

10. Show that the covariance between y_{1t} and $y_{2,t-k}$ defined by (23) and (30) equals

$$\sum_j E[\alpha_1(\omega_j)\alpha_2(\omega_j)]\cos \omega_j k + \sum_j E[\beta_1(\omega_j)\alpha_2(\omega_j)]\sin \omega_j k.$$

11. Calculate the gain $G_{12}(\omega)$ for the two time series defined by (7) at $\omega = 0$ and $\omega = \pi/2$. Interpret your answer.

12. Calculate the coherence for the two time series defined by (7) at $\omega = 0$ and $\omega = \pi/2$. Interpret your answer.

13. Calculate the gain $G_{21}(\omega)$ for the two time series defined by (7) at $\omega = 0$ and $\omega = \pi/2$. Compare your answer with that of Problem 11.

14. What is the relation between the gains $G_{12}(\omega)$ and $G_{21}(\omega)$ and the coherence $R_{12}^2(\omega)$?

15. What are the gains $G_{12}(\omega)$ and $G_{21}(\omega)$ and the coherence $R_{12}^2(\omega)$ for the two time series defined by Problem 3.

16. Specify the canonical variables for the system $y_t = .9y_{t-1} - .8y_{t-2} + u_t$. Find their cross-covariance function and from that their cross-spectral density function.

17. Specify the canonical variables for the system of (7). Find their cross-covariance function and from that their cross-spectral density function.

18. Provide numerically a bivariate first-order system of stochastic difference equations from which the spectral density function of (62) for y_{1t} can be derived. Evaluate the spectral density of y_{2t} from this system at several values of ω. Does it also have a peak? Explain.

CHAPTER 5

Dynamic Analysis
of a Simple
Macroeconomic Model

5.1 STOCHASTIC DYNAMIC ECONOMIC ANALYSIS

The science of economics has progressed through the continual process of observation, hypothesis formulation, deduction, and further observation. A hypothesis is an assumption or a set of assumptions. Deduction is the process of drawing logical implications from the hypotheses. Observation involves the process of comparing the implications with the data from the real world. The term observation is also used to denote a piece of datum that is observed rather than the act of observing, as just described. A hypothesis is said to be confirmed if its implications agree with the data observed. The agreement increases our confidence in the hypothesis, but, of course, a hypothesis can be confirmed by one set of data and later found to be in disagreement with another set of data. The disagreement can lead to rejection or modification of the hypothesis. Therefore we can never be absolutely certain of a hypothesis about the real world. When a hypothesis passes the test of many observations, it is called a theory. In economics the term theory is used more loosely than in some natural sciences and may refer to a hypothesis not well confirmed.

The process of *induction* draws conclusions concerning the hypothesis from the data observed, whereas the process of *deduction* draws implications from a set of assumptions. How much disagreement with the data is required to reject a hypothesis? How much confidence should we attach to a hypothesis from a given set of data? These are the questions of induction.

What justifies induction is a question the philosophers have pondered for many years. If we are willing to make assumptions (and the philosophers will ask "why" here) concerning the probability distribution of the data, given the hypothesis, we can draw conclusions about the hypothesis from the data in a precise manner by the method of statistical inference. Thus statistical inference is a special case of inductive inference in general. It is applicable when we are willing to assume something about the probability mechanism that generates the data from the hypothesis. We need not necessarily specify the mathematical form of the distribution but only that the joint probability distribution of the observations, given the hypothesis, has a certain structure, such as being identically and independently distributed. We can go further with the process of induction if, in addition, we are willing to assign some probability structure to the hypothesis itself before the given data are observed, as readers familiar with Bayesian statistics will realize. However much we may appreciate the contribution of statistical methods to induction, we will admit that most scientists do not always follow mathematically describable rules in drawing conclusions about hypotheses from data. More supporting evidence will ordinarily lead to a higher degree of confidence in a hypothesis, but how the degree of confidence is revised in most practical situations is more complicated than the statisticians can describe.

    I have inserted this digression on induction in order to place the subject matter of this book in perspective. In the process of observation, hypothesis formulation, and deduction in economics the subject of economic analysis is concerned mainly with the second and third parts. In static economic analysis the hypotheses lead to propositions about economic variables without regard to their timing. In dynamic economic analysis they lead to propositions about the time paths of economic variables. In stochastic dynamic economic analysis they lead to statements about the joint probability distribution of economic variables at different points of time. In all three forms we are concerned with the statements of the hypotheses and the deduction of useful propositions from them. Mathematics is often employed for deduction. In static economic analysis demand functions are deduced from assumptions about consumer maximizing utility. These functions explain the quantities of different commodities to be demanded by using prices and the income of the consumer as arguments. The supply function of a firm's output and the demand functions for its inputs are deduced from assumptions about a business firm maximizing profit. After a demand function for and a supply function of a commodity are derived they can, in turn, be used to deduce the effects of income, wage, and other explanatory variables on the price and quantity of the commodity in question.

In dynamic economic analysis we deduce the time paths of economic variables; for example, the time path for price can be deduced from the cobweb model of demand and supply of a farm product. The time path for national income can be deduced from a multiplier-accelerator model in macroeconomics. In stochastic dynamic economics we deduce propositions concerning the stochastic time series from a dynamic system of stochastic equations. In all three cases of economic analysis two types of activity are involved: the statement of the hypothesis and the deduction from the hypothesis. In this book methods of deduction from hypotheses are emphasized. The subject matter of discussion is confined mainly to macroeconomic as opposed to microeconomic hypotheses, although the methods of optimization over time developed in the later chapters on optimal control can be applied to the formulation of microeconomic hypotheses as well.

What is to be deduced from stochastic dynamic macroeconomic hypotheses? The mean or expected time paths of the variables are certainly of interest. It was shown in Chapter 3 that if the hypotheses are formulated as a system of linear stochastic difference equations, the mean time paths are identical with the solution of the nonstochastic difference equations. One interesting question concerning the means is whether they will settle down to equilibrium values if the values of the exogenous variables remain constant through time. The answer for a linear system depends on the absolute values of the characteristic roots of the system. If all roots are smaller than 1 in absolute value, the means will reach steady-state values. The remaining characteristics of the stochastic time series are related to the variances and covariances of the variables at different points in time. The variances and covariances can be used to obtain interval estimates of the time series and to measure the magnitudes of their fluctuations. Again, if the roots of a linear system are all less than 1 in absolute value, the autocovariances and cross covariances will reach equilibrium values over time and become functions only of the time lag between the variables and not of the particular time in question. When this happens we can deduce the spectral and cross-spectral density functions for the time series to characterize their cyclical properties and the lead-lag relationships of their periodic components.

We can also deduce the expected time between crossings of the mean path and between relative maxima for an individual time series. Thus much of what the economist considers important about the trends and cycles of economic time series can be deduced analytically if the system is linear.

In Section 5.2 we state and discuss briefly three of the most important hypotheses in macroeconomics. These hypotheses are incorporated in a small system of linear stochastic difference equations in Section 5.3. Because some equations are formulated in the first differences of the variables, the implications of first-differencing are noted in Section 5.4. The mean time paths and the multipliers relating them are discussed in Section 5.5. Section 5.6 analyzes the dynamic properties of the time series via the autocovariance functions, and Sections 5.7 and 5.8 deal, respectively, with the spectral and the cross-spectral properties of the system. Section 5.9 draws some conclusions about business cycle theory and dynamic economics in general from our analysis.

5.2 THREE IMPORTANT MACROECONOMIC HYPOTHESES

Three of the most important macroeconomic hypotheses are the multiplier, the acceleration principle, and the liquidity preference relation. The first is due to Keynes (1936) in his *General Theory of Employment, Interest and Money*. The acceleration principle dates back at least to J.M. Clark (1917). Keynes (1936) was also responsible for bringing the liquidity preference relation to the forefront of a theory of national income determination, although the idea behind this relation was not new at the time.

Possibly as a misnomer, the term multiplier has been used to refer to a hypothesis about the relation between consumption expenditures and income. From the relation is deduced a multiplying or many-rounded effect of an increase in autonomous expenditures on income. The term thus denotes the hypothesis when strictly speaking it describes an implication from the hypothesis about consumption behavior. Be that as it may, the original Keynesian hypothesis states that consumption expenditures depend on income but increase proportionally less than income. Being assumed a linear form, the consumption function has a positive intercept and a slope less than 1. Later authors have introduced time delays into the consumption function. Friedman (1957) assumed aggregate consumption to be a linear function of a weighted average of past incomes with geometrically declining weights. Friedman called this average "permanent income." This assumption, excluding random disturbances and stated in discrete time, is equivalent to the following consumption function:

$$C_t = \alpha_1 Y_t + \alpha_2 C_{t-1}$$
$$= \alpha_1 Y_t + \alpha_2 \alpha_1 Y_{t-1} + \alpha_2^2 \alpha_1 Y_{t-2} + \alpha_2^3 \alpha_1 Y_{t-3} + \cdots, \tag{1}$$

where C_t denotes consumption and Y_t is an appropriate concept of income such as disposable personal income. A distinction can be made between consumption and consumption expenditures. The first includes the portion consumed from durable goods, the second, the purchases of new durable goods. We assume that this relation holds approximately for consumption expenditures. Besides Y_t, the stock of money M_t may be included as an additional explanatory variable, as it is often hypothesized.

According to one hypothesis of the demand for capital goods, the ultimate demand is for the services from the total stock of capital goods available, and new additions to capital stock are made to fill the gap between the desired demand for total stock and the quantity of the stock already in existence. Denote the actual stock and the desired stock of capital goods, respectively, by K_t and K_t^*. If a constant fraction θ of the difference between K_t^* and K_{t-1} is actually added to capital stock, the hypothesis can be stated as

$$K_t - K_{t-1} = \theta(K_t^* - K_{t-1}),$$
$$K_t = \theta K_t^* + (1-\theta)K_{t-1}. \tag{2}$$

Let K_t^* be a linear function of output Y_t and the rate of interest R_t (with a negative coefficient); Y_t is measured by the same after-tax output variable used in the consumption function (1), although in a more detailed econometric study the two variables can be different. Incorporating this assumption in (2), we have

$$K_t = b_0 + b_1 Y_t + b_2 R_t + b_3 K_{t-1}. \tag{3}$$

Equation 3 has the same form as (1). Each explains a dependent variable by a linear function of the variables determining desired demand and the lagged value of the dependent variable.

Gross investment expenditures I_t is the sum of net investment ($K_t - K_{t-1}$) and replacement for existing capital stock. As an approximation, assume that replacement demand is a constant fraction δ of K_{t-1}. Then

$$I_t = (K_t - K_{t-1}) + \delta K_{t-1} = K_t - (1-\delta)K_{t-1}, \tag{4}$$

or gross investment expenditures can be considered a quasi-first-difference of capital stock, meaning the difference between K_t and $(1-\delta)K_{t-1}$. A quasi-differencing operation on both sides of (3) yields the investment function

$$I_t = \delta b_0 + b_1[Y_t - (1-\delta)Y_{t-1}] + b_2[R_t - (1-\delta)R_{t-1}]$$
$$+ b_3[K_{t-1} - (1-\delta)K_{t-1}]$$
$$= \beta_0 + \beta_1 Y_t + \beta_2 Y_{t-1} + \beta_3 R_t + \beta_4 R_{t-1} + \beta_5 I_{t-1}. \tag{5}$$

Investment function (5) differs from consumption function (1) in that the dependent variable is explained by the lagged dependent variable, together with the *quasi-differences*, and not the levels of the variables affecting desired demand. This formulation is termed the acceleration relation because it is not the flow of income or output that determines investment but the rate of change in the flow of income. If investment were defined as net investment, the explanatory variable would be exactly the rate of change rather than the quasi-difference of income.

The demand-for-money equation can be formulated along the same line as the consumption function (1) or the demand-for-capital equation (3). Using income Y_t and the rate of interest R_t as the explanatory variables of desired demand, we have the following demand equation for the stock of money:

$$M_t = \gamma_0 + \gamma_1 Y_t + \gamma_2 R_t + \gamma_3 M_{t-1}. \tag{6}$$

This equation is based on the liquidity preference hypothesis. It states that, given income or output, the higher the foregone interest cost of holding money or keeping asset in liquid form, the less money will people hold. The coefficient γ_2 in (6) is therefore assumed to be negative.

To complete the model two more equations are introduced. One is the identity

$$\text{GNP}_t = C_t + I_t + G_t \tag{7}$$

where G_t is government purchases of goods and services. For simplicity we have ignored the excess of exports over imports as a component of GNP because it is small. The second equation results from combining a definition of Y_t as $C_t + I_t + G_t - T_t$, where T_t is government taxes net of transfers and the assumption that T_t equals $g_0 + g(C_t + I_t + G_t)$.

$$Y_t = C_t + I_t + G_t - T_t$$

$$= -g_0 + (1-g)(C_t + I_t + G_t). \tag{8}$$

Equations 1, 5, 6, and 8 constitute a system of four structural equations for the four endogenous variables C_t, I_t, R_t, and Y_t; G_t and M_t are regarded as exogenous. A model based essentially on these equations was estimated, using data for the United States.

5.3 A SIMPLE MACROECONOMIC MODEL

Before this model was estimated statistically, as reported in Chow (1967), several additional decisions had to be made. First, annual data were used instead of quarterly or monthly data. The model was formulated with annual data in mind. For observations taken at shorter time intervals the distributed lag relation between capital stock and its determinants would have to be more complicated than that formulated in (2) and (3). The influence of the variables included in K_t^* in (2) would not decline geometrically as the time lag increased. Because of the delays between investment decision and actual investment expenditures, the effect of the most current values of explanatory variables may be smaller than that of the variables dated two or three quarters ago. Quarterly national income data before 1948 were not available, whereas annual data were, beginning at least in 1929. A model of annual data could be tested for a longer historical period. The period finally selected covers the years 1931–1940 and 1948–1963, a total of 26 annual observations.

It was also decided to treat two components of I_t, one for investment in producers' durable equipment plus change in business inventories and the second for new construction. Further decomposition of the first component into its two parts was tried, but the combined variable was found to be satisfactory. Each component was assumed to satisfy an equation in the form of (5). An attempt was also made to separate consumption expenditures into at least two parts, one for durables, the other for nondurables plus services. The separation was found to be unnecessary for the purpose of obtaining stable relationships.

When random disturbances were added to form a stochastic system, the assumption that they were serially uncorrelated may be questionable, at least for some equations. If the residuals u_{it} in equation (i) are assumed to follow a first-order autoregression $u_{it} = r_i u_{i,t-1} + \epsilon_{it}$, where ϵ_{it} are assumed to be identically and independently distributed, a quasi-differencing operation can be applied to the equation to yield serially independent residuals ϵ_{it}. By quasi-differencing we mean forming the difference $y_{it} - r_i y_{i,t-1}$ for both the dependent variable y_{it} and its explanatory variables. This transformation is known as the Cochrane-Orcutt transformation as it was used by Cochrane and Orcutt (1947) to deal with a regression problem in which residuals satisfied a first-order autoregression. If r_i is close to 1, an approximation simply takes the first difference Δy_{it}. First-differencing was applied to the consumption function and the demand function for money, but not to the investment equations, because these equations are already the result of quasi-differencing operations and their residuals did not appear to have high serial correlations. Since 1965, when the reported study was made, methods of estimating parameters in a system of simultaneous econometric equations with serially correlated residuals have become available. References to some of these estimation methods can be found in Chow and Fair (1973) and Chow (February 1973). They would make the approximation by use of first differences less appealing than before.

One last issue should be addressed before the statistical results are reported. Should the components of GNP be measured in constant or current dollars? This depends on whether economic agents make their consumption and investment decisions in real or nominal terms. If money illusion, that is, making a decision according to money values, is not complete, some form of deflation by a price index P would be required. It was decided to use the expenditure variables in current dollars and to add the general price index measured by the GNP deflator as an additional explanatory variable in the consumption and investment equations. Because the model does not explain the general price level, P_{t-1} was used. It serves to take out some of the effect of pure inflation on the changes in consumption and investment expenditures measured in current dollars. A more rigorous argument for this deflation device is given in Chow (1967).

The estimated equations, obtained by the method of two-stage least squares, are reported below. The estimated standard error of each coefficient appears in parentheses below the coefficient. The standard error of each equation is denoted by s^2, and DW represents the Durbin-Watson statistic for the estimated regression residuals:

$$\Delta C_t = .3083 \Delta Y_{1t} + .1938 \Delta C_{t-1} + .4079 \Delta M_t + .0783 \Delta G_t + 87.7 P_{t-1} - 4797,$$
$$\qquad (.0944) \qquad (.1301) \qquad (.4096) \qquad (.1592) \qquad (46.4)$$
$$\qquad R^2 = .850 \qquad s^2 = 13.5(10)^6 \qquad DW = 2.26. \tag{9}$$

$$\Delta I_{1t} = .2806 \Delta Y_{1t} - .6625 I_{1,t-1} + .0090 Y_{t-1} + .1683 \Delta G_t + 158.6 P_{t-1} - 6125,$$
$$\qquad (.0816) \qquad (.2907) \qquad (.0333) \qquad (.1855) \qquad (187.0)$$
$$\qquad R^2 = .682 \qquad s^2 = 12.7(10)^6 \qquad DW = 1.91. \tag{10}$$

$$\Delta I_{2t} = .1046 \Delta Y_{1t} - .2198 \Delta R_t - .5099 I_{2,t-1} + .0410 Y_{t-1} + 92.8 P_{t-1} - 5996,$$
$$\qquad (.0213) \qquad (.1440) \qquad (.2071) \qquad (.0266) \qquad (51.3)$$
$$\qquad R^2 = .752 \qquad s^2 = 1.26(10)^6 \qquad DW = 2.14. \tag{11}$$

$$\Delta R_t = .1109 \Delta Y_{1t} - .7389 \Delta M_t + .1872 \Delta G_t + .3178 \Delta M_{t-1} - 973,$$
$$\qquad (.0620) \qquad (.3048) \qquad (.1138) \qquad (.2478)$$
$$\qquad R^2 = .336 \qquad s^2 = 9.03(10)^6 \qquad DW = 2.54. \tag{12}$$

$$Y_{1t} = C_t + I_{1t} + I_{2t}. \tag{13}$$

$$Y_t = .79(Y_{1t} + G_t). \tag{14}$$

The variables are

C = total personal consumption expenditures in millions of current dollars;

I_1 = gross private domestic investment in producers' durable equipment plus change in business inventories in millions of current dollars;

I_2 = new construction in millions of current dollars;

R = yield of 20-year corporate bonds, annual percentage rate times $(10)^4$;

M = currency and demand deposits adjusted in middle of the year in millions of current dollars;

G = government purchases of goods and services in millions of current dollars;

P = GNP deflator (1954 = 100).

For sources of data and for a fuller discussion of this model and its economic implications the reader may refer to Chow (1967). A few comments on the estimated equations will suffice here.

Recall that the after-tax income variable Y is $C + I + G - T$. Given a linear tax function for T, Y becomes $-g_0 + (1 - g)(C + I + G)$. Taking first difference gives $\Delta Y = (1 - g)\Delta(C + I + G)$. For the post-World War II period $(1 - g)$ was crudely estimated to be .79, and this parameter was taken as given, as in (14). For estimation purposes the ΔY variable was separated into two parts, $(1 - g)\Delta(C + I)$ and $(1 - g)\Delta G$. Expenditures by the private sector $C + I$, denoted by Y_1, were separated from government expenditures G to determine whether their effects differ. This accounts for the presence of both ΔY_1 and ΔG in the equations instead of ΔY alone. In (11) ΔG was dropped because its coefficient had a negative sign and was not significant.

A most striking feature of the statistical results was the strong confirmation of the acceleration principle. According to this principle, investment expenditures I_t depends on Y_t and Y_{t-1} with coefficients in opposite signs but similar in magnitude. This is exactly what we found. Note that we wrote $Y_t - (1 - \delta)Y_{t-1}$ as $\Delta Y_t + \delta Y_{t-1}$ and found that the coefficient of ΔY_t was positive and the coefficient of Y_{t-1}, positive and smaller. The result is equivalent to a positive coefficient for Y_t and a negative coefficient for Y_{t-1} with a smaller absolute value. The signs of all the coefficients in these equations are correct. Some coefficients have large standard errors but they serve useful purposes: P_{t-1} is purely a deflation device and should not be discarded; ΔM in the consumption function (9) and ΔG in the investment function (10) were kept to allow for their possible influences on national income because the existence of these effects is maintained by many economists. The small coefficient of Y_{t-1} in (10) should also be kept, for if Y_t and Y_{t-1} were used instead of ΔY_t and Y_{t-1}, as just explained, the coefficient of Y_{t-1} would be larger and significant.

5.4 FIRST-DIFFERENCING AND STABILITY

Partly to eliminate serial correlation in the residuals, we took first differences of the consumption function (9) and the demand for money equation (12). Note that in spite of the appearance of ΔI_{1t} and ΔI_{2t} as dependent variables in (10) and (11) no differencing operation was applied to the formulation (5) for the investment equations; $I_{1,t-1}$ or $I_{2,t-1}$ was simply subtracted from both sides of (5). Adding back $I_{1,t-1}$ on both sides of (10) will give an equation that explains the levels I_{1t} and, similarly, I_{2t}. The coefficient of $I_{1,t-1}$ or $I_{2,t-1}$ will be increased by 1, but all other coefficients will remain unchanged. R^2 will be affected also. Thus (10) explains I_{1t} by $I_{1,t-1}$, in contrast to (9) which explains ΔC_t by ΔC_{t-1}. This is the major difference between the consumption and investment equations. The differencing operation, applied to the consumption and money demand equations, will have one important implication on the dynamic properties of the system. Let us turn to this implication.

Consider a simple stochastic difference equation $\Delta y_t = .8\Delta y_{t-1} + \epsilon_t$, where ϵ_t has mean 0 and is independently and indentically distributed through time. Δy_t is certainly a covariance stationary time series. Its properties have been studied thoroughly in Chapters 3 and 4. What can be said about y_t itself? Because y_t is the sum of past Δy_s ($s = 1,\ldots,t$) and y_0, its variance will increase through time even if the variances of Δy_s are constant. The variance of a sum of random variables equals the sum of the variances plus twice the sum of the covariances. In this example the covariances between Δy_s and Δy_{s-k} are all positive, making the variance of the y_t even larger than the sum of the variances of Δy_s. There is no question that the variance of y_t will increase without bound as t increases and accordingly that y_t is not covariance stationary.

A convenient way to examine the dynamic behavior of y_t, given that $\Delta y_t = \alpha \Delta y_{t-1} + \epsilon_t (|\alpha| < 1)$, is to study the system of two stochastic difference equations

$$\begin{bmatrix} \Delta y_{t+1} \\ y_t \end{bmatrix} = \begin{bmatrix} \alpha & 0 \\ 1 & 1 \end{bmatrix} \begin{bmatrix} \Delta y_t \\ y_{t-1} \end{bmatrix} + \begin{bmatrix} \epsilon_{t+1} \\ 0 \end{bmatrix}, \qquad (15)$$

which was obtained by combining the identity

$$y_t = y_{t-1} + \Delta y_t \tag{16}$$

with the original equation for Δy_{t+1}. The characteristic roots of this system, found by solving the characteristic equation

$$(\alpha - \lambda)(1 - \lambda) = 0.$$

are α and 1. There are two canonical variables in the system, one with coefficient α, the other with coefficient 1. Being a linear combination of these canonical variables, y_t is nonstationary. For a closer examination of the canonical variables and the linear combination that form y_t it is recommended that the reader do Problem 1 at the end of this chapter. The identity (16) for the variable y_t gives rise to one characteristic root of unity, as the coefficient of y_{t-1} in this equation equals 1. Although Δy_t is a stationary process, y_t will not be because of this root.

In general, if a system of stochastic difference equations contains the first-differences Δy_{it} of some variables, the behavior of the variables y_{it} themselves can be deduced by incorporating identities like (16) into the system. Even if the original system is covariance stationary, the time series y_{it} generated by these additional identities will not be. Thus by formulating some equations in the first differences of the endogenous variables we automatically impart nonstationarity to the variables themselves. If the formulation of these variables is regarded as *perfect*, that is, $y_{it} - .99999 y_{i,t-1}$ cannot, for example, be substituted for $y_{it} - y_{i,t-1}$, the methods introduced in Chapters 3 and 4 are not applicable to those variables whose first differences are explained by the equations. No formulation of an econometric model is perfect, however. All econometric models are approximations; the differencing operations in the model under discussion were performed as a crude method to eliminate first-order serial correlations in the consumption and money demand equations. As it turned out, the Durbin-Watson statistics for the residuals of (9) and (12) are slightly larger than 2, which suggests that the residuals may have a small negative first-order serial correlation. A quasi-differencing operation like $y_{it} - .95 y_{i,t-1}$ would probably fit the data better. If such an operation were used, the source of nonstationarity would be eliminated. Hence the nonstationarity of the system should not be regarded as reflecting the facts perfectly. If, however, the first differencing operations employed in the model are approximately correct, we would still expect the roots resulting from the more accurate quasi-differencing operations to be near 1. It will be seen that the two roots of the above system as they appeared in the computer output are .9999725 and .9999064. It would be interesting to analyze the dynamic properties of the system by using these roots as a stringent test for the applicability of the methods of analysis.

5.5 MEAN PATHS AND MULTIPLIERS

The mean or expected paths of the linear stochastic system are identical to the solution of the deterministic system obtained by omitting the random disturbances. To derive the mean paths a set of reduced form equations is required. These equations are obtained by solving the structural equations (9) to (12) for the endogenous variables $\Delta C_t, I_{1t}, I_{2,t}$, and ΔR_t. The variable Y_{1t} was eliminated by using the identity (13). Because Y_{t-1} appears in (10) and (11), an equation for Y_t due to (13) and (14), that is, $Y_t = Y_{t-1} + .79(\Delta C_t + I_{1t} + I_{2t} + \Delta G_t) - .79(I_{1,t-1} + I_{2,t-1})$, was used in addition to (9) to (12). The resulting reduced-form equations are shown in Table 5.1. The intercepts and the coefficients of P_{t-1} are omitted because they are not needed for later analysis.

Table 5.1 Reduced Form Equations

Endogenous Variables	ΔC_{-1}	$I_{1,-1}$	$I_{2,-1}$	ΔR_{-1}	Y_{-1}	ΔM	ΔG	ΔM_{-1}
(1) ΔC	.3744	−.6173	−.4751	0	.0466	.9393	.2697	−.0651
(2) I_1	.1644	−.2243	−.4324	0	.0514	.4837	.3425	−.0592
(3) I_2	.0470	−.1606	.3665	0	.0531	.3007	.0087	−.0868
(4) ΔR	.0650	−.2220	−.1709	0	.0168	−.5477	.2561	.2943
(5) Y	.4627	−1.5818	−1.2174	0	1.1194	1.3617	1.2805	−.1668

By denoting the vector of endogenous variables in Table 5.1 by y_t we can write the reduced form in matrix notation as

$$y_t = A_1 y_{t-1} + C_0 x_t + C_1 x_{t-1} + d_t, \tag{18}$$

where x_t is a vector with ΔM_t and ΔG_t as elements and d_t is a 5×1 vector whose elements combine the intercepts and the effects of P_{t-1} in the corresponding reduced-form equations. Because ΔM_{t-1} appears in the system, x_{t-1} is needed. The second column of the C_1 matrix is 0 because ΔG_{t-1} is absent. To derive the mean paths and the various multipliers it is convenient to rewrite the system (18) in order to eliminate all *lagged* exogenous variables as

$$\begin{bmatrix} y_t \\ x_t \end{bmatrix} = \begin{bmatrix} A_1 & C_1 \\ 0 & 0 \end{bmatrix} \begin{bmatrix} y_{t-1} \\ x_{t-1} \end{bmatrix} + \begin{bmatrix} C_0 \\ I \end{bmatrix} x_t + \begin{bmatrix} d_t \\ 0 \end{bmatrix} \qquad (19)$$

or simply as

$$\tilde{y}_t = A\tilde{y}_{t-1} + Cx_t + \tilde{d}_t. \qquad (20)$$

Each vector x_{t-k} of exogenous variables lagged k periods, if it appears, can be eliminated by attaching x_{t-k+1} to the newly defined vector $\tilde{y}_t$ of endogenous variables; for example, if $C_2 x_{t-2}$ were added to the right-hand side of (18), x_{t-1} should appear as the third component of $\tilde{y}_t$. Using the form (20), we can easily express the solution or the (vector) mean path of the system as

$$\tilde{y}_t = A^t \tilde{y}_0 + Cx_t + ACx_{t-1} + \cdots + A^{t-1}Cx_1 + d_t + Ad_{t-1} + \cdots + A^{t-1}d_1.$$

$$(21)$$

The original endogenous variables y_t are contained in the first subvector of $\tilde{y}_t$ and are therefore of special interest. The mean path can be calculated from (20) or (21). The properties of the path $A^t \tilde{y}_0$, due to interactions of the endogenous variables alone, can be studied by the methods in Chapter 2.

Equation 21 is called the *final form* of the econometric model. It is obtained from the reduced form after lagged endogenous variables are replaced by lagged exogenous variables. Various multipliers can be defined and calculated by (21). The *impact multiplier* of the jth exogenous variable on the ith endogenous variable is given by the i-j element of the matrix C. In the model of Table 5.1 the impact multipliers of ΔM and ΔG on Y are, respectively, 1.3617 and 1.2805, for example. The *delayed multipliers* are the derivatives of the endogenous variables with respect to lagged exogenous variables. From (21) the delayed multiplier of the jth exogenous variable lagged k periods on the ith endogenous variable is the i-j element of $A^k C$. The *intermediate-run multipliers* measure the combined effects of changes in the exogenous variables that have persisted over several periods. The intermediate-run multiplier of the jth exogenous variable persisting over m periods on the ith current endogenous variable is the i-j element of the sum

$$C + AC + A^2 C + \cdots + A^{m-1}C.$$

The (very) *long-run multiplier* is the limit, if it exists, of the corresponding intermediate-run multiplier as m approaches infinity. Using (29) in Chapter 3 and assuming $(I - A)$ to be nonsingular, we rewrite the above matrix of intermediate-run multipliers as

$$(I + A + A^2 + \cdots + A^{m-1})C = (I - A)^{-1}(I - A^m)C. \qquad (22)$$

The limit of (22) exists as m approaches infinity if and only if the limit of A^m exists, which is the case if the characteristic roots of A are all smaller than 1 in absolute value. From the result in Section 3.6, following (23) in Chapter 3, this last condition is also a necessary and sufficient condition for the mean path to reach a steady state. Of course, if the mean path never reaches a steady state in the first place, how can we speak of the derivative of the steady state with respect to permanent values of the exogenous variables? If a steady state for the mean path exists, the long-run multipliers are given by the matrix $(I - A)^{-1}C$, for A^m will approach the 0 matrix in (22). The mean paths of some economic models, however, have no steady-state solutions. Models that explain economic growth endogenously are obvious examples. In such models long-run multipliers with m approaching infinity are meaningless and useless concepts.

5.6 AUTOCOVARIANCE PROPERTIES OF THE SYSTEM

Different sets of endogenous variables can be selected from the structural equations (9) to (14) for the purpose of stochastic dynamic analysis. We have selected the five variables $C, I_1, I_2, R,$ and Y_1, denoted, respectively, by $y_1, y_2, y_3, y_4,$ and y_5. Thus all first differences Δy_{it} are replaced by $y_{it} - y_{i,t-1}$ in these equations. Let y_t denote the vector of these five endogenous variables and let x_t denote the vector $(M_t, G_t, P_{t-1}, 1)'$. The structural equations can be written as

$$B_0 y_t = B_1 y_{t-1} + B_2 y_{t-2} + B_3 x_t + B_4 x_{t-1} + B_5 x_{t-2} + \epsilon_t. \qquad (23)$$

Because of ΔC_{t-1} and ΔM_{t-1} in the original equations, y_{t-2} and x_{t-2} are required in (23). Note that except for the first column the elements of B_2 are 0; except for the first column, the elements of B_5 are 0; ϵ_t stands for a vector of random residuals whose variances were estimated by s^2 in (9) to (12). The reduced form of (23) is

$$y_t = B_0^{-1}B_1 y_{t-1} + B_0^{-1}B_2 y_{t-2} + B_0^{-1}B_3 x_t + B_0^{-1}B_4 x_{t-1} + B_0^{-1}B_5 x_{t-2} + B_0^{-1}\epsilon_t$$

$$= A_1 y_{t-1} + A_2 y_{t-2} + C_0 x_t + C_1 x_{t-1} + C_2 x_{t-2} + B_0^{-1}\epsilon_t. \qquad (24)$$

The effects of x_t and lagged x_{t-k} on the mean paths of the endogenous variables have already been discussed in Section 5.5. (See also Problem 8.) We shall study from here on the dynamic properties of the stochastic system:

$$y_t = A_1 y_{t-1} + A_2 y_{t-2} + B_0^{-1} \epsilon_t. \tag{25}$$

To convert this system into first order we need to add only one variable, C_{t-1}, to the vector y_t. The coefficients of the reduced form that results are presented in Table 5.2. They are the elements of the matrix A in $y_t = A y_{t-1}$. The characteristic roots of this matrix are

λ_1	λ_2	λ_3	λ_4 and λ_5	λ_6
.9999725	.9999064	.4838	$.0761 \pm .1125i$	$-.00004142$

The corresponding right characteristic vectors are

$$
\begin{matrix}
-.008 & 1.143 & .320 & .283 & \pm & .591i & .000 \\
-.000 & .013 & -.586 & -2.151 & \pm & .742i & 2.241 \\
-.001 & .078 & .889 & -.215 & \pm & .135i & .270 \\
1.024 & .271 & .069 & -.231 & \pm & .163i & .307 \\
-.009 & 1.235 & .623 & -2.082 & \pm & 1.468i & 2.766 \\
-.008 & 1.143 & .662 & 4.772 & \pm & .714i & -4.399
\end{matrix}
\tag{26}
$$

Note the two roots λ_1 and λ_2, which are very close to 1, as a result of the first-differencing of (9) and (12); λ_6 is nearly 0, thus reflecting the identity for C_{t-1}. A pair of complex roots, λ_4 and λ_5, has absolute value 0.1358 and angle 55.92°, giving a highly damped cycle of 6.44 years to a component of the deterministic time path and of the autocovariance function for each endogenous variable. Let us find the autocovariance functions.

Table 5.2 Reduced Form For Five Selected Variables

Endogenous Variables	Coefficient of					
	C_{-1}	$I_{1,-1}$	$I_{2,-1}$	R_{-1}	$Y_{1,-1}$	C_{-2}
C	2.306146	.314472	.456660	.0	$-.894965$	$-.374377$
I_1	1.012405	.623718	.415630	.0	$-.807454$	$-.164353$
I_2	.289449	.081830	.608930	.0	$-.200483$	$-.046989$
R	.400127	.113120	.164267	1.0	$-.321932$	$-.064956$
Y_1	3.608000	1.020020	1.481220	.0	-1.902902	$-.585718$
C_{-1}	1.000000	.000000	.000000	.0	.000000	.000000

We apply (41) in Chapter 3 for this purpose. To do so the covariance matrix of the residuals of the reduced form (25) is needed. It is obtained from the estimated covariance matrix of the residuals of the structural equations by

$$V = E\left[(B_0^{-1}\epsilon_t)(B_0^{-1}\epsilon_t)' \right] = E\left[B_0^{-1}\epsilon_t\epsilon_t' B_0^{-1'} \right]$$

$$= B_0^{-1}(E\epsilon_t\epsilon_t')B_0^{-1'}. \tag{27}$$

After the identity for $y_{6,t} = C_{t-1}$ is added to the reduced form the covariance matrix of the residuals is calculated as

$$V = \begin{bmatrix} 8.250 & 7.290 & 2.137 & 2.277 & 17.68 & 0 \\ 7.290 & 7.135 & 1.992 & 2.165 & 16.42 & 0 \\ 2.137 & 1.992 & .618 & .451 & 4.746 & 0 \\ 2.277 & 2.165 & .451 & 1.511 & 4.895 & 0 \\ 17.68 & 16.42 & 4.746 & 4.895 & 38.84 & 0 \\ 0 & 0 & 0 & 0 & 0 & 0 \end{bmatrix}. \tag{28}$$

All information is now available for the application of (41) in Chapter 3, where w_{ij} is defined by (35) in that chapter.

The variances and selected áutocorrelations are given in Table 5.3. Because of the two roots very close to 1, the system is very close to being nonstationary. The variances are very large, compared with the diagonal elements of V. The autocorrelations decrease very slowly as the lag k increases. The numerical accuracies of these calculations are subject to question, but are presented to show what will happen when the system is very close to being nonstationary and to find out whether certain dynamic characteristics can still be deduced in such circumstances.

Table 5.3 Variances and Selected Autocorrelations

Variable	$\gamma_{ii,0}$	$\rho_{ii,1}$	$\rho_{ii,2}$	$\rho_{ii,3}$	$\rho_{ii,10}$	$\rho_{ii,35}$
1. C	$2.487680(10)^4$	.999820	.999705	.999608	.998953	.996619
2. I_1	$1.054430(10)$	.261258	.291833	.313463	.319593	.318871
3. I_2	$1.178579(10)^2$	.996139	.995414	.995068	.994150	.991825
4. R	$1.668905(10)^4$	.999948	.999917	.999886	.999669	.998891
5. Y_1	$2.905648(10)^4$	.999027	.998907	.998821	.998164	.995832

By applying the formula from Section 3.4,

$$2\pi/\cos^{-1}[(1-2\rho_1+\rho_2)/(2\rho_1-2)],$$

for the expected time between relative maxima to each of the five time series, we obtain

C	I_1	I_2	R	Y_1
3.59	2.97	3.16	3.54	3.10

These numbers are not unreasonable in view of the approximately four-year cycles reported and the fact that minor down-turns are sometimes not counted in the business cycle chronology. (See also Problem 11.) It would be interesting to note the expected time between relative maxima for the first-order autoregressive time series $y_t = ay_{t-1} + u_t$ for $a = .9999$ and $a = 0$. The first is 3.99987 and the second is 3. Thus for first-order autoregression the expected time between maxima lies between three and four time units. (See Problem 12.) For higher order systems other outcomes are possible. We turn to Section 5.7 to determine whether a model like $y_t = .9999 \times y_{t-1} + u_t$ is at all a reasonable first approximation for many economic time series.

5.7 SPECTRAL PROPERTIES OF INDIVIDUAL TIME SERIES

Using Table 5.2, the characteristic roots, the associated characteristic vectors in (26), and the covariance matrix in (28), we can apply Formula 51 in Chapter 4 to compute all the spectral and cross-spectral densities. Selected values of the (normalized) spectral densities of the five variables are presented in Table 5.4; the first two are graphed in Figures 5.1 and 5.2. Several comments are in order.

Table 5.4 Selected Values of Normalized Spectral Densities

$\dfrac{\omega}{2\pi}$	0	$\dfrac{1}{360}$	$\dfrac{1}{36}$	$\dfrac{1}{18}$	$\dfrac{1}{9}$	$\dfrac{1}{6}$	$\dfrac{1}{4}$	$\dfrac{1}{3}$	$\dfrac{1}{2}$
1. C	6814	.7819	.0020	$.55(10^{-3})$	$.18(10^{-3})$	$.10(10^{-3})$	$.63(10^{-4})$	$.47(10^{-4})$	$.36(10^{-4})$
2. I_1	2180	.4050	.1585	.1661	.1886	.2108	.2333	.2407	.2391
3. I_2	6781	.7810	.0048	$.31(10^{-2})$	$.21(10^{-2})$	$.17(10^{-2})$	$.13(10^{-2})$	$.12(10^{-2})$	$.10(10^{-2})$
4. R	22235	.2601	.0007	$.17(10^{-3})$	$.49(10^{-4})$	$.26(10^{-4})$	$.16(10^{-4})$	$.13(10^{-4})$	$.11(10^{-4})$
5. Y_1	6808	.7815	.0023	$.79(10^{-3})$	$.42(10^{-3})$	$.35(10^{-3})$	$.32(10^{-3})$	$.30(10^{-3})$	$.28(10^{-3})$

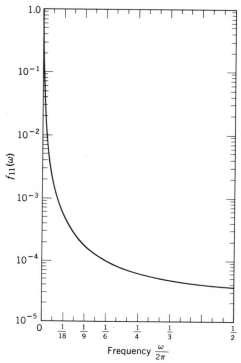

Figure 5.1   Spectral density of consumption expenditures.

First, the results of Table 5.4 show highest densities for the lowest frequencies (or longest cycles). This finding agrees with the results obtained by spectral analyses of economic data, reported by Nerlove (1964, p. 255), Adelman (1965), Granger and Hatanaka (1964), Granger (1966, p. 150), and Fand (1966, p. 364). Generally speaking, many economic time series have high positive correlations between successive observations that cause smooth time paths or reflect the importance of long cycles. As pointed out by various authors, including Nerlove (1964, p. 271) and Granger (1966), a time series that satisfies a first-order stochastic difference equation $y_t = a y_{t-1} + u_t$ with the coefficient a slightly below 1 will have a spectral density that gives the highest values at the lowest frequencies. Such a spectral density function was studied in Section 4.1. In Section 4.9 it was stated that the spectral density function of a time series generated by a system of stochastic difference equations like (9) through (14) is a linear combination of the spectral density functions of the

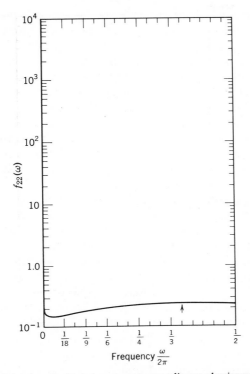

Figure 5.2 Spectral density of private investment expenditures plus inventory change.

first-order canonical variables. If some roots are slightly below unity, the spectral properties of the corresponding canonical variables can be preserved in the linear combination. This explains the results of Table 5.4.

Second, if some roots are extremely close to unity, as in our model, the components in the above linear combination will have extremely high densities at the lowest frequencies. We observe, however, that the magnitudes of these extremely high densities shown in Table 5.4 are not different from those found elsewhere by the use of spectral analysis; for example, in Nerlove (1964, p. 255) and Adelman (1965, pp. 457–58). Thus there is no reason to abandon our pragmatic view of using those roots very close to unity that are due to equations formulated in first differences. Indeed, such formulation and the resulting roots can explain the extremely high spectral densities observed at very low frequencies.

Third, are there local peaks in the spectral densities showing distinct cycles? From spectral analyses of many economic time series Granger

(1966) has concluded that the "typical spectral shape" valid for most economic time series resembles the spectral density of a first-order autoregression with coefficient very close to 1. This density has no relative maximum for ω between 0 and π. On the other hand, spectral densities estimated by other authors, including Adelman (1965) and Fand (1966), often show local peaks. As we pointed out in Chapter 4, a linear stochastic system like the one being analyzed can produce local peaks in the spectral densities. Complex roots can, but do not necessarily, produce local maxima. Even positive and real roots alone in a bivariate first-order system can do so, as illustrated by (61) in Chapter 4. Thus the linear stochastic system is quite flexible in generating spectral density functions of various shapes.

Fourth, our empirical results in Table 5.4 show that consumption, new construction, 20-year-bond yield, and GNP (less government and net foreign expenditures) have spectral densities monotonically decreasing with frequency, thus exemplifying Granger's "typical spectral shape" just mentioned. This is so in spite of the pair of complex roots λ_4 and λ_5 because they are small in absolute value. By comparison Adelman (1965) has found a local peak for output and a second maximum at $\omega = \pi$ for consumption by spectral analysis. We will not settle the difference here. As shown in Figure 5.2, the spectral density for durable equipment expenditure plus inventory change has a local maximum at ω near $.72\pi$, corresponding to cycles of about 2.8 years, but the density is almost constant for ω from $.5\pi$ to π (or for cycles of four to two years in duration). This result is consistent with the importance of short cycles in that time series.

5.8 CROSS-SPECTRAL PROPERTIES

In this section we report on the cross-spectral densities between the first four variables C, I_1, I_2, and R and total private expenditures on goods and services Y_1. We present first the regression coefficient and the (squared) correlation tnat relate the periodic components of each of the four variables to the corresponding components in Y_1. They are, respectively, the gain

$$G_{i5}(\omega) = \frac{|f_{i5}(\omega)|}{f_{55}(\omega)} \qquad (i = 1,\ldots,4),$$

and the coherence

$$R_{i5}^2(\omega) = \frac{|f_{i5}(\omega)|^2}{f_{ii}(\omega)f_{55}(\omega)} \qquad (i = 1,\ldots,4),$$

as defined in Section 4.8, where $f_{ij}(\omega)$ stands for the complex-value cross-spectral densities functions.

Figures 5.3 to 5.6 show, respectively, the gains of the first four variables on the fifth variable (GNP less government and net foreign expenditures) and the corresponding coherence. The gain (regression coefficient) of consumption on private-domestic GNP is about .92 for very low frequencies and reduces monotonically to below .4 for frequencies higher than $\frac{1}{4}$. This indicates that short-period fluctuations in income have smaller effects on short-period fluctuations in consumption expenditures than long-period fluctuations, a result consistent with the permanent income hypothesis, as noted by Howrey (1967). Observe, however, that although the regression coefficients are small for short-period fluctuations, the squared correlation coefficient (coherence) remains high. This may be partly the result of using consumption expenditures, which include expenditures on, rather than consumption of, durable goods. Previous studies of demand for consumer durable goods reported in Chow (1957) and Chow (1960) have pointed out the dependence of durable-good *expenditures* on current income, even though *consumption* of durable goods may depend mainly on permanent income.

The gain (Figure 5.4) of durable equipment expenditures plus inventory change on private GNP is small for very long cycles, but increases

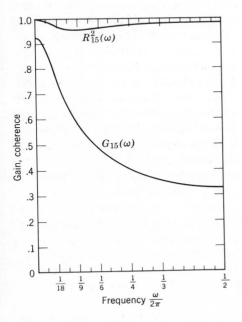

Figure 5.3 Gain and coherence of consumption on total private expenditures.

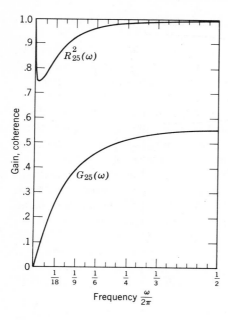

Figure 5.4 Gain and coherence of investment on total private expenditures.

monotonically to more than .5 for short cycles. This result agrees with the acceleration principle, also noted by Howrey (1967). A regression coefficient as high as .5 does show that the short-period fluctuations of inventory change are quite violent. The regression coefficient (Figure 5.5) of new construction expenditures also increases as the cycle shortens to about nine years but decreases slightly for shorter cycles. For these short cycles the coefficient is about .12, much less than the coefficient for durable equipment plus inventory change—a reasonable result. Note that the squared correlation coefficients (coherence) for both types of investment expenditure remain fairly high at all frequencies, thus confirming the dependence of investment expenditures on income.

The regression coefficient of 20-year-corporate bond yield in tenths of a percentage point per billion dollars (Figure 5.6) is more nearly constant throughout all frequencies, with some tendency to decrease as frequency increases, thus showing no acceleration and perhaps a little dampening effect of income. More interesting is the small squared correlation coefficient between interest rate and income—lower than .5 for four-year cycles. It has been pointed out by Turvey (1965) that inverting a liquidity-preference demand equation for money is a poor way of explaining the rate of interest.

Figure 5.7 shows the phase difference $\psi_{i5}(\omega)/2\pi$ in cycles between the

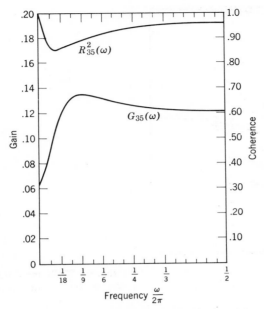

Figure 5.5 Gain and coherence of new construction on total private expenditures.

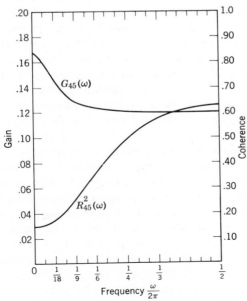

Figure 5.6 Gain and coherence of long-term interest rates on total private expenditures.

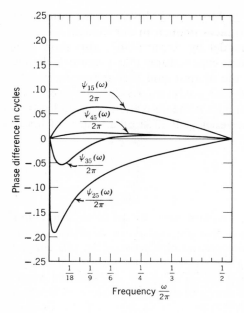

Figure 5.7 Phase-difference cross spectral densities.

first four variables and private-domestic GNP. For consumption expenditures $\psi_{15}(\omega)$ is always positive, which indicates that income leads consumption by .06 of a cycle, or .36 year, for six-year cycles and by .036 of a cycle, or .108 year, for three-year cycles. It is interesting to compare our results with those of Moore and Shiskin (1967), in which leads and lags at turning points of economic time series are compared. Some of the figures quoted below have been kindly supplied by Geoffrey Moore. The study by Moore and Shiskin reveals that the same consumption variable has a mean lag of .8 month at turning points. Of course, comparisons with the Moore-Shiskin figures provide only rough checks, for they deal with mean leads or lags at turning points.

Durable equipment expenditures plus inventory change lead GNP at all frequencies—by .066 of a cycle, or .4 year, for six-year cycles and by .025 of a cycle, or .075 year, for three-year cycles. Moore and Shiskin have found that change in business inventory has a mean lead of 3.8 months, but durable equipment expenditures have a slight (0.3 month) mean lag. New construction, by Figure 5.7, leads at low frequencies and lags very slightly at high frequencies (for cycles shorter than six years). Moore and Shiskin have found that the same variable has a mean lead of 3.4 months. By Figure 5.7 the 20-year corporate bond yield is lagging GNP slightly. Moore and Shiskin show a small lag of 0.9 month for corporate bond yields. These rough comparisons with Moore and Shiskin do not contradict the findings concerning phase differences implicit in our model.

The results of simulating this model by Arzac (1967) also show that consumption lags, durable equipment expenditures plus inventory change lead, new construction leads, and the interest rate lags at turning points, thus providing a rough check with our analytical results. An interesting unsolved problem in stochastic dynamics is the derivation of mean leads or lags at turning points between time series generated by a system of linear stochastic difference equations.

5.9 ECONOMIC IMPLICATIONS

In economics, as in other sciences, we construct a model to say something about observations not used in its construction, but the reliability of the model itself has to be tested against unused observations. In this chapter we perform the dual role of testing a model and at the same time using it to say something about the nature of business cycles. Because the model has not been sufficiently tested, its implications on the nature of business cycles are not conclusive. We have found, however, that the cyclical properties of the model, judged from the individual time series and from the relations between time series, are broadly consistent with other observations.

In the five aggregative economic time series included very long cycles dominate shorter cycles, but the spectral density may have another peak at a higher frequency. Although the spectral density functions for consumption expenditures, new construction, private-domestic GNP, and long-term interest rate may be monotone-decreasing functions of frequency, expenditures on durable equipment plus inventory change may have another peak. The importance of short-period fluctuations in this last variable is in agreement with other observations.

The regression coefficients of the other four variables on GNP, computed by the corresponding cyclical components, show larger effects on consumption expenditures at low than at high frequencies, the opposite effects on investment expenditures, and only slightly larger effects on the interest rate at low frequencies. Compared with GNP, consumption expenditures lag, expenditures on durable equipment plus inventory change lead, new construction expenditures (except for high frequencies) lead, and the rate of interest lags so little that it is almost coincidental.

As a contribution to the theory of business cycles, we have found that the acceleration principle, although capable of generating complex roots in the system, does not necessarily produce local peaks in the spectral densities; we have found only one (very flat) local peak in durable equipment expenditures plus inventory change. Although the acceleration principle is important, and empirically well supported, its importance cannot be fully assessed by a nonstochastic model. For that matter, a nonstochastic linear model may be inadequate for the study of business cycles, but the present investigation has not uncovered any evidence against the use of a linear stochastic model, even a small one, in business-cycle theory. The importance of incorporating stochastic disturbances in the study of macroeconomic models, as stated at the beginning of Chapter 3, has now been documented. Without random disturbances no dynamic economic model will fit observed time series data. By incorporating random disturbances even simple linear models can do fairly well. The methods introduced so far are certainly applicable to comparative stochastic dynamic analysis in economics. By changing the parameters in the model we can deduce the consequences in terms of the various measures of the dynamic characteristics of the system.

Many references have already been mentioned in the text of this chapter. The material presented in Sections 5.7, 5.8, and 5.9 has been drawn mainly from Chow and Levitan (August 1969).

PROBLEMS

1. Specify the stochastic difference equations (including the residuals) satisfied by the canonical variables of system (15). What is the matrix B of the two right characteristic vectors in this example? Express Δy_{t+1} and y_t as linear combinations of the canonical variables.

2. It has been pointed out that first differences were taken for (9) and (12) to obtain serially uncorrelated residuals in these equations. Would some quasi-differencing operations be preferable? How would you determine the quasi-differencing operators by statistical analysis of the data?

3. If all first differences in (9) and (12) were replaced by the quasi-differences $y_{it} - .95 y_{i,t-1}$ and all other aspects of (9) to (12) were to remain the same, what would happen to the mean time paths of the system and to the multipliers?

4. Under the assumptions in Problem 3, what would happen to the autocovariances?

5. Under the assumptions in Problem 3, what would happen to the spectral and cross-spectral densities?

6. Using the coefficients in Table 5.1, set up a system of reduced-form equations in which the lagged exogenous variables are eliminated. What are the delayed multipliers of ΔG_{t-1} and ΔM_{t-1} on Y_t? What are the two-period ($m=2$) intermediate-run multipliers of ΔG and ΔM on Y_t?

7. Using Table 5.1, find the multipliers of ΔG_{t-3} and ΔM_{t-3} on Y_t.

8. Rewrite the system

$$y_t = A_1 y_{t-1} + A_2 y_{t-2} + C_0 x_t + C_1 x_{t-1} + C_2 x_{t-2} + d_t$$

in the form of (20),

$$\tilde{y}_t = A\tilde{y}_{t-1} + Cx_t + \tilde{d}_t.$$

9. Let G_t and M_t instead of ΔG_t and ΔM_t be chosen as the exogeneous variables in the system in Table 5.1. Write out the reduced form in terms of these exogeneous variables. Find the multipliers of M_{t-k} and G_{t-k} $(k=1,2,3,4)$ on Y_t.

10. Obtain the first row in Table 5.2 from the first row in Table 5.1, using four significant figures.

11. Calculate the expected time between maxima for the time series y_{1t} specified by (7) in Chapter 4. Compare your answer with the length of the damped cycles in the autocorrelation function and the length given by the peak of the spectral density. Comment on the differences.

12. Show that the expected time between relative maxima for a first-order autoregressive time series $y_t = ay_{t-1} + u_t$ with a nonnegative coefficient ranges between 3 and 4. What would the range be if the coefficient a could be negative? State the implicit assumptions underlying your calculations.

13. Using a computer program, perform all necessary calculations to answer the questions in Problems 3, 4, and 5. Do they agree with your theoretical reasoning?

14. Using a computer program, sutdy the mean paths and selected multipliers (of your own choice) pertaining to the time series ΔC and ΔR specified by Table 5.1.

15. Using a computer program, study the autocovariance characteristics of the time series ΔC and ΔR specified by Table 5.1.

16. Using a computer program, study the spectral densities of the time series ΔC and ΔR specified by Table 5.1.

17. Using a computer program, study the cross-spectral density of ΔC and ΔR specified by Table 5.1. Include in the analysis the phase-difference cross-spectral density, the gain of the components of ΔC on the components of ΔR, and the coherence. Comment briefly on the in-phase and out-of-phase cross-spectral densities. Discuss the economic meanings of your results.

18. Refer to the model in (9) to (14). If the coefficient $(1-g) = .79$ in (14) were changed to .81 by government lowering the marginal rate of taxation g, how would you study the economic consequences in terms of the mean rate of growth and the nature of fluctuations? Assume that government expenditures take the same course as before.

CHAPTER 6

Analysis of Nonstationary and Nonlinear Models

In this chapter we first extend the concepts of autocovariance and spectral properties to linear stochastic difference equations with roots greater than 1 in absolute value. The remainder of the chapter deals with the analysis of nonlinear systems. By linearizing a nonlinear system, or by using some other analytical device, the methods developed for deducing dynamic properties analytically from linear systems can be applied. Computer simulations can be used to generate samples from a nonlinear system, and the simulation results can be analyzed by spectral and other methods. Nonlinearities in the model will produce nonstationary time series, and spectral methods will have to be modified for the study of nonstationary time series.

6.1* ANALYSIS OF AN EXPLOSIVE FIRST-ORDER AUTOREGRESSION

Consider first the simple case of a univariate first-order linear stochastic difference equation,

$$z_t = \lambda z_{t-1} + \epsilon_t, \qquad (t = 1, \dots, T), \tag{1}$$

where ϵ_t are mutually independent random variables with mean 0 and variance w and λ *is greater than 1*. Given z_0, the autocovariance function $\text{cov}(z_t, z_{t-k})$ for such a series is a function of both t and k and will not reach a steady state as t increases, as (6) in Chapter 3 has revealed. The

122

variance of this series according to that equation is

$$w(\lambda^2 - 1)^{-1}(\lambda^{2t} - 1), \qquad (\lambda > 1),$$

and it increases with t. Can we define a time-invariant autocovariance function or spectral density function that is useful for the study of dynamic properties of this time series?

The answer is yes, provided that an exponential trend is taken from the sample $(z_1, \ldots, z_T)$ of this time series and the autocovariance properties of the deviations from the trend are examined. Note the new meaning given to the variance and covariances. In the last paragraph the variance of z_t means that from the vantage point of time 0 a mathematical expectation is taken of the squared deviations of the random variable z_t from its mean $z_0\lambda^t$. This variance will increase with t. In the present situation a sample z_1, $z_2, \ldots, z_T$ is generated by the stochastic difference equation (1) and an exponential trend is fitted *after* observing the sample. Now consider the autocovariances of the deviations of the sample observations from this fitted trend. A main difference between the two cases is that the mean $z_0\lambda^t$ in the first instance is announced *before* the sample is taken, whereas in the second instance the trend is determined *after* the sample is taken. Although, as t gets large, z_t may deviate a great deal from its pre-announced trend $z_0\lambda^t$, the deviation of z_t from a fitted trend can be controlled by bringing the trend close to the actual observations of the time series. The idea of taking the trend from a univariate first-order autoregression by the method described below and of studying the auto-covariance properties of the resulting deviations is due to Quenouille (1957). The study of spectral properties and the generalization to higher order systems of stochastic difference equations are contained in Chow and Levitan (June 1969).

Quenouille (1957, pp. 57–58) pointed out that an exponential trend can be found such that the deviations $z_1^d, \ldots, z_T^d$ of any sample $z_1, \ldots, z_T$ of (1) from the trend values constitute a covariance-stationary series. To show this, let us first impose the specification

$$z_t^d = \lambda z_{t-1}^d + \epsilon_t, \qquad (t = 1, \ldots, T), \qquad (2)$$

on the deviations z_t^d, where the ϵ_t are identical to those in (1). Note first that z_t^d in (2) can differ from z_t in (1) because z_0^d can differ from z_0. Second, define the trend x_t as the difference $z_t - z_t^d$. It will satisfy

$$x_t \equiv z_t - z_t^d = \lambda(z_{t-1} - z_{t-1}^d) \equiv \lambda x_{t-1} \qquad (3)$$

and is therefore an exponential trend. To produce a series z_t^d that satisfies

(2) and is also a sample from a stationary process we select the terminal value z_T^d from a distribution that has 0 mean and is independent of ϵ_T and works backward by (2), that is,

$$z_{t-1}^d = \lambda^{-1} z_t^d - \lambda^{-1} \epsilon_t, \qquad (t = T, T-1, \dots, 1). \tag{4}$$

The z_t^d so obtained are a sample, going backward in time, from a stationary first-order autoregression with coefficient λ^{-1} and independent residuals $-\lambda^{-1}\epsilon_t$. The steady-state variance of this series is, by (8) in Chapter 3,

$$\operatorname{var}(-\lambda^{-1}\epsilon_t) \cdot (1 - \lambda^{-2})^{-1} = w(\lambda^2 - 1)^{-1}. \tag{5}$$

To study the dynamic behavior of (4) in the steady state we can choose z_T^d from a distribution with variance equal to (5). Alternatively, we can simply let $z_T^d = 0$ if T is sufficiently large.

To show that the z_t^d in (4) can also be interpreted as a sample of a stationary first-order autoregression going forward in time we specify the equation

$$z_t^d = \lambda^{-1} z_{t-1}^d + \eta_t, \qquad (t = 1, \dots, T), \tag{6}$$

and show that the residuals η_t indeed have 0 mean and constant variance and are serially uncorrelated. By (4) and (6) these residuals satisfy

$$\eta_t = z_t^d - \lambda^{-1} z_{t-1}^d = (1 - \lambda^{-2}) z_t^d + \lambda^{-2} \epsilon_t. \tag{7}$$

To rid (7) of z_t^d we can apply (4) repeatedly and obtain

$$z_{t-1}^d = -\lambda^{-1}(\epsilon_t + \lambda^{-1}\epsilon_{t+1} + \cdots + \lambda^{-(T-t)}\epsilon_T - \lambda^{-(T-t)} z_T^d). \tag{8}$$

Thus z_{t-1}^d can be regarded as a linear function of the future ϵ's if the end point z_T^d is chosen by either method stated in the last paragraph. The second method of setting $z_T^d = 0$ ensures that the trend value x_T will equal z_T. When (8) is used to substitute for z_t^d in (7), we have

$$\eta_t = -(1 - \lambda^2)\lambda^{-1}(\epsilon_{t+1} + \lambda^{-1}\epsilon_{t+2} + \lambda^{-2}\epsilon_{t+3} + \cdots) + \lambda^{-2}\epsilon_t. \tag{9}$$

From (9) it follows that the residuals η_t of the forward scheme (6) will have the same autocovariance function as the residuals $-\lambda^{-1}\epsilon_t$ of the backward scheme (4). The expectation of η_t is 0. Its variance is

$$E\eta_t^2 = (1 - \lambda^{-2})^2\lambda^{-2}(1 + \lambda^{-2} + \lambda^{-4} + \cdots)w + \lambda^{-4}w = \lambda^{-2}w, \tag{10}$$

the same as the variance of $-\lambda^{-1}\epsilon_t$. The covariance between η_t and η_{t-k} $(k>0)$ is

$$E\eta_t\eta_{t-k} = E\left[-(1-\lambda^{-2})\lambda^{-1}(\epsilon_{t+1}+\lambda^{-1}\epsilon_{t+2}+\cdots)+\lambda^{-2}\epsilon_t\right]$$

$$\times\left[-(1-\lambda^{-2})\lambda^{-1}(\epsilon_{t-k+1}+\lambda^{-1}\epsilon_{t-k+2}+\cdots)+\lambda^{-2}\epsilon_{t-k}\right]$$

$$=(1-\lambda^{-2})^2\lambda^{-2-k}(1-\lambda^{-2})^{-1}w-(1-\lambda^{-2})\lambda^{-2-k}w=0. \qquad (11)$$

Equation 8 can be used to derive the autocovariance function and the spectral density function of the detrended series z_t^d of (6). It will also be used in Section 6.2 to obtain the autocovariance and spectral properties of explosive systems of linear stochastic difference equations. In this section we have found that *any sample from the nonstationary process* (1), *with ϵ_t as residuals, can be decomposed into an exponential trend plus a sample from the stationary series* (6) *whose residuals η_t are each the sum of $\lambda^{-2}\epsilon_t$ and a term independent of it*. These residuals have mean 0, variance $\lambda^{-2}w$, and are uncorrelated. (See Problem 1.)

6.2* DYNAMIC PROPERTIES OF EXPLOSIVE LINEAR SYSTEMS

In this section we decompose a sample from the system

$$y_t = Ay_{t-1}+u_t \qquad (12)$$

into a sample from a stationary system plus exponential trends and derive the autocovariance matrix and spectral density matrix of the stationary system. The $p\times p$ matrix A in (12) is assumed to be real and nonsingular. The covariance matrix of u_t is denoted by V, which is assumed to be positive semidefinite, with some equations in (12) that are possibly identities.

A set of canonical variables z_t, which are linear combinations of y_t, is defined by diagonalizing the matrix A,

$$A = BD_\lambda B^{-1}, \qquad (13)$$

where D_λ is the diagonal matrix consisting of the characteristic roots of A, and the columns of B are the corresponding (right) characteristic vectors. Equations 12 and 13 imply

$$B^{-1}y_t = D_\lambda B^{-1}y_{t-1}+B^{-1}u_t \qquad (14)$$

or

$$z_t = D_\lambda z_{t-1}+\epsilon_t.$$

The canonical variables z_t are defined as $B^{-1}y_t$, and the residuals ϵ_t $= B^{-1}u_t$ in the transformed system have a covariance matrix

$$E\epsilon_t\epsilon_t' = B^{-1}VB^{-1'} = W = (w_{ij}). \tag{15}$$

If some roots are greater than 1 in absolute value, the system will be nonstationary. We first treat a single real root greater than 1, then, later on in this section, many real roots greater than 1, leaving the complex roots greater than 1 in absolute value to Section 6.3. Let $\lambda_1 > 1$. By the result in Section 6.1 a sample of the first canonical variable

$$z_{1t} = \lambda_1 z_{1,t-1} + \epsilon_{1t}, \qquad (t = 1, \ldots, T), \tag{16}$$

can be decomposed into an exponential trend

$$x_{1t} = \lambda_1 x_{1,t-1}, \qquad (t = 1, \ldots, T), \tag{17}$$

plus a sample from a stationary series

$$z_{1t}^d = \lambda_1^{-1} z_{1,t-1}^d + \eta_{1t}, \qquad (t = 1, \ldots, T), \tag{18}$$

where η_{1t} have 0 mean and are mutually uncorrelated.

A sample from the multivariate time series (12) can be similarly decomposed. Observations on the first canonical variable are assumed to be decomposed as in the preceding paragraph. Observations on all other canonical variables will be unchanged. Let b_{ij} be the ijth element of B and $b_{i\cdot}$ the ith row of B. The sample of the ith variable is partitioned as

$$y_{it} = b_{i\cdot}z_t^d + b_{i1}x_{1t}, \qquad (t = 1, \ldots, T), \tag{19}$$

where z_t^d denotes the vector z_t with the first component replaced by z_{1t}^d. We are concerned with the dynamic properties of the detrended observations

$$y_{it}^d = b_{i\cdot}z_t^d, \qquad (t = 1, \ldots, T), \tag{20}$$

after the exponential trends $b_{i1}x_{1t}$ $(i = 1, \ldots, p)$ have been eliminated.

Consider first the autocovariance matrix of the (detrended) canonical variables

$$\Gamma_k^{zd} = E(z_t^d z_{t-k}^{d'}) = E(z_{it}^d z_{j,t-k}^d) = (\gamma_{ij,k}^{zd}). \tag{21}$$

Here the superscript z indicates that the variables are canonical rather than the original variables y, and the superscript d signifies "detrended." Expectations here and elsewhere are taken under the assumption that the correct value of λ, and not an estimate of it, is used for detrending. The covariances $\gamma_{ij,k}^{zd}$ for the three cases $(i,j \neq 1)$, $(i,j = 1)$, and $(i = 1, j \neq 1$ or

$i \neq 1, j = 1$) are given, respectively, by expressions (22), (23), and (24).

$$
\begin{aligned}
\gamma_{ij,k}^{zd} = \gamma_{ij,k}^{z} &= E z_{it} z_{j,t-k} \\
&= E\left(\epsilon_{it} + \lambda_i \epsilon_{i,t-1} + \lambda_i^2 \epsilon_{i,t-2} + \cdots\right)\left(\epsilon_{j,t-k} + \lambda_j \epsilon_{j,t-k-1} + \cdots\right) \\
&= \lambda_i^k\left(1 + \lambda_i \lambda_j + \lambda_i^2 \lambda_j^2 + \cdots\right) E\epsilon_{it}\epsilon_{jt} \\
&= \frac{\lambda_i^k w_{ij}}{1 - \lambda_i \lambda_j} = \gamma_{ji,-k}^{z}, \qquad (i,j \neq 1; \ k \geqslant 0).
\end{aligned} \tag{22}
$$

Using (18) and the properties of η_{1t} given in (10) and (11), we have

$$
\lambda_{11,k}^{zd} = \gamma_{11,-k}^{zd} = \frac{\lambda_1^{-k}\left(\lambda_1^{-2} w_{11}\right)}{1 - \lambda_1^{-2}}, \qquad (k \geqslant 0). \tag{23}
$$

$$
\begin{aligned}
\gamma_{1j,k}^{zd} &= E\left(z_{1,t}^d z_{j,t-k}\right) \\
&= -E\left(\lambda_1^{-1} \epsilon_{1,t+1} + \lambda_1^{-2} \epsilon_{1,t+2} + \cdots\right)\left(\epsilon_{j,t-k} + \lambda_j \epsilon_{j,t-k-1} + \cdots\right) \\
&= 0 = \gamma_{j1,-k}^{zd}, \qquad (j \neq 1; \ k \geqslant 0),
\end{aligned} \tag{24a}
$$

and

$$
\begin{aligned}
\gamma_{1j,-k}^{zd} &= -E\left(\lambda_1^{-1} \epsilon_{1,t+1} + \lambda_1^{-2} \epsilon_{1,t+2} + \cdots\right)\left(\epsilon_{j,t+k} + \lambda_j \epsilon_{j,t+k-1} + \cdots\right) \\
&= -\left(\lambda_1^{-k} + \lambda_1^{-k+1}\lambda_j + \lambda_1^{-k+2}\lambda_j^2 + \cdots + \lambda_1^{-1}\lambda_j^{k-1}\right) w_{1j} \\
&= \frac{\left(\lambda_j^k - \lambda_1^{-k}\right) w_{1j}}{1 - \lambda_1 \lambda_j} = \gamma_{j1,k}^{zd}, \qquad (j \neq 1; \ k \geqslant 1).
\end{aligned} \tag{24b}
$$

Here we ignore the possibility that $\lambda_1 \lambda_j = 1$, in which case $(1 - \lambda_1^k \lambda_j^k)/(1 - \lambda_1 \lambda_j)$ should be replaced by k and (24b) would equal $-\lambda_1^{-k} k w_{1j}$.

Using these covariances, we can obtain the corresponding spectral densities by

$$
f_{ij}^{zd}(\omega) = \frac{1}{\pi} \sum_{k=-\infty}^{\infty} \gamma_{ij,k}^{zd} e^{-i\omega k}, \qquad (i = \sqrt{-1}\,). \tag{25}
$$

Because the key term in the covariance functions is λ_1^{-k} or λ_j^k, the weighted sum (25) can be performed by using

$$
\sum_{k=0}^{\infty} \lambda_j^k e^{-i\omega k} = \frac{1}{1 - \lambda_j e^{-i\omega}}. \tag{26}
$$

The spectral densities are (27), (28), and (29).

$$f_{ij}^{zd}(\omega) = \frac{1}{\pi}\left(\sum_{k=0}^{\infty} \gamma_{ij,k}^{zd} e^{-i\omega k} + \sum_{k=0}^{\infty} \gamma_{ij,-k}^{zd} e^{i\omega k} - \gamma_{ij,0}^{zd} \right)$$

$$= \frac{1}{\pi}\cdot\frac{w_{ij}}{1-\lambda_i\lambda_j}\left(\frac{1}{1-\lambda_i e^{-i\omega}} + \frac{1}{1-\lambda_j e^{i\omega}} - 1 \right)$$

$$= \frac{1}{\pi}\cdot\frac{w_{ij}}{\left(1-\lambda_i e^{-i\omega}\right)\left(1-\lambda_j e^{i\omega}\right)}, \qquad (i,j\neq 1), \tag{27}$$

$$f_{11}^{zd}(\omega) = \frac{1}{\pi}\cdot\frac{\lambda_1^{-2}w_{11}}{\left(1-\lambda_1^{-1}e^{-i\omega}\right)\left(1-\lambda_1^{-1}e^{i\omega}\right)}$$

$$= \frac{1}{\pi}\cdot\frac{w_{11}}{\left(1-\lambda_1 e^{-i\omega}\right)\left(1-\lambda_1 e^{i\omega}\right)} \tag{28}$$

$$f_{1j}^{zd}(\omega) = \frac{1}{\pi} \sum_{k=1}^{\infty} \gamma_{1j,-k}^{zd} e^{i\omega k}$$

$$= \frac{1}{\pi}\cdot\frac{w_{1j}}{1-\lambda_1\lambda_j}\left(\frac{\lambda_j e^{i\omega}}{1-\lambda_j e^{i\omega}} - \frac{\lambda_1^{-1}e^{i\omega}}{1-\lambda_1^{-1}e^{i\omega}} \right)$$

$$= \frac{1}{\pi}\cdot\frac{w_{1j}}{\left(1-\lambda_1 e^{-i\omega}\right)\left(1-\lambda_j e^{i\omega}\right)} \tag{29a}$$

and

$$f_{j1}^{zd}(\omega) = \frac{1}{\pi}\cdot\frac{w_{j1}}{\left(1-\lambda_j e^{-i\omega}\right)\left(1-\lambda_1 e^{i\omega}\right)}. \tag{29b}$$

The results (28) and (29) show that even when a characteristic root λ_1 is greater than unity, or a canonical variable z_{1t} is nonstationary, the spectral or cross-spectral density functions $f_{ij}^{zd}(\omega)$ of the canonical variables, once detrended, will have the same form as the function (27) for the stationary canonical variables. The autocovariance or cross-covariance functions (23) and (24) involving the detrended canonical variable z_{1t}^d differ in form from the function (22) for the originally stationary canonical variables; the corresponding spectral density functions have the same form, however. This result does not lead to mathematical inconsistency because the covariance functions (23) and (24) are defined and valid only for $\lambda_1 > 1$, whereas the covariance functions (22) are defined for $\lambda_i < 1$ and $\lambda_j < 1$.

It is easy to show that the formula (27) applies to two or more real roots greater than unity. For $\lambda_1 > 1$ and $\lambda_2 > 1$, say, we need to check only $f_{12}^{zd}(\omega)$. Using (8), we obtain

$$
\begin{aligned}
\gamma_{12,k}^{zd} &= E\left(z_{1,t}^d z_{2,t-k}^d\right) \\
&= E\left(\lambda_1^{-1}\epsilon_{1,t+1} + \lambda_1^{-2}\epsilon_{1,t+2} + \cdots\right)\left(\lambda_2^{-1}\epsilon_{2,t-k+1} + \lambda_2^{-2}\epsilon_{2,t-k+2} + \cdots\right) \\
&= \frac{\lambda_2^{-k}\left(\lambda_1^{-1}\lambda_2^{-1}w_{12}\right)}{1 - \lambda_1^{-1}\lambda_2^{-1}} = \gamma_{21,-k}^{zd}, \qquad (k \geqslant 0).
\end{aligned} \tag{30}
$$

The corresponding cross-spectral density function is

$$
\begin{aligned}
f_{12}^{zd}(\omega) &= \frac{1}{\pi}\left(\sum_{k=0}^{\infty}\gamma_{12,k}^{zd}e^{-i\omega k} + \sum_{k=0}^{\infty}\gamma_{12,-k}^{z}e^{i\omega k} - \gamma_{12,0}^{z}\right) \\
&= \frac{1}{\pi}\frac{\lambda_1^{-1}\lambda_2^{-1}w_{12}}{1 - \lambda_1^{-1}\lambda_2^{-1}}\left(\frac{1}{1 - \lambda_2^{-1}e^{-i\omega}} + \frac{1}{1 - \lambda_1^{-1}e^{i\omega}} - 1\right) \\
&= \frac{1}{\pi}\frac{w_{12}}{(\lambda_2 - e^{-i\omega})(\lambda_1 - e^{i\omega})} \\
&= \frac{1}{\pi}\frac{w_{12}}{(1 - \lambda_1 e^{-i\omega})(1 - \lambda_2 e^{i\omega})},
\end{aligned} \tag{31}
$$

which has the same form as (27).

In fact, this analysis can be applied when there are complex roots. Obviously, if the complex roots are smaller than 1 in absolute value, all of our derivations in this section remain valid. For complex roots greater than 1 in absolute value the reader is referred to Section 6.3.

We have derived the autocovariance matrix $\Gamma_k^{zd} = [\gamma_{ij,k}^{zd}]$ and the spectral density matrix $F^{zd}(\omega) = [f_{ij}^{zd}(\omega)]$ of the detrended canonical variables. The corresponding matrices for the detrended original variables y_t^d, defined by (20), are simply

$$
\Gamma_k^d = B\Gamma_k^{zd}B' \tag{32}
$$

and

$$
F^d(\omega) = BF^{zd}(\omega)B' = \frac{1}{\pi}B\left(\frac{w_{ij}}{(1 - \lambda_i e^{-i\omega})(1 - \lambda_j e^{i\omega})}\right)B'. \tag{33}
$$

The main result of this section is summarized by the following theorem:

Theorem. Given system (12), with $A = BD_\lambda B^{-1}$, D_λ a diagonal matrix consisting of the roots λ_j of A, $Eu_t u_t' = V$, $(w_{ij}) = B^{-1}VB^{-1'}$, (33) is the spectral density matrix of

$$y_{it}^d = b_i. z_t^d,$$

where $b_i.$ is the ith row of B, $z_t = B^{-1}y_t$, $z_{jt}^d = z_{jt}$ for $|\lambda_j| < 1$, $z_{jt}^d = z_{jt} - x_{jt}$ for $|\lambda_j| > 1$ and for the exponential trend $x_{jt} = \lambda_j x_{j,t-1}$.

Strictly speaking, our method breaks down when a root is *exactly* 1, but this rarely happens. Even when certain theorizing, such as assuming linear equations for the first differences of some variables, could lead to roots of unity if the theory were strictly adhered to, we would still recommend using (33), for the theory itself is only an approximation. This issue was discussed in Chapter 5. The study of spectral properties after trends have been taken from the time series generated by a stochastic model is appropriate for comparison with spectral densities estimated directly from economic data because trend elimination has also preceded spectral analysis in practice. Our method also produces the trends implicit in the model which can be compared with the observed trends.

6.3* THE CASE OF COMPLEX ROOTS GREATER THAN ONE IN ABSOLUTE VALUE

We now examine the applicability of the method in Section 6.2 to complex roots greater than 1 in absolute value. Because complex roots come in conjugate pairs, let λ_1 and λ_2 be a pair. As pointed out in Section 2.8, the corresponding right characteristic vectors b_1 and b_2 are also complex conjugates. Define the left characteristic vectors as the row vectors r_i which satisfy $r_i A = \lambda_i r_i$. It is easy to see that the rows of B^{-1}, denoted by b^i, are the left characteristic vectors of A. The vectors b^1 and b^2 corresponding to λ_1 and λ_2 are also conjugate complex. (See Problem 2.) Therefore $z_{1t} = b^1 y_t$ and $z_{2t} = b^2 y_t$ are conjugate complex, as are $\epsilon_{1t} = b^1 u_t$ and $\epsilon_{2t} = b^2 u_t$. From any sample of the pair of complex canonical time series

$$z_{1t} = \lambda_1 z_{1,t-1} + \epsilon_{1t}$$

and

$$z_{2t} = \lambda_2 z_{2,t-1} + \epsilon_{2t}, \qquad (t = 1, \ldots, T),$$

we subtract the sample of the complex stationary series

$$z_{1t}^d = \lambda_1^{-1} z_{1,t-1}^d + \eta_{1t}$$

and

$$z_{2t}^d = \lambda_2^{-1} z_{2,t-1}^d + \eta_{2t}, \qquad (t = 1, \ldots, T),$$

where, as before, z_{1t}^d and z_{2t}^d are generated backward in time, with z_{1T}^d and z_{2T}^d selected at random from their conjugate bivariate stationary process or set equal to 0 for sufficiently large T. The remaining "trends"

$$x_{1t} = z_{1t} - z_{1t}^d = \lambda_1^t x_{10}$$

and

$$x_{2t} = z_{2t} - z_{2t}^d = \lambda_2^t x_{20}, \qquad (t = 1, \ldots, T),$$

will be conjugate complex.

For the partition of the sample of the original variables we replace (19) with

$$y_{it} = b_{i \cdot} z_t^d + b_{i1} x_{1t} + b_{i2} x_{2t}, \qquad (t = 1, \ldots, T). \tag{34}$$

The contribution of λ_1 and λ_2 to the stationary part $b_{i \cdot} z_t^d$ is real because the sum $b_{i1} z_{1t}^d + b_{i2} z_{2t}^d$ is real. The nonstationary part is

$$b_{i1} x_{10} \lambda_1^t + b_{i2} x_{20} \lambda_2^t.$$

Letting r and θ be the absolute value and the angle of the roots and s_i and p_i, the absolute value and the angle of the coefficients, we can write the nonstationary part as (superscript $i = \sqrt{-1}$)

$$s_i r^t (e^{i(\theta t + p_i)} + e^{-i(\theta t + p_i)}) = 2 s_i r^t \cos(\theta t + p_i), \tag{35}$$

which is an explosive cosine function of time. Although our method of decomposition is formally applicable to a pair of complex (canonical) time series, we have to keep in mind the cyclical nature of the nonstationary part eliminated from the original time series.

6.4 LINEARIZING A NONLINEAR SYSTEM

The autocovariance matrix and the spectral density matrix of a nonlinear stochastic system will in general be (matrix) functions of time. Therefore, for a nonlinear system, we cannot expect the autocovariance and spectral properties to be stationary. However, if the system is not too far from being linear or the above dynamic properties change only slowly through time, it will be useful to deduce these properties for a particular time interval. Even if the system is highly nonlinear, it will still be interesting to examine and compare these properties at different time periods. Such a comparison will reveal how the nonlinearities affect the spectral properties and will provide some measure of the extent of nonlinearity in the system. Once methods are available to estimate spectral and cross-spectral densities valid for economic time series during particular time periods the time-dependent spectral properties deduced from a nonlinear model can be compared with the empirically estimated counterparts.

The first method to be discussed is based on linearizing a nonlinear system of econometric equations. Ordinarily the econometric equations to be analyzed take the form of structural equations. Several endogenous variables appear simultaneously in each structural equation, but one endogenous variable is usually separated out. With additive residuals the equations can be written as

$$y_{it} - \Phi_i(y_{1t},\ldots,y_{pt}; y_{1,t-1},\ldots,y_{p,t-1}; x_{1t},\ldots,x_{qt}) = \epsilon_{it} \tag{36}$$

or

$$y_{it} - \Phi_i(y_t, y_{t-1}, x_t) = \epsilon_{it}, \qquad (i=1,\ldots,p).$$

Here all lagged endogenous variables dated $t-k$ for $k>1$ and all lagged exogenous variables are assumed to have been eliminated by identities. The residuals $\epsilon_{it}(i=1,\ldots,p)$ are assumed to have a joint distribution with mean 0 and a covariance matrix unchanging through time, and the vector $\epsilon_t = (\epsilon_{1t},\ldots,\epsilon_{pt})'$ is assumed to be independent of ϵ_s for $t \neq s$. The values of x_t are assumed to be fixed and given. The aim is to study the dynamic properties of the deviations of the vector time series y_t generated by these equations from some trend values.

Denote by $\bar{y}_t$ the solution to the deterministic model obtained by ignoring the random disturbances, that is,

$$\bar{y}_{it} = \Phi_i(\bar{y}_t, \bar{y}_{t-1}, x_t), \tag{37}$$

and denote the deviation $y_t - \bar{y}_t$ by y_t^*. We can approximate (36) by a second-order Taylor expansion around $\bar{y}_t$ and $\bar{y}_{t-1}$ as follows:

$$y_{it}^* + \bar{y}_{it} - \Phi_i(\bar{y}_t, \bar{y}_{t-1}, x_t) - \left(\frac{\partial \Phi_i}{\partial y_t}\right)' y_t^* - \left(\frac{\partial \Phi_i}{\partial y_{t-1}}\right)' y_{t-1}^*$$

$$- \tfrac{1}{2} y_t^{*\prime} \left(\frac{\partial^2 \Phi_i}{\partial y_t \partial y_t'}\right) y_t^* - \tfrac{1}{2} y_{t-1}^{*\prime} \left(\frac{\partial^2 \Phi_i}{\partial y_{t-1} \partial y_{t-1}'}\right) y_{t-1}^*$$

$$- y_t^{*\prime} \left(\frac{\partial^2 \Phi_i}{\partial y_t \partial y_{t-1}'}\right) y_{t-1}^* \cong \epsilon_{it}, \qquad (38)$$

where $\partial \Phi_i / \partial y_t$ is the $p \times 1$ vector of partial derivatives of Φ_i with respect to the elements of y_t and $\partial^2 \Phi_i / \partial y_t \partial y_{t-1}'$ is the $p \times p$ matrix whose $m - n$ element is the second partial of Φ_i with respect to y_{mt} and $y_{n,t-1}$, all derivatives being evaluated at $\bar{y}_t$ and $\bar{y}_{t-1}$. If the terms involving second and higher order derivatives can be ignored, and this can be a poor assumption if the model is highly nonlinear, then (38) can be approximated by

$$y_{it}^* - \left(\frac{\partial \Phi_i}{\partial y_t}\right)' y_t^* - \left(\frac{\partial \Phi_i}{\partial y_{t-1}}\right)' y_{t-1}^* = \epsilon_{it}. \qquad (39)$$

Defining the matrices

$$B_{1t} \equiv \left(\frac{\partial \Phi_1}{\partial y_t} \cdots \frac{\partial \Phi_p}{\partial y_t}\right)' \quad \text{and} \quad B_{2t} \equiv \left(\frac{\partial \Phi_1}{\partial y_{t-1}} \cdots \frac{\partial \Phi_p}{\partial y_{t-1}}\right)' \qquad (40)$$

and combining (39) for $i = 1, \ldots, p$ we can write the linearized system for the vector y_t^* of deviations as

$$y_t^* - B_{1t} y_t^* - B_{2t} y_{t-1}^* = \epsilon_t. \qquad (41)$$

The matrices B_{1t} and B_{2t} of the coefficients in this system are time-dependent, but we can select a particular time period to fix the coefficients B_1 and B_2 and analyze the dynamic behavior of the system

$$y_t^* = (I - B_1)^{-1} B_2 y_{t-1}^* + (I - B_1)^{-1} \epsilon_t \qquad (42)$$

by using the methods set forth in this book.

Use of a linearized model in the manner described above for the purpose of studying spectral properties was suggested by Bowden (1972), who applied the method to the Klein-Goldberger model. Howrey (1971) also linearized the same model for the purpose of studying its stochastic properties. Howrey evaluated its linearized version about the vector of sample means. Goldberger (1959) had performed the same linearization and derived various multipliers for the Klein-Goldberger model.

The following procedure can be applied to linearize a model around the sample means. Let the vector of sample means be $\bar{y}_{s-1}$ and define the vector $\bar{y}_s$ as the solution of

$$\bar{y}_{is} = \Phi_i(\bar{y}_s, \bar{y}_{s-1}, \bar{x}), \qquad (i = 1, \ldots, p), \qquad (43)$$

where $\bar{x}$ is the vector of sample means of the exogenous variables. Then linearizing the left-hand side of

$$y_{is} - \Phi_i(y_s, y_{s-1}, \bar{x}) = \epsilon_{is} \qquad (44)$$

as a function of y_s and y_{s-1} by the method in (38) and (39) will give

$$y_{is}^* - \left(\frac{\partial \Phi_i}{\partial y_s}\right)' y_s^* - \left(\frac{\partial \Phi_i}{\partial y_{s-1}}\right)' y_{s-1}^* = \epsilon_{is}, \qquad (45)$$

where $y_s^* = y_s - \bar{y}_s$ and $y_{s-1}^* = y_{s-1} - \bar{y}_{s-1}$ and the vectors $\partial \Phi_i/\partial y_s$ and $\partial \Phi_i/\partial y_{s-1}$ of partial derivatives are evaluated, respectively, at $\bar{y}_s$ and $\bar{y}_{s-1}$.

6.5* SPECTRAL PROPERTIES WITHOUT LINEARIZING THE MODEL

A second approach to deducing spectral properties from a nonlinear model is to consider one definition of the spectral density matrix for a linear system and to apply the same to a nonlinear system. This approach was used by Howrey and Klein (1972).

The definition chosen to compute the spectral density matrix of a linear system with 0 mean is as follows:

In our notation a linear system with serially uncorrelated residuals is written as

$$y_t = Ay_{t-1} + u_t = u_t + Au_{t-1} + A^2 u_{t-2} + A^3 u_{t-3} + \cdots, \qquad (46)$$

where t is sufficiently large for $A^t y_0$ to be ignored; but let us consider a more general linear system with possibly moving average residuals:

$$y_t = Ay_{t-1} + u_t + Gu_{t-1}$$

$$= u_t + (A + G)u_{t-1} + A(A + G)u_{t-2} + A^2(A + G)u_{t-3} + \cdots$$

$$= D_0 u_t + D_1 u_{t-1} + D_2 u_{t-2} + D_3 u_{t-3} + \cdots. \qquad (47)$$

Thus y_t can be written as a weighted sum of $u_t, u_{t-1}, u_{t-2}, \ldots$, which are uncorrelated in time and have the same contemporaneous covariance matrix V. (See Problem 10.)

The autocovariance matrix of y_t at the steady state is, for $k \geq 0$,

$$\Gamma_k = E(D_0 u_t + D_1 u_{t-1} + D_2 u_{t-2} + \cdots)(u'_{t-k}D'_0 + u'_{t-k-1}D'_1 + u'_{t-k-2}D'_2 + \cdots$$

$$= D_k VD'_0 + D_{k+1} VD'_1 + D_{k+2} VD'_2 + D_{k+3} VD'_3 + \cdots. \tag{48}$$

The spectral density matrix is obtained by summing (48) term by term:

$$F(\omega) = \frac{1}{\pi}\left(\sum_{k=0}^{\infty} \Gamma_k e^{-i\omega k} + \sum_{k=0}^{\infty} \Gamma'_k e^{i\omega k} - \Gamma_0 \right)$$

$$= \frac{1}{\pi}\left[\left(\sum_{k=0}^{\infty} D_k e^{-i\omega k} \right) VD'_0 + \left(\sum_{k=0}^{\infty} D_{k+1} e^{-i\omega k} \right) VD'_1 + \cdots \right.$$

$$\left. + D_0 V\left(\sum_{k=0}^{\infty} D'_k e^{i\omega k} \right) + D_1 V\left(\sum_{k=0}^{\infty} D'_{k+1} e^{i\omega k} \right) + \cdots - \Gamma_0 \right]. \tag{49}$$

The point to note in (49) is that the spectral density matrix of a linear system can be calculated by taking the Fourier transform $\sum_{k=0}^{\infty} D_k e^{-i\omega k}$ of the matrices D_k in (47).

For a nonlinear system Howrey and Klein (1972) suggest obtaining a deterministic solution corresponding to a given time path of the exogenous variables of interest by setting the random disturbances of the model equal to 0. The vector y_t^* of the deviations of the stochastic time series y_t generated by the model from the above solution is expressed as a weighted sum $\sum_{s=0}^{N} D_s u_{t-s}$ of the random disturbances. Note that the i-j element $d_{ij,s}$ of the matrix D_s is the partial derivative of y_{it}^* with respect to $u_{j,t-s}$. To estimate $d_{ij,s}$ let $u_{j,1} = \epsilon$ and $u_{j,t} = 0$ for $t \neq 1$, where ϵ is some small number. Further, let $u_{it} = 0$ for $i \neq j$. Thus the only residual activated is $u_{j,1}$. Now compute $y_{i,s}^*$ as the difference of the solution from the deterministic solution mentioned for $s = 1, 2, \ldots, N$. The resulting $y_{i,s}^*$ is divided by ϵ to form an estimate of the derivative $d_{ij,s}$. The choice of the size of ϵ is a problem in calculating numerical derivatives; ϵ should be small compared with the variation in the variables $y_{i,s}$. When the chosen value of ϵ is allowed to vary somewhat, say, to be a factor from 4 to $\frac{1}{4}$, the resulting estimates of the derivatives should remain approximately the same. In any case, by using the estimates of $d_{ij,s}$ we can apply (49) to calculate the

136   ANALYSIS OF NONSTATIONARY AND NONLINEAR MODELS

spectral density matrix. The summations to infinity have to be replaced by summations to N. Howrey and Klein (1972) have applied this method to calculate spectral properties from the Wharton model. They have pointed out, however (p. 600), that "no attempt is made at this point to justify this approach to the analysis of nonlinear systems beyond the analogy with linear systems." It seems reasonable to suppose that the method will work well only if the system, net of the effects of random disturbances and of the exogenous variables, has a stable equilibrium. Otherwise the definition of spectral density used in its construction will not work even for a linear system.

6.6 TWO METHODS FOR SOLVING NONLINEAR STRUCTURAL EQUATIONS

The methods of both Sections 6.4 and 6.5 require that a nonlinear system like (36) be solved for the current endogenous variables. Two extremely useful iterative methods for solving such a system are the Gauss-Siedel and the Newton-Raphson. An iterative method computes a vector $y^{(r+1)}$ at the $r+1$st iteration or trial from the vector $y^{(r)}$ at the preceding iteration. This method is said to converge at the rth iteration if $y^{(r+1)}$ equals $y^{(r)}$ to as many significant digits as specified.

Perhaps the most popular method for solving nonlinear structural equations in econometrics is the Gauss-Siedel. Let the structural equations be (36). The random disturbances ϵ_{it} may be the result of random drawings from a joint distribution with mean 0 and covariance matrix Σ as the model stipulates or set equal to 0. The first occurs when stochastic simulations are performed, the second when nonstochastic simulations are applied. From the viewpoint of solving the equations for the vector y_t, given y_{t-1} and x_t, we treat the vector ϵ_t as given in either case. Let the values of the unknowns in the rth iteration be $y_t^{(r)}$. The Gauss-Siedel method computes the values for the next iteration by the equations

$$y_{1t}^{(r+1)} = \Phi_1\left(y_{1t}^{(r+1)}, y_{2t}^{(r)}, \ldots, y_{pt}^{(r)}; y_{t-1}; x_t\right) + \epsilon_{1t}$$

$$y_{2t}^{(r+1)} = \Phi_2\left(y_{1t}^{(r+1)}, y_{2t}^{(r+1)}, y_{3t}^{(r)}, \ldots; y_{t-1}; x_t\right) + \epsilon_{2t}$$

$$y_{3t}^{(r+1)} = \Phi_3\left(y_{1t}^{(r+1)}, y_{2t}^{(r+1)}, y_{3t}^{(r+1)}, \ldots; y_{t-1}; x_t\right) + \epsilon_{3t} \tag{50}$$

$$y_{pt}^{(r+1)} = \Phi_p\left(y_{1t}^{(r+1)}, \ldots, y_{p-1,t}^{(r+1)}, y_{pt}^{(r+1)}; y_{t-1}; x_t\right) + \epsilon_{pt}.$$

Several points should be noted. First, the unknowns in each iteration are computed one at a time. In (50) $y_{1t}^{(r+1)}$ is computed first, then used to compute $y_{2t}^{(r+1)}$, and both, in turn, are used to compute $y_3^{(r+1)}$, and so on. Thus the p equations (36) are solved one at a time rather than simultaneously in each iteration. Second, in nearly all econometric applications the variable y_{it} is *not* an argument of the function Φ_i. This special feature makes the solution of each equation simple; it amounts to the evaluation of the function Φ_i. Otherwise, we may choose to solve for $y_{it}^{(r+1)}$ in each equation by some iterative method, which may itself be a difficult problem. Of course, we may also use $y_{it}^{(r)}$ as an argument in Φ_i to obtain $y_{it}^{(r+1)}$. It is this feature that makes the Gauss-Siedel method so attractive for solving nonlinear econometric equations. Third, it is obvious from the description of the method that it matters how the equations are ordered. There are $p!$ ways of ordering the p equations and some will give faster convergence than others. Fourth, given any ordering, there is no guarantee that the iterations will converge to the solution; for example, consider the following two equations for y_1 and y_2, with the subscript t omitted:

$$y_1 = 2y_2 + 1$$
$$y_2 = y_1 + 3. \tag{51}$$

Starting with $y_1^{(1)} = 0$ and $y_2^{(1)} = 0$, say, the iterations by the Gauss-Siedel method will not converge. In practice, however, when damping factors are introduced to inhibit oscillations and when the ordering and other convergence problems are properly dealt with, the Gauss-Siedel method has been successively applied to the solution of nonlinear structural equations in econometrics.

The second method to be briefly stated is the Newton-Raphson method. Let us write the system of equations (36) in vector form by defining Φ as a column vector of functions with Φ_i as elements and f as another vector of functions satisfying the vector equation

$$f(y) = y - \Phi(y) - \epsilon = 0. \tag{52}$$

In (52) the subscript t of y_t has been suppressed and so have the known parameters y_{t-1} and x_t of the functions. To solve $f(y) = 0$ the Newton-Raphson method iterates by the vector equation

$$y^{(r+1)} = y^{(r)} - \left[\frac{\partial f_i(y)}{\partial y_j} \right]^{-1} f(y^{(r)}) \tag{53}$$

where the $i-j$ element of the matrix in squared brackets is the partial derivative of $f_i(y)$ with respect to y_j evaluated at $y^{(r)}$. One property of the Newton-Raphson method is that if the system is linear, that is,

$$f(y) = Ay - b = 0,   \tag{54}$$

convergence occurs in one iteration beginning with any initial trial $y^{(0)}$, provided, of course, that the matrix A is nonsingular. In the linear case the first iteration is

$$y^{(1)} = y^{(0)} - A^{-1}[Ay^{(0)} - b] = A^{-1}b.   \tag{55}$$

Modifications to the method are available to speed up convergence, but we shall not go into them here.

6.7 DYNAMIC PROPERTIES THROUGH STOCHASTIC SIMULATIONS

Once the nonlinear equations are solved, y_t can be computed, given y_{t-1} and x_t, together with values of ϵ_t in stochastic simulations. Samples of y_t $(t = 1, \ldots, T)$ can be generated by using the computer. With n samples or replications we can study various dynamic characteristics of the time series generated by the nonlinear system.

The analysis of simulation results from nonlinear stochastic models in economics can be conveniently classified in two types—short-run and long-run analyses. In the short-run analysis the behavior of the system in a short period is examined. Usually not more than several years are covered. Such a study can throw light on whether the econometric model can satisfactorily explain certain historical episodes, such as a recession, and on whether a certain choice for some policy variables included in the exogenous variables x_t could have improved the performance of the economy during particular historical periods. In general, given certain initial conditions y_0, historical or otherwise, we can study the short-run behavior of the system and the short-run impacts of certain policies on the system. The topic of short-run policy analysis is studied later on in this book. Examples of short-run dynamic analyses are Duesenberry, Eckstein, and Fromm (1960), Liu (1963), and De Leeuw (1964). Among the important characteristics to be examined are the expected time paths and the variances around these paths, which can be estimated by using the corresponding sample statistics obtained from repeated stochastic simulation experiments. By designing the experiments properly, as suggested by Hammersley and Morton (1956) and Hammersley and Handscomb (1964), for instance, we can improve the efficiency of the sample statistics. Of course, a short-run analysis can also be applied to a future period. This is the subject of econometric forecasting. Like the subject of efficient design of simulation experiments, it is one of the statistical problems in econometrics and is not treated in this book.

For the purpose of a long-run analysis other characteristics besides the mean paths and the variances are often studied. They include the covariances of different time series observed contemporaneously or at different times, the mean time interval from peak to peak or from trough to trough for each series, and the mean time interval between the adjacent peaks of two time series. These characteristics can be estimated by averaging over n samples. Under appropriate conditions moments computed across samples will reach the same limits as moments computed over time by using one sample of many observations. In this case estimates can also be obtained by averaging the appropriate observations from one large sample over time. Averages over time have been used to study the dynamic properties of econometric models because many of the characteristics recorded by the researchers at the National Bureau of Economic Research for comparison with the model characteristics were obtained by taking averages of historical data over time. Long-run analyses are examplified by Adelman and Adelman (1959), Adelman (1963), and Arzac (1967). Besides the comparison of the characteristics mentioned, one set of interesting questions to be answered is the extent to which different forces included in the dynamic model can explain the characteristics of economic fluctuations. Can we simply rely on the deterministic system by using no random disturbances and only constant values or constant growth rates for the exogenous variables? To what extent will the random disturbances contribute to a more realistic picture? To what extent are fluctuations in the exogenous variables also needed? Can certain rules applied to the policy variables or certain institutional arrangements contained, for example, in the Social Security Act reduce fluctuations and/or promote growth in the long run? The last question belongs to the second half of this book dealing with optimal control.

Readers of this book up to this point will realize that spectral methods can be applied to study long-run properties of the system. In addition to the methods of linear approximation described in Sections 6.4 and 6.5, we can apply spectral and cross-spectral analysis to data obtained by stochastic simulations of the model. Naylor, Wertz, and Wonnacott (1969) discuss several aspects of this analysis, using a simple linear model as an example. If the model is nonlinear, the spectral properties of the time series generated will not be constant through time. It may be hoped that by taking out certain trends from the data the resulting series will be approximately covariance-stationary. If so, the methods of spectral and cross-spectral analysis can be applied. To take out the trend from a series of data we may fit a polynomial of an appropriate degree. The residuals from the polynomial may also be divided by an increasing function of time to prevent an explosive variance, if necessary. Alternatively, we may use quasi-differences $y_{it} - ky_{i,t-1}$, where k is some positive number smaller than or equal to 1. If taking the quasi-difference of the data once does not produce reasonable results, we can take the quasi-difference of the quasi-difference or even the third quasi-difference. Such operations were applied by Nerlove (1964) before the estimation of spectral densities, in which the recommended value of k is .75.

6.8* SLOWLY CHANGING SPECTRAL DENSITIES

If the methods suggested fail to produce a set of covariance-stationary time series, we may accept the lack of stationarity as a reality and try to devise spectral methods to deal with it. Two such attempts are briefly indicated in this section. Both are generalizations of the concepts of spectral and cross-spectral densities for nonstationary time series. Both approaches assume that the change in spectral properties is slow through time.

The first approach was used by Granger and Hatanaka (1964) to analyze economic time series thought to have slowly changing spectral properties. The method was also summarized in Granger (1967). In (23) and (24) in Section 4.5 a stationary time series was defined as a sum (integral) of sine and cosine functions with random coefficients, and the spectral density at a given frequency was defined as the variance of the coefficients of the periodic functions at that frequency. To generalize these concepts it seems natural to allow the random coefficients $\alpha(\omega_j)$ and $\beta(\omega_j)$ at each frequency to have an additional parameter t. Thus (23) in Chapter 4 is replaced by

$$y_{1t} = \sum_j \left[\alpha_1(\omega_j, t)\cos\omega_j t + \beta_1(\omega_j, t)\sin\omega_j t \right] \tag{56}$$

to model a time series y_{1t}. As before, all $\alpha_1(\omega_j, t)$ and $\beta_1(\omega_j, t)$ are assumed to be statistically independent and the expectations $E[\alpha_1^2(\omega_j, t)]$ and $E[\beta_1^2(\omega_j, t)]$ are assumed to be equal and are used to define the spectral density function. Similarly, we can model a second time series y_{2t} with random coefficients $\alpha_2(\omega_j, t)$ and $\beta_2(\omega_j, t)$. The specifications for these coefficients as well as the cross-spectral density function and related concepts can be defined as in Section 4.7.

To estimate the spectral density for y_{1t}, as just defined, we have to modify the computation of the regression coefficients a_{1j} and b_{1j} as given in (21) in Chapter 4. Instead of using just an average of $y_{1t}\cos\omega_j t$ over t for

a_{1j}, we compute $a_{1j}(t)$ as a properly chosen moving average of $y_{1s} \cos \omega_j s$:

$$a_{1j}(t) = \frac{2}{t} \sum_{s=1}^{t} w_{t-s} y_{1s} \cos \omega_j s,$$

$$b_{1j}(t) = \frac{2}{t} \sum_{s=1}^{t} w_{t-s} y_{1s} \sin \omega_j s,$$

(57)

where w_{t-s} are the appropriate weights to be specified and $b_{1j}(t)$ is similarly defined as a moving average. Then $\frac{1}{2}[a_{1j}^2(t) + b_{1j}^2(t)]$ can be used to estimate the spectral density near ω_j at time t. The method is called demodulation. Further details can be found in Granger and Hatanaka (1964).

A second approach to calculating spectral properties of nonstationary econometric models was suggested by Hatanaka and Suzuki (1967). This approach is similar in motivation to the approach of Howrey and Klein for generalizing the concept of a spectral density matrix to nonlinear systems described in Section 6.5. Both start from a definition of spectral density designed for a stationary linear system and then apply the same definition (except for taking a certain sum to a finite but large number N rather than infinity) to the new situation.

To obtain a definition of spectral density that can be generalized, let us go back to (63) in Chapter 4. It was shown there that

$$\frac{1}{\pi} \sum_{k=-N+1}^{N-1} c_{11,k} \cdot \cos \omega_j k = \frac{1}{\pi N}\left[\left(\sum_{t=1}^{N} y_{1t} \cos \omega_j t\right)^2 + \left(\sum_{t=1}^{N} y_{1t} \sin \omega_j t\right)^2\right],$$

(58)

where

$$c_{11,k} = c_{11,-k} = \frac{1}{N} \sum_{t=1}^{N-k} y_{1t} y_{1,t+k}.$$

(59)

Taking expectations on both sides of (58) yields

$$\frac{1}{\pi} \sum_{k=-N+1}^{N-1} \left(\frac{1}{N} \sum_{t=1}^{N-k} E y_{1t} y_{1,t+k}\right) \cdot \cos \omega_j k$$

$$= \frac{1}{\pi N} E\left[\left(\sum_{t=1}^{N} y_{1t} \cos \omega_j t\right)^2 + \left(\sum_{t=1}^{N} y_{1t} \sin \omega_j t\right)^2\right].$$

(60)

If y_{1t} is a covariance-stationary time series with 0 mean, then taking the limit of the left-hand side of (60) as N goes to infinity will give the spectral density function

$$\lim_{N \to \infty} \frac{1}{\pi} \sum_{k=-N+1}^{N-1} \gamma_{11,k} \cos \omega_j k = f_{11}(\omega_j). \tag{61}$$

For a nonstationary time series y_{1t} Hatanaka and Suzuki define the right-hand side of (60) as the "pseudospectrum."

These authors have pointed out several desirable properties of the pseudospectrum. It is the frequency decomposition of the arithmetic mean of the time-changing variances over the sample period, just as the spectrum of a stationary time series is the frequency decomposition of the constant variance. A time-changing autocovariance can be defined such that its mean and the pseudospectrum are a Fourier transform pair. If the time series is defined over an infinitely long time domain and it is stationary, the pseudospectrum is identical with the spectrum. Further details are contained in the reference cited. (See also Problems 16 and 17.)

Although spectral concepts were originally defined for covariance-stationary time series, we have shown that they are also applicable to nonstationary time series. To apply spectral methods to nonstationary time series several methods have been summarized in this chapter. They include taking out trends from the data generated by explosive linear systems, linearizing a nonlinear system, and revising the spectral concepts to make them dependent on time. Two of the methods, those by Howrey and Klein and by Hatanaka and Suzuki, are essentially applications of a definition designed for linear systems to the study of nonlinear systems. It should be noted that applying linear concepts to the study of nonlinear phenomena has been a fairly common practice among research workers. The world is never strictly linear, so that any application of a linear concept to the real world is in a sense an example of such a practice. More than this, econometricians and statisticians have computed correlation coefficients and linear regression coefficients for bivariate relationships that are not linear. Strictly speaking, these coefficients are defined for linear relationships only, but they have been useful also for the nonlinear. The methods of Howrey-Klein and Hatanaka-Suzuki are merely further examples of applying linear concepts to nonlinear situations. The whole idea of covariances (and of spectral and cross-spectral densities which are variances and covariances) is basically a linear concept. They are applicable, as we have suggested in this chapter, to models that are not too nonlinear.

6.9 CONCLUDING REMARKS

Here ends the first part of this book which has dealt with the analysis of stochastic dynamic systems in economics. The mathematical details aside, it is hoped that the conception of an economy as a dynamic and stochastic entity has been successfully communicated. We should allow for models that show the economic variables as forever changing. A good theory or model that can explain the time paths of many economic variables may not be one that will ensure the existence of equilibrium values for the variables. It seems desirable to free ourselves from thinking carried over from comparative static analysis which places much emphasis on the existence of equilibrium values. The importance of stochastic elements in a dynamic system should also be stressed. Concepts and tools for studying the impact of stochastic elements on the system have been suggested and studied. Spectral methods are useful in characterizing certain dynamic properties of an econometric model. A model can be confronted with historical experience or compared with another model in terms of these characteristics. Its validity can thus be effectively scrutinized.

Once the dynamic properties of a system can be studied, it seems natural to consider the design of policies for the improvement of the dynamic performance, as we propose to do from Chapter 7 on.

PROBLEMS

1. Assuming that $\lambda = 1.2$, find the autocovariance function and the spectral density function of z_t^d as specified by (6).

2. Show that the left characteristic (row) vectors r_1 and r_2 corresponding to a pair of conjugate complex roots λ_1 and λ_2 are conjugate complex. *Hint.* See Problem 14 in Chapter 2.

3. Consider a stochastic system

$$y_{1t} = 5y_{1,t-1} - y_{2,t-2} + u_{1t},$$

$$y_{2t} = 2y_{1,t-1} + 2y_{2,t-1} + u_{2t},$$

where u_{1t} and u_{2t} are random variables with mean 0 and covariance matrix I. What are the autoregressive equations for the canonical variables? How are the residuals in these equations distributed?

4. Specify the ex post trends of the canonical variables in Problem 3. Find the autocovariance matrix and the spectral density matrix of the deviations of the canonical variables from these trends.

5. Specify the ex post trends of y_{1t} and y_{2t} in Problem 3. Find the autocovariance matrix and the spectral density matrix of the deviations of y_{1t} and y_{2t} from these trends.

6. Let a system be

$$y_{1t} = 1 + 2e^{-.05y_{2t}}$$

$$y_{2t} = .4e^{0.2y_{1t}}.$$

Linearize the system around $y_{1t} = 3$ and $y_{2t} = 1$.

7. Find the second-order Taylor expansion of the system given in Problem 6 around $y_{1t} = 3$ and $y_{2t} = 1$.

8. Starting with initial values $y_{1t} = 3$ and $y_{2t} = 1$, solve the system of equations in Problem 6 by the Gauss-Siedel method.

9. Starting with initial values $y_{1t} = 3$ and $y_{2t} = 1$, solve the system of equations in Problem 6 by the Newton-Raphson method. Show each iteration.

10. Express the vector time series y_t given by

$$y_t = A_1 y_{t-1} + A_2 y_{t-2} + u_t + G_1 u_{t-1} + G_2 u_{t-2}$$

as a weighted sum of u_t and past u_{t-k}. Write out the coefficients of u_{t-k} for $k = 0, 1, \ldots, 5$ explicitly.

11. Describe how you would compute the spectral density at $\omega = \pi/2$ by using the Howrey and Klein method for the time series y_{1t} as specified by the system

$$y_{1t} = .8y_{1,t-1} + .4e^{-.05y_{2,t-1}} + .2 + u_{1t},$$

$$y_{2t} = .1e^{.2y_{1,t-1}} + .75y_{2,t-1} + 0 + u_{2t},$$

where u_{1t} and u_{2t} are statistically independent, having 0 means and standard deviations .1. The spectral density is to be evaluated at the steady state of the deterministic system obtained by setting $u_{1t} = u_{2t} = 0$. First find the steady-state values of the two variables. Then write out your answer in the form of a set of instructions to a computer programmer.

12. Describe how you would compute the spectral density required in Problem 11 by linearizing the model. Write out the linearized model and provide a set of instructions to a computer programmer.

13. Write a critical essay (including a summary and a critical evaluation) of no more than 1000 words on one of the following stochastic dynamic analyses:

    a. Duesenberry, Eckstein, and Fromm (1960);
    b. Liu (1963);
    c. De Leeuw (1964);
    d. Adelman and Adelman (1959);
    e. Adelman (1963);
    f. Morishima and Saito (1964);
    g. Arzac (1967);
    h. Chow and Levitan (June 1969);
    i. Howrey (1971);
    j. Bowden (1972);
    k. Howrey and Klein (1972).

14. List the most important questions to be raised in a long-run stochastic dynamic analysis of an econometric system and discuss their economic significance.

15. List the most important questions to be raised in a short-run stochastic dynamic analysis of an econometric system and discuss their economic significance.

16. Show that the pseudospectrum defined by Hatanaka and Suzuki is the mathematical expectation of sample estimates of spectra obtained by using $(1 - |k|/N)$ as the lag window for the estimation.

17. Show that the integral of the pseudospectrum defined by Hatanaka and Suzuki is the arithmetic mean of the variances (about 0) of the time series over the sample period.

18. Payments to the unemployed under the Social Security Act are said to be an automatic stabilizer in the economy. Provide a simple formulation of this institution in terms of the relevant structural equations for the economy of the United States. There is no need to specify a complete model but only the part that is affected. Describe the methods that can be used to study the impact of this institution on economic fluctuations and growth.

Part 2

Control of Dynamic Economic Systems

CHAPTER 7

Optimal Control of Known Linear Systems: Lagrange Multipliers

The subject of optimal control in economics was described on page 74 of the May 9, 1973, issue of *Business Week* in an article entitled, "Optimal Control: A Mathematical Supertool," partly as follows.

> How hard and how soon do you put on the brakes in an overheating economy without causing a skid into a catastropic recession? And what policy tools do you use in what combination? The questions are more than hypothetical. U.S. policymakers are facing just such a situation today, and they are getting very little precise guidance from the economists who advise them. The econometric models that the advisers use to forecast the path of the economy can provide rough guidelines. But, by themselves, they cannot determine the best possible timing and dosage for each available policy remedy.
>
> Control theory has swept into the economic profession so rapidly in the past two or three years that most economists are only dimly aware that it is around. But for econometricians and mathematical economists, and for the companies and government agencies that use their skills, it promises an improved ability to manage short-run economic growth, investment portfolios, and corporate cash positions.

In this chapter we develop optimal control rules for the setting of the policy variables in the context of a linear system of stochastic difference

149

equations with known parameters. The method employed is that of Lagrange multipliers. Chapter 8 solves the same problem by the method of dynamic programming. In this development we point out the relations between methods of optimization for static, dynamic, and stochastic dynamic models. The nature of our solution is also elucidated. Analyses of economic policy problems are reserved for Chapter 9.

7.1 RELATION BETWEEN STOCHASTIC DYNAMIC ANALYSIS AND OPTIMAL CONTROL

Techniques for the analysis of dynamic properties of econometric models can be applied to the selection of government policies. Given a policy, they can be used to deduce the dynamic characteristics of the system and alternative policies can be compared. A policy in the context of macroeconomics is a strategy concerning the choice of the *policy variables* in the system. Policy variables are those variables that can be controlled or manipulated by government authorities. They are also known as *control variables* or *instruments*. In most applications the policy variables are a subset of the exogenous variables in an econometric model. The model in such an instance does not explain how the values of these variables are determined. If we specify a rule or an equation for a policy variable, it becomes endogenous.

Two kinds of policy should be distinguished. One is a specification of the time paths of the policy variables at the beginning of a planning period. These paths are to be followed by the policy maker without regard to future events. The second kind is a specification of the policy variables as functions of observations yet to be made. This function is called a *control rule* or a *control equation*. The adjective *feedback* is added to indicate that results of current policy will in turn determine future policy. An example of a linear feedback control equation is

$$x_t = Gy_{t-1} + g, \tag{1}$$

where $x_t =$ a $q \times 1$ vector of policy variables,

       $G =$ a $q \times p$ matrix of coefficients in the equation,

       $g =$ a vector of intercepts.

When a feedback control rule is used, the values of the policy variables in the future will depend on future observations. The first kind is called an *open loop*, and the second, a *closed loop* policy.

Both kinds of policy can be incorporated into an econometric model for dynamic analysis. The first amounts to specifying a time path for the vector x_t. Using a linear model, we have

$$y_t = Ay_{t-1} + Cx_t + b + u_t$$

$$= A^t y_0 + u_t + Au_{t-1} + A^2 u_{t-2} + \cdots + A^{t-1} u_1$$

$$+ b + Ab + A^2 b + \cdots + A^{t-1} b + Cx_t + ACx_{t-1} + \cdots + A^{t-1} Cx_1. \quad (2)$$

The effects of x_{t-k} $(k = 0, \ldots, t-1)$ can be ascertained from the last part of (2). Second, a control rule in the form of (1) can be combined with the linear model (2) to form

$$y_t = (A + CG)y_{t-1} + (b + Cg) + u_t. \quad (3)$$

The dynamic properties of the system (3) can be analyzed by the methods developed in the preceding chapters.

What criteria should be used to judge the outcome of a given policy? If the means, variances, and covariances are important characteristics of the time series for dynamic analysis, perhaps they should be included in the criterion function. A *criterion* or *objective function* is a scalar function that measures the desirability of the variables or their characteristics. When the variables are deterministic, they can enter as arguments in the objective function. In this case the term *welfare function* is also used in the context of macroeconomics. When the variables are random, however, a welfare function is also random. Some parameter in the distribution of this random function, such as the mean, has to be selected as the criterion. One mathematically convenient objective function for the present purpose is the expectation of a quadratic function of the economic variables generated by the stochastic model. As can easily be shown, this function has the means, variances, and covariances of the time series as its arguments. (See Problem 1.)

Two observations should be made concerning the use of an objective function for policy analysis. First, any objective function is only an approximation of the preference of the policy maker, just as an econometric model is only an approximation of the real world. If the objective function proposed is found to be imperfect in some respect, the relevant question is whether the ensuing analysis will provide useful results that are otherwise unavailable. An alternative is to make the objective function mathematically more complicated. This may make the computation of an optimal solution for the policy variables more difficult or even impossible. There will then be a choice between an exact solution to an approximate formulation of the problem and an approximate solution to an exact formulation of the problem. Which is preferable depends on the accuracies of the approximations. If it is economical, probably both solutions should be tried and examined. To appreciate the usefulness of having an objective function, however crude, for the purpose of calculating some desirable strategy for the policy variables, imagine the enormous number of possibilities available. In the context of (3) the choice of G and g offers an immense number of possibilities. It is difficult to tell from simple calculations what control equation in the form of (1) will produce desirable results. Thus some explicit formulation of an objective function, however imperfect, is a necessity for a systematic search of good policies. The tools of optimal control are used in the systematic search.

The second observation, somewhat related to the first, is that the recommendations from an optimal control calculation may not be the end of an analysis. Many dynamic characteristics of the econometric system may be of interest, but only some of them can be conveniently incorporated in the objective function. After an optimal control equation is obtained, it will be useful to deduce the important characteristics of the resulting system (3) by the methods of stochastic dynamic analysis; for example, if it is desirable to reduce the short-period fluctuations of a certain time series, we may wish to calculate not only the variance of the time series but also that part of the total variance contributed by the high-frequency components. The area under the power spectrum of the time series for ω larger than a specified value may be important to measure. Methods of stochastic dynamic analysis can be applied after optimal control calculations are made.

We have just pointed out that the tools of stochastic dynamics are not sufficient for the search for good policies. The tools of optimal control are necessary in this regard. However, optimal control solutions are often based on an abstraction from the important dynamic characteristics of the system. While the methods of optimal control are efficient ways of obtaining some reasonable good policies, if such policies exist, they should in turn be supplemented by the methods of stochastic dynamics after the optimal solutions are obtained.

7.2 CONTROL PROBLEM FOR A LINEAR SYSTEM AND A QUADRATIC WELFARE FUNCTION

Much of the remainder of this book is concerned with a control problem that employs a linear econometric model and an objective function which is the expectation of a quadratic welfare function of the relevant variables. Generalizations to nonlinear systems or nonquadratic welfare functions are discussed in Chapter 12. In this chapter and the following the parameters of the linear system are assumed to be known constants.

The system employed for policy analysis is the reduced form of a linear econometric model:

$$y_t = A_{1t}y_{t-1} + \cdots + A_{mt}y_{t-m} + C_{0t}x_t + \cdots + C_{nt}x_{t-n} + b_t + u_t, \qquad (4)$$

where y_t is a vector of endogenous or dependent variables, x_t is a vector of variables subject to control, A_{it} and C_{it} are given constant matrices, and u_t is a serially uncorrelated vector with mean 0 and covariance matrix V. Time subscripts are used for the matrices A_{it} and C_{it} to make the treatment more general. Readers of Section 6.4 can anticipate that one way to study the control of a nonlinear system is to approximate it by a linear system with time-varying coefficients. Exogenous variables in the system that are not subject to control are treated either as part of b_t (also assumed to be given constants) or as a part of u_t.

To simplify analysis (4) is rewritten as a first-order system,

$$
\begin{bmatrix} y_t \\ y_{t-1} \\ \vdots \\ y_{t-m+1} \\ x_t \\ x_{t-1} \\ \vdots \\ x_{t-n+1} \end{bmatrix}
=
\begin{bmatrix}
A_{1t} & A_{2t} & \cdots & A_{mt} & C_{1t} & \cdots & C_{nt} \\
I & 0 & \cdots & 0 & 0 & \cdots & 0 \\
 & & \cdots & & & & \\
0 & 0 & \cdots I & 0 & 0 & \cdots & 0 \\
0 & 0 & \cdots & 0 & 0 & \cdots & 0 \\
0 & 0 & \cdots & 0 & I\cdot & \cdots 0 & 0 \\
 & & \cdots & & & \cdots & \\
0 & 0 & \cdots & 0 & 0 & \cdots I & 0
\end{bmatrix}
\begin{bmatrix} y_{t-1} \\ y_{t-2} \\ \vdots \\ y_{t-m} \\ x_{t-1} \\ x_{t-2} \\ \vdots \\ x_{t-n} \end{bmatrix}
$$

$$
+
\begin{bmatrix} C_{0t} \\ 0 \\ \vdots \\ 0 \\ I \\ 0 \\ \vdots \\ 0 \end{bmatrix} x_t
+
\begin{bmatrix} b_t \\ 0 \\ \vdots \\ 0 \\ 0 \\ 0 \\ \vdots \\ 0 \end{bmatrix}
+
\begin{bmatrix} u_t \\ 0 \\ \vdots \\ 0 \\ 0 \\ 0 \\ \vdots \\ 0 \end{bmatrix},
\qquad (5)
$$

which is redesignated as

$$y_t = A_t y_{t-1} + C_t x_t + b_t + u_t. \tag{6}$$

Note that the newly defined y_t includes current and (possibly) lagged dependent variables as well as current and (possibly) lagged control variables, whereas x_t remains the same as before. The purposes of writing the original system (4) in the form (5) are twofold. First, the system is converted to first order to avoid cumbersome algebra; for example, the control equation does not have to include y_{t-2} and x_{t-1} as arguments because they are already included in the redefined y_{t-1}. Second, the control variables themselves are imbedded in the vector y_t of dependent variables so that the welfare function has only y_t as arguments, even when some control variables x_t are, in fact, included. This also simplifes the derivation. Note also that autocorrelations in the residuals u_t can be eliminated by the method in Section 3.11.

The performance of the system is measured by the deviations of y_t, as defined in (6), from the target vectors a_t $(t = 1, \ldots, T)$. The vectors a_t have the same dimension as y_t, and because the latter include lagged variables only selected elements of a_t corresponding to the current variables are relevant. Specifically, the objective is to minimize

$$E_0 W = E_0 \sum_{t=1}^{T} (y_t - a_t)' K_t (y_t - a_t), \tag{7}$$

where the expectation E_0 is conditional on the initial condition y_0, again in the notation of (6), and K_t are known, symmetric (usually diagonal), positive semidefinite matrices, with 0 elements normally corresponding to lagged (endogenous and control) variables. Often, if the econometric model contains a large number of endogenous variables, only a small number will be included in the welfare cost function; that is, they will be multiplied by nonzero elements in the diagonal of the matrix K_t. *An optimal control problem is to minimize the expected welfare loss (7), given the econometric model (6).*

Two of the favorite candidates for inclusion in the welfare function are the changes in the price level and the rate of unemployment. There are various ways to incorporate these two aspects of the economy into the welfare loss function. If the first difference of a general price index and the percentage rate of unemployment are among the variables in the vector y_t, we can simply assign appropriate values such as 2.5 points per year and 4 percentage points for the corresponding elements in the target vector a_t. If only the price level but not its first difference is included in the list of dependent variables, there are at least two ways to specify the welfare function. One is to convert the desired changes in the price level to a time path for the price index. This may be a path of a constant annual growth of 2.5 points in the index or a constant percentage growth from the actual value at period 0 or from a desired value at period 0. The corresponding elements in the vectors a_t $(t = 1, \ldots, T)$ are thus specified. The second way is to introduce a new endogenous variable Δy_{it} by using the identity $\Delta y_{it} = y_{it} - y_{i,t-1}$, assuming that y_{it} is the general price index. A target path can be specified for this new variable. Note, however, that controlling the level of a variable is different from controlling its first difference. The first does not penalize the period-to-period changes and the second does. Steering y_{it} near a constant growth path will not discriminate between a smooth time path of y_{it} and a highly oscillating one, as long as $\sum_{t=1}^{T} k_{ii,t}(y_{it} - a_{it})^2$ is the same. Controlling Δy_{it} to make it as close to a constant as possible will attach a high cost to the oscillating time path. It is also feasible to use both the level and the first differences of the same variable in the welfare function.

Two possible defects of the quadratic welfare function (7) should be mentioned. First, a positive deviation from target is assigned the same cost as a negative deviation of the same magnitude. If 2.5 is the target for price change, an achievement of 2.0 will be treated the same way as 3.0. This feature, however, may not seriously damage the usefulness of the quadratic function. In a study of the inflation-unemployment tradeoff, for example, we can make the targets for both variables so idealistic and unlikely to be achieved that the actual solution may be above both targets. If so, the chance of getting to the low side of either target is small, and the errors introduced in assessing the negative deviations will be small. As we have emphasized, a quadratic welfare cost function is a means for systematic search for good policies. The welfare weights as given in the diagonal elements of the matrices K_t as well as the targets a_t should not be regarded as accurate for all ranges of the dependent variables. They are only approximate values of the parameters in the welfare function applicable to a certain range. In practice, the values of K_t and a_t can be changed in successive optimal control calculations. If one set of a_t is found to be too far from the achievable path, we may modify them to improve the quadratic approximation near the neighborhood of the solution. Viewed in this light, the quadratic welfare function can be very useful.

A second possible shortcoming is the *additive* form of function (7), namely, that an expectation is taken for the *sum* of functions of variables in different periods. The function cannot reflect a greater or smaller variance in the total welfare loss from all periods. To illustrate this point consider the case of one scalar variable y_t for two periods. Let the expected loss function be

$$EW = E(y_1^2 + y_2^2). \tag{8}$$

Let y_1 be a random variable taking only values 0 and 1 with equal probabilities and let y_2 be distributed likewise and statistically independent of y_1; EW will be 1. The sum $y_1 + y_2$ can take values 0, 1, and 2 with probabilities .25, .5, and .25, respectively. Now change the joint distribution to make $y_1 = y_2$, so that the sum $y_1 + y_2$ will take values 0 and 2 with probabilities .5 and .5. The latter appears to be a more risky situation, but it has the same expected loss as before. This shortcoming can be overcome by incorporating the cross-product term $y_1 y_2$ in the loss to penalize positive covariance between the outcomes in the two periods. As an approximation, however, the additive form is probably accurate enough for most purposes because the mean of a sum is more important than its variance. In Chapter 11 a more general loss function incorporating cross products of variables in different periods is employed and the additional mathematical difficulties can be appreciated.

7.3 TWO-PART DECOMPOSITION OF THE OPTIMAL CONTROL PROBLEM

It will be fruitful to decompose the optimal control problem in Section 7.2 into two parts. The first is a deterministic control problem that uses the deterministic model

$$\bar{y}_t = A_t \bar{y}_{t-1} + C_t \bar{x}_t + b_t, \qquad (\bar{y}_0 = y_0), \tag{9}$$

which is obtained by setting the random disturbance u_t in the stochastic model (6) equal to its mean value 0. We use bars to denote the endogenous and control variables in this deterministic model. The second part is a stochastic control problem that uses the stochastic model

$$y_t^* = A_t y_{t-1}^* + C x_t^* + u_t, \qquad (y_0^* = 0), \tag{10}$$

for the deviation $y_t^* = (y_t - \bar{y}_t)$ of the random vector y_t from the path given by the above deterministic model; (10) is obtained by subtracting (9) from (6), using the definition $x_t^* = x_t - \bar{x}_t$. If the random disturbance u_t were 0, the control problem in Section 7.2 would be reduced to one of minimizing $W_1 = \sum_{t=1}^{T}(\bar{y}_t - a_t)' K_t(\bar{y}_t - a_t)$ with respect to $\bar{x}_t$, subject to the constraint of

(9). The optimal path for $\bar{x}_t$ would be deterministic because it can be announced at the beginning of period 1. The existence of u_t makes the actual y_t deviate from the path $\bar{y}_t$ given by (9) and using the optimal path $\bar{x}_t$ for the deterministic control problem. It is therefore necessary to modify the optimal setting $\bar{x}_t$ by x_t^* in order to control the deviation $y_t^* = y_t - \bar{y}_t$ given by (10). Corresponding to the decomposition of the model into (9) and (10) is the decomposition of the expected welfare cost

$$EW = \sum_{t=1}^{T} (\bar{y}_t - a_t)' K_t (\bar{y}_t - a_t) + E \sum_{t=1}^{T} y_t^{*'} K_t y_t^* = W_1 + EW_2. \quad (11)$$

The two parts of the resulting control problem are then the following. First, a deterministic control problem is to minimize W_1, defined by (11), with respect to $\bar{x}_t$, given the model (9). Second, a stochastic control problem is to minimize EW_2 as defined by (11) with respect to x_t^*, given the stochastic model (10). These two problems are solved by the method of Lagrange multipliers. When the separate solutions $\bar{x}_t$ and x_t^* are obtained, the policy variables x_t will be set equal to their sum. Much of the material in this discussion is taken from Chow (February 1972 and October 1972).

7.4 SOLUTION OF DETERMINISTIC CONTROL BY LAGRANGE MULTIPLIERS

One elementary way to solve the deterministic control problem is to introduce the vectors λ_t of Lagrange multipliers and differentiate the Lagrangian expression

$$L_1 = \frac{1}{2} \sum_{t=1}^{T} (\bar{y}_t - a_t)' K_t (\bar{y}_t - a_t) - \sum_{t=1}^{T} \lambda_t' (\bar{y}_t - A_t \bar{y}_{t-1} - C_t \bar{x}_t - b_t). \quad (12)$$

Using the differentiation rules,

$$\frac{\partial}{\partial x}(x'a) = a \quad \text{and} \quad \frac{\partial}{\partial x}(\tfrac{1}{2}x'Ax) = Ax, \quad (13)$$

where $a =$ a vector of constants,
$\quad A =$ a symmetric matrix of constants,
$\quad x =$ the variable vector,

we obtain the following vectors of derivatives.

$$\frac{\partial L_1}{\partial \bar{y}_t} = K_t(\bar{y}_t - a_t) - \lambda_t + A'_{t+1}\lambda_{t+1} = 0, \qquad (t = 1, \ldots, T; \lambda_{T+1} = 0), \quad (14)$$

$$\frac{\partial L_1}{\partial \bar{x}_t} = C'_t\lambda_t = 0, \qquad (t = 1, \ldots, T), \tag{15}$$

$$\frac{\partial L_1}{\partial \lambda_t} = -(\bar{y}_t - A_t\bar{y}_{t-1} - C_t\bar{x}_t - b_t) = 0, \qquad (t = 1, \ldots, T). \tag{16}$$

Equations 14, 15, and 16 constitute a set for the unknowns $\bar{y}_t$, $\bar{x}_t$, and λ_t, $t = 1, \ldots, T$. The method of Lagrange multipliers is applied here to minimize a function of variables at different points of time subject to the constraints of a system of dynamic equations in the same way that a function of many variables is minimized subject to several constraints that do not involve time. The dynamic situation is treated simply by defining the variables $\bar{y}_t$, $\bar{x}_t$, and λ_t at different points in time as separate variables. The dynamic nature of the problem, however, gives a special structure to the simultaneous equations (14), (15), and (16) and requires special methods for their efficient solution.

One efficient way of solving the system (14) to (16) is to start with $t = T$ and repeat the following three steps backward in time for $t = T-1, \ldots, 1$. First, (14) is used to express λ_t as a function of $\bar{y}_t$. Second, the result, together with (15) and (16), is used to solve for $\bar{x}_t$. Third, the results of the first two steps together with (16) are used to express $\bar{y}_t$ and λ_t as linear functions of $\bar{y}_{t-1}$. Using the last linear function, express λ_{t-1} as a linear function of $\bar{y}_{t-1}$, and we are in step one of the next round. For $t = T$ step one, using (14), gives

$$\lambda_T = K_T\bar{y}_T - K_Ta_T + A'_{T+1}\lambda_{T+1} = H_T\bar{y}_T - h_T, \tag{17}$$

where, anticipating generalization to $t < T$, we have set

$$H_T = K_T \tag{18}$$

and

$$h_T = K_Ta_T. \tag{19}$$

By (15), (16), and (17) in the second step

$$C'_T\lambda_T = 0 = C'_T(H_T\bar{y}_T - h_T) = C'_T(H_TA_T\bar{y}_{T-1} + H_TC_T\bar{x}_T + H_Tb_T - h_T), \tag{20}$$

which implies

$$\bar{x}_T = G_T \bar{y}_{T-1} + g_T, \tag{21}$$

where

$$G_T = -(C_T' H_T C_T)^{-1} C_T' H_T A_T, \tag{22}$$

$$g_T = -(C_T' H_T C_T)^{-1} C_T' (H_T b_T - h_T). \tag{23}$$

The matrix $C_T' H_T C_T$ is assumed to be nonsingular. If the rank r of $C_T' H_T C_T$ is smaller than the number q of control variables, we can arbitrarily set $(q - r)$ elements of $\bar{x}_T$ in (20) equal to any desired values and solve (20) for the remaining elements of $\bar{x}_T$. More on this point appears in Section 7.7. In the third step we use (16) and (21) to solve for $\bar{y}_T$ as a function of $\bar{y}_{T-1}$:

$$\bar{y}_T = (A_T + C_T G_T)\bar{y}_{T-1} + b_T + C_T g_T; \tag{24}$$

this result can be applied to (17) to express λ_T also as a function of $\bar{y}_{T-1}$,

$$\lambda_T = H_T(A_T + C_T G_T)\bar{y}_{T-1} + H_T(b_T + C_T g_T) - h_T. \tag{25}$$

Having solved for λ_T in terms of $\bar{y}_{T-1}$, we substitute (25) into (14) to obtain an equation analogous to (17) in the first step:

$$\lambda_{T-1} = K_{T-1}\bar{y}_{T-1} - K_{T-1}a_{T-1} + A_T'\lambda_T = H_{T-1}\bar{y}_{T-1} - h_{T-1}, \tag{26}$$

where

$$H_{T-1} = K_{T-1} + A_T' H_T(A_T + C_T G_T), \tag{27}$$

$$h_{T-1} = K_{T-1}a_{T-1} - A_T' H_T(b_T + C_T g_T) + A_T' h_T. \tag{28}$$

The development from (20) on can now be followed, with $T - 1$ replacing T, and so on.

To apply this solution to the deterministic control problem use the pair of equations (22) and (27) to obtain $G_T, H_{T-1}, G_{T-1}, \ldots$, consecutively backward in time with (18) as the initial condition. Then, given H_t, use the pair of equations (23) and (28) to obtain $g_T, h_{T-1}, g_{T-1}, \ldots$, consecutively backward in time with (19) as the initial condition. Having obtained G_t and g_t, we set the optimal $\bar{x}_t$ by the linear feedback control rule (21) on $\bar{y}_{t-1}$.

Several features of this solution should be noted. First, the optimal $\bar{x}_t$ is a linear function of $\bar{y}_{t-1}$. A linear feedback control equation results from the minimization. Second, the coefficients G_t and g_t in the optimal feedback control equation $\bar{x}_t = G_t y_{t-1} + g_t$ are obtained by solving two sets of difference equations backward in time from $t = T$ to $t = 1$. Let there be p dependent variables in $\bar{y}_t$, including lagged variables and control variables as specified in the notation of (6). Let there be q control variables. The matrices G_t are $q \times p$. They are obtained sequentially by (22) and (27), which can be combined to form a matrix difference equation in the $p \times p$ matrix H_t:

$$H_{t-1} = K_{t-1} + A_t' H_t (A_t + C_t G_t)$$

$$= K_{t-1} + A_t' H_t A_t - A_t' H_t C_t (C_t' H_t C_t)^{-1} C_t' H_t A_t. \tag{29}$$

Equation 29 is known as a matrix Riccati equation and is an example of a nonlinear difference equation. The matrices H_t are symmetric because H_T is symmetric; given symmetric K_{T-1} and H_T, H_{T-1} is also symmetric by (29), and so on. The difference equation (29) is easy to solve. Each step for one time period involves essentially matrix multiplications. The matrix $C_t' H_t C_t$ to be inverted is of order q only. In practice, the number q of control variables is probably quite small even if the number p of dependent variables in the model is large. Similarly, the intercepts g_t in the optimal control equations are obtained by solving (23) and (28), which can be combined to form a vector difference equation in the $p \times 1$ vector h_t:

$$h_{t-1} = K_{t-1} a_{t-1} - A_t' H_t (b_t + C_t g_t) + A_t' h_t$$

$$= K_{t-1} a_{t-1} - A_t' (H_t b_t - h_t) + A_t' H_t C_t (C_t' H_t C_t)^{-1} C_t' (H_t b_t - h_t)$$

$$= K_{t-1} a_{t-1} + (A_t + C_t G_t)' (h_t - H_t b_t). \tag{30}$$

Third, to appreciate the efficiency of this method of solution, consider the admittedly inefficient method of solving equations (14), (15) and (16) simultaneously as one set of linear equations in the unknowns $\bar{x}_t, \bar{y}_t,$ and $\lambda_t,$ $t = 1, \ldots, T$. There are $T \times (q + 2p)$ of these equations and the problem is much bigger. (See Problem 2.)

7.5 ROLE OF MATHEMATICAL PROGRAMMING AND THE MINIMUM PRINCIPLE

In this section the possible uses of mathematical programming and Pontryagin's minimum principle for solving the deterministic control problem are briefly indicated.

Mathematical programming is a set of methods for maximizing or minimizing an objective function subject to equality or inequality constraints in the unknowns. There are two possible differences between most mathematical programming problems and the present problem. First, the former often have no temporal structure in either the objective function or the constraints as the control problem has. Second, their constraints often include inequality constraints that are absent in the control problem. If we wish to impose inequality restrictions on the control variables x_t, we will find methods of mathematical programming useful. If restrictions on the policy variables are considered important in our setup, they are incorporated in the form of penalties on deviations of the control variables from specified targets. This treatment may be adequate for many practical applications. The methods of mathematical programming are varied and are not studied here. It should be mentioned, however, that a theorem of Kuhn and Tucker (1951) for minimizing a nonlinear function subject to inequalities is often applied. For the application of mathematical programming to deterministic control problems the reader is referred to Canon, Cullum, and Polak (1970).

The maximum or minimum principle of Pontryagin et al. (1962) was developed for solving deterministic control problems when time is continuous rather than discrete, as assumed in this book. Instead of difference equations, differential equations are used to represent the dynamics of the system. Random disturbances are assumed to be absent in the treatment of Pontryagin, although there have been recent attempts to modify the minimum principle for application to stochastic control problems. For the treatment of deterministic control problems in discrete time attempts have also been made to state the minimum principle in discrete form, but the principle does not accomplish any more than the method of Lagrange multipliers in this situation. It may be useful, however, to point out some resemblance between the solution given above and certain analogous elements in a discrete version of the minimum principle.

A discrete version of the minimum principle is usually stated in three parts. These parts constitute a set of necessary conditions for minimizing an objective function which is a sum of functions for different periods subject to a set of difference equations. For the problem formulated in

Section 7.4 the analogous conditions are the following:
First, a *Hamiltonian* defined as

$$H(\bar{y}_t, \bar{x}_t, \lambda_t) = \tfrac{1}{2}(\bar{y}_t - a_t)' K_t(\bar{y}_t - a_t) - \lambda_t'(\bar{y}_t - A_t\bar{y}_{t-1} - C_t\bar{x}_t - b_t) \quad (31)$$

is minimized by the optimal value of the control vector $\bar{x}_t$, given the optimal values of $\bar{y}_t$ and λ_t. Second, the dependent variables $\bar{y}_t$ (known as the *state variables* in the control literature) corresponding to the optimal values of the control variables and a set of variables λ_t called the *costate variables* satisfy, respectively, the following difference equations obtained by differentiating the Hamiltonian with respect to λ_t and $\bar{y}_t$:

$$\frac{\partial H}{\partial \lambda_t} = -\bar{y}_t + A_t\bar{y}_{t-1} + C_t\bar{x}_t + b_t = 0 \quad (32)$$

and

$$\frac{\partial H}{\partial \bar{y}_t} = K_t(\bar{y}_t - a_t) - \lambda_t + A_{t+1}'\lambda_{t+1} = 0. \quad (33)$$

Third, if the vector $\bar{y}_T$ in the terminal period has to satisfy $k \leqslant p$ restrictions $g_1(\bar{y}_T) = 0, \ldots, g_k(\bar{y}_T) = 0$, then λ_T must satisfy

$$\lambda_T = \sum_{i=1}^{k} \mu_i \frac{\partial g_i}{\partial \bar{y}_T} \quad (34)$$

for some numbers $\mu_1, \ldots, \mu_k$. This is known as the *transversality condition*. It does not apply to the problem formulated in Section 7.4 because the problem does not impose any restriction on the terminal state $\bar{y}_T$.

This description of the minimum principle applied to our problem differs from descriptions in the control literature; for example, Chang (1961) and Katz (1962). In the control literature the lagged control variable $\bar{x}_{t-1}$ rather than $\bar{x}_t$ is used to determine $\bar{y}_t$ in the difference equation (9). Furthermore, the difference equations are sometimes written with $\Delta\bar{y}_t$ rather than $\bar{y}_t$ as a dependent variable. These differences aside, the above description gives the essential features of the minimum principle for our problem. As the reader has undoubtedly noted the Hamiltonian is simply the scalar function obtained by taking one part of sum, corresponding to t, from the Lagrange expression L_1 defined by (12). The minimization of the Hamiltonian (31) with respect to $\bar{x}_t$ corresponds to our differentiation in (15). Equations 32 and 33 correspond, respectively, to (16) and (14). Finally, if we replace $\tfrac{1}{2}(\bar{y}_T - a_T)' K_T(\bar{y}_T - a_T)$ in (12) by $\Sigma_{i=1}^{k} \mu_i g_i(\bar{y}_T)$, where $\mu_1, \ldots, \mu_k$ are new Lagrange multipliers and $g_1, \ldots, g_k$ are functions restricting the terminal y_T, then differentiating the Lagrangian expression with respect to $\bar{y}_T$ will give (34).

7.6 SOLUTION OF STOCHASTIC CONTROL BY LAGRANGE MULTIPLIERS

The second part of the control problem is to minimize EW_2 from (11), given the stochastic system (10). We seek the optimal *linear* feedback equation for the control variable x_t^* without justifying the proposition that the optimal feedback equation is linear; that is, we search for the optimal matrix G_t in the linear equation

$$x_t^* = G_t y_{t-1}^*, \qquad (t = 1, \ldots, T). \tag{35}$$

The problem of finding the best linear decision rule is an interesting one even if the truly optimal rule is not linear. Furthermore, the optimal rule is indeed linear for this problem, as shown in Chapter 8.

To solve the problem by the method of Lagrange multipliers we first rewrite the objective function in terms of the (nonstationary) covariance matrices $Ey_t^* y_t^{*\prime}$ of the vectors y_t^* and utilize a set of restrictions on these covariance matrices to form a Lagrange expression. The objective function is written as

$$EW_2 = E \sum_{t=1}^{T} \mathrm{tr}(y_t^{*\prime} K_t y_t^*) = E \sum_{t=1}^{T} \mathrm{tr}(K_t y_t^* y_t^{*\prime}) = \sum_{t=1}^{T} \mathrm{tr}\, K_t (Ey_t^* y_t^{*\prime}), \tag{36}$$

where tr, or trace, is the sum of the diagonal elements of a matrix and use is made of the property $\mathrm{tr}(BG) = \mathrm{tr}(GB)$. In the second term $\mathrm{tr}(y_t^{*\prime} K_t y_t^*)$ is simply the scalar $(y_t^{*\prime} K_t y_t^*)$ itself. (See Problem 4.) To find a set of restrictions on the covariance matrices when the control rule (35) is enforced we substitute (35) into (10),

$$y_t^* = (A_t + C_t G_t) y_{t-1}^* + u_t$$

$$= R_t y_{t-1}^* + u_t, \qquad (y_0^* = 0), \tag{37}$$

where R_t denotes $A_t + C_t G_t$. Postmultiply (37) by $y_t^{*\prime}$ and take expectation

$$Ey_t^* y_t^{*\prime} = R_t Ey_{t-1}^* y_t^{*\prime} + Eu_t y_t^{*\prime}$$

$$= R_t Ey_{t-1}^* y_t^{*\prime} + V, \tag{38}$$

because by (37) $Eu_t y_t^{*\prime} = Eu_t u_t' = V$. Similarly, transpose equation (37), premultiply the result by y_{t-1}^*, and take expectation

$$Ey_{t-1}^* y_t^{*\prime} = (Ey_{t-1}^* y_{t-1}^{*\prime}) R_t'. \tag{39}$$

Substitution of (39) into (38) gives the desired difference equation in $(Ey_t^* y_t^{*\prime}) = \Gamma(t,0) = \Gamma_{.t}$:

or
$$Ey_t^* y_t^{*\prime} = V + R_t (Ey_{t-1}^* y_{t-1}^{*\prime}) R_t' \tag{40}$$

$$\Gamma_{.t} = V + (A_t + C_t G_t) \Gamma_{.t-1} (A_t + C_t G_t)', \qquad t = 1, \dots, T.$$

To incorporate (40) as a set of constraints for $t = 1, \dots, T$ in a Lagrange expression for minimizing (36) a $p \times p$ matrix H_t of Lagrange multipliers is introduced for each constraint. If each element of a $p \times p$ matrix X is constrained to be zero, the sum of p^2 constraints in the form $\sum_{i,j} h_{ij} x_{ij}$ has to be added to the Lagrange expression, where x_{ij} is the $i - j$ element of X and h_{ij} is the corresponding Lagrange multiplier. The sum of these constraints can be written as the trace of the product HX'. Because the ith diagonal element of the product HX' is $\sum_j h_{ij} x_{ij}$, the trace of HX', denoted by tr(HX'), is $\sum_{i,j} h_{ij} x_{ij}$, a term to be included in the Lagrange expression if the matrix X is constrained to be the 0 matrix. Therefore for our problem of minimizing (36), subject to the matrix constraints (40), we write the Lagrange expression as

$$L_2 = \sum_{t=1}^{T} \text{tr}(K_t \Gamma_{.t}) - \sum_{t=1}^{T} \text{tr}\{H_t[\Gamma_{.t} - V - (A_t + C_t G_t)\Gamma_{.t-1}(A_t + C_t G_t)']\}.$$

$$\tag{41}$$

The variables are the elements of G_t, $\Gamma_{.t}$, and H_t.
  The differentiation rule

$$\frac{\partial}{\partial G} \text{tr}(BG) = \frac{\partial}{\partial G} \text{tr}(GB) = B' \tag{42}$$

is applied to (41). (See Problems 4 and 5 concerning this rule.) Consider differentiating

$$\text{tr}(H_t C_t G_t \Gamma_{.t-1} A_t') = \text{tr}(G_t \Gamma_{.t-1} A_t' H_t C_t)$$

with respect to G_t, for example. Application of this rule yields

$$(\Gamma_{.t-1} A_t' H_t C_t)' = C_t' H_t A_t \Gamma_{.t-1}$$

as the matrix of derivatives. Note that the matrix H_t of Lagrange multipliers is symmetric because the corresponding constraints

$$\Gamma_{\cdot t} - V - (A_t + C_t G_t)\Gamma_{\cdot t-1}(A_t + C_t G_t)' = 0$$

constitute a symmetric matrix. The derivatives of L_2 are

$$\frac{\partial L_2}{\partial G_t} = 2C_t' H_t A_t \Gamma_{\cdot t-1} + 2C_t' H_t C_t G_t \Gamma_{\cdot t-1} = 0, \qquad t = 1, \ldots, T, \tag{43}$$

$$\frac{\partial L_2}{\partial \Gamma_{\cdot t}} = K_t - H_t + (A_{t+1} + C_{t+1}' G_{t+1})' H_{t+1}(A_{t+1} + C_{t+1} G_{t+1}) = 0,$$

$$t = 1, \ldots, T-1, \tag{44a}$$

$$\frac{\partial L_2}{\partial \Gamma_{\cdot T}} = K_T - H_T = 0, \tag{44b}$$

$$\frac{\partial L_2}{\partial H_t} = -[\Gamma_{\cdot t} - V - (A_t + C_t G_t)\Gamma_{\cdot t-1}(A_t + C_t G_t)'] = 0, \quad t = 1, \ldots, T. \tag{45}$$

Equations 43, 44, and 45 provide a set of necessary conditions for the unknowns G_t, $\Gamma_{\cdot t}$ and H_t. Equation 43 can be satisfied by choosing

$$G_t = -(C_t' H_t C_t)^{-1} C_t' H_t A_t. \tag{46}$$

Equation 44 is equivalent to

$$H_t = K_t + (A_{t+1} + C_{t+1} G_{t+1})' H_{t+1}(A_{t+1} + C_{t+1} G_{t+1}). \tag{47}$$

Therefore the unknowns G_t and H_t can be found by using the initial condition $H_T = K_T$ from (44b) and solving (46) and (47) alternately backward in time for $t = T, T-1, \ldots, 1$. It is interesting to observe that the coefficients G_t in the optimal feedback control equations $x_t^* = G_t y_{t-1}^*$ for controlling the variances of the random deviations y_t^* are identical with the coefficients G_t in the optimal feedback control equations $\bar{x}_t = G_t \bar{y}_{t-1} + g_t$ for steering the means $\bar{y}_t$ to the targets a_t obtained previously for the deterministic control problem. Equation 47 is easily shown to be the same as (27). Having solved for the optimum G_t and H_t, we can find the covariance matrices $\Gamma_{\cdot t}$ by (45), namely,

$$\Gamma_{\cdot t} = V + (A_t + C_t G_t)\Gamma_{\cdot t-1}(A_t + C_t G_t); \qquad t = 1, \ldots, T. \tag{48}$$

Equation 48 is solved forward in time from $t = 1$ to $t = T$, using the initial condition $\Gamma_{\cdot 0} = Ey_0^* y_0^{*\prime} = 0$ which is due to the definition $\bar{y}_0 = y_0$ or $y_0^* = 0$.

7.7 THE COMBINED SOLUTION AND THE MINIMUM EXPECTED LOSS

Combining the solutions of Sections 7.4 and 7.6, respectively, for the deterministic and stochastic parts of the control problem, we obtain the optimal control equation for x_t:

$$x_t = \bar{x}_t + x_t^* = G_t \bar{y}_{t-1} + g_t + G_t y_{t-1}^*$$

$$= G_t y_{t-1} + g_t. \tag{49}$$

The matrices G_t are computed by using (22) and (27) or equivalently (46) and (47). The vectors g_t are computed by (23) and (28). In the computations the targets a_t affect only g_t but not G_t. Both G_t and g_t are computed from information already known before any observation on $(y_1, \ldots, y_T)$ is made. The optimal policy x_t is set by (49), using G_t and g_t and the observation on y_{t-1}.

The combined solution (49) could have been found by minimizing the expected welfare cost (11) of the original problem, subject to the constraints (9) and (40), using the method of Lagrange multipliers. The Lagrange expression for the combined problem is

$$L = \tfrac{1}{2} \sum_{t=1}^{T} (\bar{y}_t - a_t)' K_t (\bar{y}_t - a_t) + \tfrac{1}{2} \sum_{t=1}^{T} \text{tr}(K_t \Gamma_{\cdot t})$$

$$- \sum_{t=1}^{T} \lambda_t' (\bar{y}_t - A_t \bar{y}_{t-1} - C_t \bar{x}_t - b_t)$$

$$- \tfrac{1}{2} \sum_{t=1}^{T} \text{tr}\{ H_t [\Gamma_{\cdot t} - V - (A_t + C_t G_t) \Gamma_{\cdot t-1} (A_t + C_t G_t)'] \}. \tag{50}$$

Differentiating L with respect to $\bar{x}_t$, $\bar{y}_t$, and λ_t yields (14), (15), and (16). The solution of these equations provides a control rule for $\bar{x}_t$. Differentiating L with respect to G_t, $\Gamma_{\cdot t}$ and H_t yields (43), (44), and (45). Their solution provides a control rule for x_t^*. The combined solution for x_t is their sum, given in (49).

Using this solution, we can compute the minimum expected welfare loss. The expected loss is conveniently divided into two parts. The deterministic part

$$W_1 = \sum_{t=1}^{T} (\bar{y}_t - a_t)' K_t (\bar{y}_t - a_t)$$

is evaluated by using the solution $\bar{y}_t$ of the deterministic system under optimal control; that is,

$$\bar{y}_t = A_t \bar{y}_{t-1} + C_t \bar{x}_t + b_t$$

$$= (A_t + C_t G_t) \bar{y}_{t-1} + C_t g_t + b_t. \tag{51}$$

This solution can be calculated forward in time for $t = 1, 2, \ldots, T$, given $\bar{y}_0 = y_0$. The second part of the expected loss is

$$EW_2 = \sum_{t=1}^{T} \text{tr}(K_t \Gamma_{.t}).$$

The matrices $\Gamma_{.t}$ required in its evaluation are obtained from (48). If the dynamic system were deterministic, with u_t absent from (6), the optimal control equation would still be $x_t = G_t y_{t-1} + g_t$, the same as the combined solution for the stochastic model (6), but the minimum welfare loss would consist of only the deterministic part W_1. The second part EW_2 of the expected loss is a measure of the importance of the random disturbances in the welfare calculations.

If the number of target variables (the number of nonzero elements in the $p \times p$ diagonal matrix K_t) equals the number $q \leqslant p$ of control variables, the time path $\bar{y}_t$ generated by the deterministic system (51) under optimal control will meet the targets exactly and the deterministic part W_1 of the minimum expected welfare loss will be 0, provided that C_t is of rank q. This theorem is proved in two parts,

$$K_t(A_t + C_t G_t) = 0, \qquad (t = T, T-1, \ldots, 1), \tag{52}$$

and

$$K_t(a_t - b_t - C_t g_t) = 0, \qquad (t = T, T-1, \ldots, 1). \tag{53}$$

Equations 52 and 53, with (51), imply that $K_t \bar{y}_t = K_t a_t$; that is, the q target variables in $\bar{y}_t$ equal the specified target values.

To prove (52) we need only to prove that the q rows on the left-hand side corresponding to the q nonzero elements of K_t are 0 because the remaining $(p - q)$ rows equal the 0 row vectors of K_t postmultiplied by

$(A_t + C_t G_t)$. Form a $q \times q$ diagonal matrix K_{1t} by using the nonzero elements of K_t. Also form a $q \times p$ matrix A_{1t} and a $q \times q$ matrix C_{1t} by selecting the q rows of A_t and C_t, respectively, that correspond to the nonzero elements of K_t; (52) will be satisfied if

$$K_{1t}(A_{1t} + C_{1t}G_t) = 0. \tag{54}$$

The jth column of G_T computed by (22), with $H_T = K_T$, is a set of coefficients in the regression of the jth column of $K_T^{\frac{1}{2}} A_T$ on the explanatory variables $-K_T^{\frac{1}{2}} C_T$ obtained by the method of least squares. Because the regression coefficients are not affected by rearranging the order of the observations, let the first q observations on the dependent variable be the nonzero observations, that is, the jth column of $K_{1T}^{\frac{1}{2}} A_{1T}$. The corresponding q observations on the q explanatory variables are $-K_{1T}^{\frac{1}{2}} C_{1T}$. The remaining $(p - q)$ observations on the dependent and the explanatory variables are all 0. The matrix G_T by (22) becomes

$$G_T = -\left[\begin{pmatrix} K_{1T}^{\frac{1}{2}} C_{1T} \\ 0 \end{pmatrix}' \begin{pmatrix} K_{1T}^{\frac{1}{2}} C_{1T} \\ 0 \end{pmatrix} \right]^{-1} \begin{pmatrix} K_{1T}^{\frac{1}{2}} C_{1T} \\ 0 \end{pmatrix}' \begin{pmatrix} K_{1T}^{\frac{1}{2}} A_{1T} \\ 0 \end{pmatrix}$$

$$= -(C'_{1T} K_{1T} C_{1T})^{-1} C'_{1T} K_{1T} A_{1T}. \tag{55}$$

Each column of G_T is a set of coefficients in the regression of a column of $K_{1T}^{\frac{1}{2}} A_{1T}$ on $-K_{1T}^{\frac{1}{2}} C_{1T}$. For each regression the number of observations equals the number of explanatory variables. The regression residuals are 0, provided that the matrix C_{1T} is of rank q, which is the case if C_T has rank q, but the left-hand side of (54) consists precisely of columns of these regression residuals except for a factor $K^{\frac{1}{2}}$, and (52) is proved for $t = T$. Using (52) and (27), we have $H_{T-1} = K_{T-1}$. Repeating these arguments will prove (52) for $t = T - 1$ and so on for all t.

The proof of (53) is analogous. Observe that g_T computed by (23), with $h_T = K_T a_T$, is a set of coefficients in the regression of $K_T^{\frac{1}{2}}(a_T - b_T)$ on $K_T^{\frac{1}{2}} C_T$ obtained by the method of least squares. Agin, if K_T has only q nonzero elements, only q observations on the dependent variable and the explanatory variables are nonzero. The q regression residuals using these observations are 0. This proves (53) for $t = T$. Using (53) and (28), we have $h_{T-1} = K_{T-1} a_{T-1}$. Repetition of the above argument will prove (53) for all t.

As a corollary, we state that *if the number of target variables is less than or equal to the number of instruments the time path* $\bar{y}_t$ *generated by the deterministic system* (51) *under optimal control will meet the targets exactly and the deterministic part* W_1 *of the minimum expected welfare loss will be* 0.

The corollary is true if the number of target variables is equal to q. If it is equal to $r < q$, the matrix K_t in (52) and (53) will have only r nonzero elements. To satisfy (52) and (53) consider only the r rows on the left-hand side of each of the equations that correspond to the nonzero elements of K_t. Here we have fewer observations than the number q of explanatory variables because the matrix C_t has q columns. There are many ways to choose the regression coefficients G_t and g_t to make the regression residuals in (52) and (53) equal to 0. To obtain a unique solution for G_t and g_t we can arbitrarily fill in additional nonzero elements in the diagonal of K_t until their number equals q. The resulting G_t and g_t can be used to satisfy (52) and (53), where K_t is defined by the original problem as having only r nonzero elements. Thus more target variables are introduced without affecting the solution to the original problem with $r < q$ target variables. This proves the corollary.

In practice, however, if we use a linear model with known coefficients for optimal control calculations, the number of target variables should not be smaller than the number of instruments. If it is smaller, there is something to gain by stabilizing one more variable (possibly an instrument). This will not increase the deterministic part of the welfare loss, nor the minimum expected sum of squared deviations of the other target variables from their target values, as we show below.

When the number of target variables equals the number of instruments, the target variables in the deterministic system (51) will take their specified values and the deterministic part W_1 of the expected welfare loss will be 0. The stochastic part of the loss EW_2, according to (36), is $\sum_{t=1}^{T} E(y_t^{*\prime} K_t y_t^*)$, where by (37)

$$K_t^{\frac{1}{2}} y_t^* = K_t^{\frac{1}{2}}(A_t + C_t G_t) y_{t-1}^* + K_t^{\frac{1}{2}} u_t, \qquad (t = 1, \ldots, T). \qquad (56)$$

Using (52) and (56), we find that EW_2 is simply $\sum_{t=1}^{T} E u_t' K_t u_t$.

When the number of target variables exceeds the number of instruments, not all target values can be satisfied by the solution $\bar{y}_t$ of the deterministic system (51). In terms of the regression interpretation of (52) and (53), there are more observations than explanatory variables, and the regression residuals will not be all 0. Both W_1 and EW_2 will be positive. The jth column of G_t, given by (46), can still be interpreted as coefficients in the regression of the jth column of $H_t^{\frac{1}{2}} A_t$ on the q explanatory variables

$- H_t^{\frac{1}{2}} C_t$, where H_t is no longer equal to K_t. By (47) H_t is symmetric and in general not diagonal, but we can still write $H_t^{\frac{1}{2}}$ as the matrix whose square equals H_t. As regression coefficients G_t are chosen to obtain small residuals

$$H_t^{\frac{1}{2}} A_t + H_t^{\frac{1}{2}} C_t G_t = H_t^{\frac{1}{2}} R_t, \tag{57}$$

where $A_t + C_t G_t = R_t$. In fact, G_t minimizes $\operatorname{tr}(R_t' H_t R_t)$. (See Problem 6.) The resulting G_t will also minimize $\sum_{t=1}^T E(y_t^* K_t y_t^*)$ as we have shown in Section 7.6. An interpretation of H_t is given in Section 8.6.

In the standard treatment of the control problem of this section by control engineers, as summarized by Pindyck (1973), for example, the quadratic welfare function is written separately for the dependent variables y_t and the instruments x_t, the latter not being imbedded as a subvector of y_t. The algrithm for computing the optimal feedback control equations requires that the matrix of the quadratic form of x_t in the welfare function be nonsingular. This implies that all instruments have to be included in the set of target variables, a requirement that is avoided in the treatment given above. As long as the total number of target variables equals or exceeds the number of instruments, our method yields a unique set of optimal control equations. It is not necessary for all or even any of the instruments to be included in the set of target variables (or assigned positive weights in the welfare function), provided that there are enough dependent variables serving as targets.

7.8 THE STEADY-STATE SOLUTION

Under what circumstances will the optimal control equation (49) reach a steady-state in the sense of having G_t and g_t invariant through time? This question is answered separately for G_t and g_t.

Because G_t is obtained, together with H_t, by solving (46) and (47) backward in time and the matrices A_t and C_t are involved in these equations, we can hardly expect to obtain a steady-state solution for G_t if A_t and C_t are functions of time. It will therefore be assumed that A_t and C_t are time invariant and the t subscript is dropped. Further, let $K_t = K$ for all t. The matrix equations which the steady-state G and H have to satisfy are by (46) and (47)

$$G = -(C'HC)^{-1} C'HA, \tag{58}$$

$$H = K + (A + CG)' H (A + CG)$$

$$= K + R'HR. \tag{59}$$

If a matrix G can be found to make $R = (A + CG)$ so small that (59) can be satisfied for some H, then a steady-state exists for G_t. Consider the special case when both A and C are scalars. The scalar R can be made 0 by choosing $G = -C^{-1}A$ and $H = K$ is the solution of (59). In a somewhat more general case C is a nonsingular $p \times p$ matrix; G can be set equal to $-C^{-1}A$ to make the matrix R vanish and $H = K$ is the solution of (59). This is a case of having the same number q of control variables as the number p of dependent variables. It cannot occur if we have to imbed the instruments x_t in the vector y_t of (6) because p will then be larger than q. When the control variables are imbedded in the vector y_t as in (6), let the matrix K be diagonal with rank q. We have shown in Section 7.7 that if the number of target variables equals the number of instruments (52) will hold. We can then select $H = K$ and have $HR = 0$ by (52) so that (59) can be satisfied.

The equality between the number of target variables and the number q of instruments, together with the assumption that C is of rank q, has been found sufficient for the existence of a steady-state solution for G. It is not a necessary condition. Let the number of target variables be greater than q; G can no longer be chosen to make $R'KR$ vanish, but a pair of matrices G and H may still be found to satisfy (58) and (59) simultaneously. This is possible if and only if, for a matrix G staisfying (58), the roots of $R = A + CG$ are all smaller than 1 in absolute value. The stated condition is necessary and sufficient for a solution to exist for (59). The proof is left as an exercise in Problems 7 and 8.

We now turn to q_t in the feedback control equation. The assumption is made that A_t, C_t, b_t, and K_t are all time-invariant. The difference equations (23) and (28), with t replacing T, need a steady-state solution for g_t and h_t. Equivalently, (30) which combines these equations should have a steady-state solution for h_t. Under the stated assumption (30) becomes

$$h_{t-1} = Ka_t + (A + CG_t)'(h_t - H_t b). \tag{60}$$

From (60), it can be seen that h_t will reach a steady state if a_t, G_t, and H_t are all time-invariant. If G_t and H_t have steady-state solutions, all roots of $R = (A + CG)$ will be smaller than 1 in absolute value, as we have pointed out in the last paragraph. The infinite series

$$I + R' + R'^2 + R'^3 + \cdots$$

will converge to $(I - R')^{-1}$ or the matrix $(I - R')$ will be nonsingular. The steady state h from (60) can then be found by solving

$$h = Ka + R'(h - Hb) \tag{61}$$

or

$$(I - R')h = Ka - Hb.$$

Thus under the assumptions of time-invariant A, C, b, K, and a and a steady-state solution for G, the intercept g_t in the optimal feedback control equation will also reach a steady state.

Notice that the difference equations (27) and (28) for H_t and h_t are solved *backward* in time. If a steady state is reached in either equation, it will be for *small* values of t; that is to say, a time-invariant control rule will be applied during the early periods of the planning horizon of T periods. This result is reasonable because as we get closer to period T the solution will be affected by the assumption that the future will no longer matter. We cannot expect an optimal rule derived by due consideration of future loss to be the same as an optimal rule that ignores the future. Therefore the coefficients G_t and g_t must change as t approaches T simply because the relevant future becomes shorter. Without using any special terminal condition for the problem, it is suggested that T be made long enough so that the computed G_t and g_t for t close to T can be disregarded. To decide how many of the early G_t and g_t can be taken seriously we can observe whether G_t during these periods are almost constant. If they are, the remaining time periods are numerous enough in the optimal control calculations. The intercepts g_t during these early periods can still change because the target a_t changes, but, as demonstrated in the last paragraph, constant G_t will imply constant g_t under the assumptions of a time-invariant model and time-invariant targets. The changing g_t due to changing targets are still trustworthy.

In summary, if a time-invariant linear model is assumed, and K_t is constant, a necessary and sufficient condition for G_t to reach a steady state is that the roots of $R = A + CG$ are all smaller than 1 in absolute value. If G_t reaches a steady state, the intercept g_t will also, provided that the target a_t remains the same through time. In applications it is more reasonable to assume a constant target vector if the variables are defined as first differences or rates, such as the first difference in GNP or the rate of unemployment. The targets for the level of GNP and the number of persons unemployed will presumably change through time. For further discussion of the steady-state solution and some economic applications the reader may refer to Chow (1970).

7.9 BRIEF COMMENTS ON RELATED APPROACHES TO QUANTITATIVE ECONOMIC POLICY

Any treatise on quantitative economic policy should include an acknowledgement of debt to the pioneering work of Tinbergen (1952) and (1956).

Tinbergen's discussion, broader in scope, covers possible changes in political, legal, and economic institutions under which economic decision making takes place. It has provided much basic thinking on the subject, including the important concepts of instruments and targets to which we have referred. The theorem and corollary in Section 7.7 can be considered partly an extension of Tinbergen's treatment of the numbers of targets and instruments. Our framework is stochastic, whereas his is deterministic.

Any study of macroeconomic policy based on an econometric model should also include an expression of gratitude to the significant work of Theil (1958 and 1964). The problem formulated by Theil is almost identical to the problem in Section 7.2. His method of solution is different, however, mainly because the concept of feedback control is not used. Essentially, Theil solves the multiperiod problem by determining the *quantities* $x_1, \ldots,$ x_T simultaneously rather than the *functions* (or control equations) for calculating these quantities according to future observations. By his method only the result x_1 for the current period is applied. After the first period elapses the quantities $x_2, \ldots, x_T$ will be determined simultaneously and x_2 is applied. After the second period elapses, the quantities $x_3, \ldots, x_T$ will be determined, and so on. A recent note of Norman (1974) proves that the optimal solution for the first period by Theil's method is identical to ours. The usefulness of the two methods appears to be different, however; for example, by Theil's method it is difficult, if not impossible, to evaluate the expected minimum welfare loss analytically as we have done in Section 7.7. Furthermore, without a feedback control equation it is difficult to study the dynamic properties of the system under control by methods introduced in Chapters 3, 4, and 6 of this book.

A brief comment should be made concerning the concept of *certainty equivalence*. This concept was first introduced by Simon (1956). It was adapted by Theil (1958) to the macroeconomic policy problem. The principle of certainty equivalence, when applied to decision under uncertainty, is to adopt a policy that would be optimal if the random variables involved were equal to certain constants, such as their expected values. Consider the problem of minimizing $Ey_1'y_1$, given the model,

$$y_1 = Ay_0 + Cx_1 + u_1.$$

The solution, obtained by differentiating

$$Ey_1'y_1 = E(y_0'A' + x_1'C' + u_1')(Ay_0 + Cx_1 + u_1)$$

$$= y_0'A'Ay_0 + x_1'C'Cx_1 + 2x_1'C'Ay_0 + Eu_1'u_1$$

with respect to x_1, is

$$x_1 = -(C'C)^{-1}C'Ay_0.$$

This solution is identical to the optimal solution for the problem when the random vector u_1 is replaced by its mean 0. Thus the certainty equivalent solution is optimal for the problem under uncertainty (incorporating the residual u_1) in this case. We have shown that for a *one-period* problem of minimizing the expectation of a *quadratic* welfare loss function, subject to a *linear* model with an *additive* random disturbance, the certainty equivalence solution is optimum. Theil has tried to extend the method of certainty equivalence to the multiperiod problem in Section 7.2. He sets up the problem by considering the vectors $x_1, \ldots, x_T$ control variables for all periods simultaneously and succeeds in showing that the certainty equivalence solution *for x_1 only* is optimal. Therefore a new solution has to be found for x_2 after period 1 elapses. He has called this the principle of *first-period certainty equivalence.* The discussion in this chapter has shown that a *multiperiod certainty equivalence* exists for the problem stated if the solution is expressed in the form of *feedback control equations.* For all periods $t = 1, \ldots, T$ the optimal feedback control equations $x_t = G_t y_t + g_t$ obtained by ignoring the random disturbances u_t are indeed optimal for the stochastic problem incorporating u_t. Our solution of the multiperiod deterministic control problem in Section 7.4, when expressed in feedback form, is identical with the solution of the stochastic control problem in Section 7.7.

In the next chapter the problem in Section 7.2 is solved by the method of dynamic programming and comments are made concerning the contributions by control engineers. Comments on the literature on control in economics, however brief and incomplete, should always include a reference to A. W. Phillips (1958) in which some of the basic ideas can be found, although the model used there is only univariate and the method, being of the frequency domain, applies only to the steady-state solution and is inefficient from the present point of view. Mention should also be made of the work of Holt, Modigliani, Muth, and Simon (1963) as one of the first applications of the ideas of control to economics.

PROBLEMS

1. Express the objective function (7) as a function of the mean vectors and covariance matrices of the variables involved.

2. Letting $p = 10$, $q = 3$, and $T = 12$, set up (14), (15), and (16) as one system of simultaneous equations in the form $Ax = b$ for the unknowns $\bar{x}_t$, $\bar{y}_t$, and λ_t $(t = 1,\ldots,12)$. Comment on the size of this problem, compared with the problem of obtaining the optimal control equations by (29), (30), and related equations.

3. Simplify the system of simultaneous equations in Problem 2 by omitting the use of Lagrange multipliers in the original formulation of the control problem in Section 7.4 and substituting the dynamic equations (9) into the objective function W_1 instead. How much simpler is this formulation? How does it compare with the solution using equations (29), (30) and related equations?

4. Assuming that a matrix B is $p \times q$ and a matrix G is $q \times p$, show that $\text{tr}(BG) = \text{tr}(GB)$.

5. Defining the derivative of a scalar function with respect to (the elements of) a matrix as the matrix of the corresponding derivatives, prove the differentiation rule (42).

6. Show that the matrix G_t given by equation (46) minimizes $\text{tr}(R_t' H_t R_t)$ where R_t is defined by (57).

7. Let $G = -(C'HC)^{-1}C'HA$ and let the roots of $R = A + CG$ be smaller than 1 in absolute value. Show that a solution exists for $H = K + R'HR$, where K is a given positive semidefinite matrix. *Hint.* Write the equation as a difference equation

$$H_{t-1} = K + R'H_t R = K + R'KR + R'H_{t+1}R$$

$$= K + R'KR + R'^2 KR^2 + R'^3 KR^3 + \cdots.$$

Review Section 3.6.

8. Let $G = -(C'HC)^{-1}C'HA$ and let some roots of $R = A + CG$ be greater than 1 in absolute value. Show that no solution exists for $H = K + R'HR$, where K is a given positive semidefinite matrix. *Hint.* Same as for Problem 7.

9. Provide a numerical example of two endogenous variables and one control variable (not imbedded in the vector y_t) that satisfy the first-order equation $y_t = Ay_{t-1} + Cx_t + u_t$ such that no steady-state solution to the optimal control equation exists for minimizing $E\sum_{t=1}^{T} y_t' y_t$. Show your calculations.

10. Using the model and criterion function in Problem 9, provide a numerical example for the existence of a time-invariant optimal feedback control equation. Show your calculations.

11. Set up the model $y_t = .4x_t + .6x_{t-1} + 10 + u_t$ in the form of (6), where $Eu_t = 0$ and $Eu_t^2 = 1$. Assume that x_t does not enter the loss functions and that the purpose of control is to steer y_t to the target 0. Let the time horizon T be 5. Find the solution for G_t, H_t, g_t, and h_t. Comment on the behavior of the control variable x_t. What is the expected welfare loss?

12. How would you stabilize the control variable x_t in Problem 11? Use a quadratic welfare function of our own choice and find the solution for G_t, H_t, g_t, and h_t.

13. Compute the expected welfare loss for Problem 12.

14. Consider using the model in Table 5.1 in Chapter 5 to steer the variables ΔC_t, $I_{1,t}$, $I_{2,t}$, and ΔR_t to given targets. Set up the model in the notation of (6). What are the dimensions of the matrices A, C, and b_t? Specify a welfare function that seems reasonable and justify the parameter values chosen.

15. Set up the computations for G_T, H_{T-1}, g_T, and h_{T-1} in Problem 14.

16. Assuming that the optimal control equations $x_t = G_t y_{t-1} + g_t$ is known, set up the computations for the expected welfare loss in Problem 14.

CHAPTER 8

Optimal Control of Known Linear Systems: Dynamic Programming

The method of dynamic programming is applied to the solution of the control problem in Section 7.2. This method constitutes the principal tool employed in textbooks written by control engineers, for example, Aoki (1967), Astrom (1970), Brockettt (1970), Bryson and Ho (1969), and Kushner (1971), although the exposition in this chapter is different from theirs. Besides introducing a useful method that is applied to more difficult control problems later on in this book, this chapter serves as an elementary introduction to several topics in control that are considered important by the control scientists and of interest also to economists.

8.1 SOLUTION TO THE LINEAR-QUADRATIC CONTROL PROBLEM BY DYNAMIC PROGRAMMING

We first solve the problem in Section 7.2 by using Bellman's (1957) method of dynamic programming. Given a linear model

$$y_t = A_t y_{t-1} + C_t x_t + b_t + u_t \qquad (1)$$

and a quadratic loss function

$$W = \sum_{t=1}^{T} (y_t - a_t)' K_t (y_t - a_t) = \sum_{t=1}^{T} (y_t' K_t y_t - 2 y_t' K_t a_t + a_t' K_t a_t), \qquad (2)$$

176

the problem is to choose $x_1, \ldots, x_T$ to minimize the conditional expectation $E_0 W$, given the initial condition y_0. Recall that $Eu_t = 0$ and $Eu_t u_t' = V$ and that (1) is the result of rewriting a higher order system and imbedding the control variables x_t in the vector y_t as described in Section 7.2.

By the method of *dynamic programming* the problem for the last period T is solved first, given the initial condition y_{T-1}. Having found the best policy x_T for the last period contingent on any initial condition y_{T-1}, we solve the two-period problem for the last two periods by choosing an optimal x_{T-1}, contingent on the initial condition y_{T-2}. Having found the optimal x_T and x_{T-1}, we similarly solve the three-period problem by choosing the optimal x_{T-2}, and so on. In the last stage the optimal x_1 for the first period is found, given the information y_0 available at the beginning of period 1. Two features of this method should be noted. The problem is solved *backward in time*, step by step. At each step only *one* (vector) *unknown* x_t is determined. Thus the problem of T unknowns $x_1, \ldots, x_T$ (or unknown policies to be precise) is transformed to T problems, each involving one unknown x_t. By the *principle of optimality* the solution by dynamic programming is optimal because at each time t, whatever the initial condition, all future policies are optimal. At the beginning of time T the solution for x_T is optimal by construction. At the beginning of $T-1$ the pair of strategies x_T and x_{T-1} is optimal because, whatever the outcome at the end of period $T-1$, x_T is optimal and because x_{T-1} is optimal by construction. This argument extends to any period t.

Let us begin by considering the problem for the last period T. It is to minimize

$$V_T = E_{T-1}(y_T - a_T)' K_T (y_T - a_T) = E_{T-1}(y_T' H_T y_T - 2y_T' h_T + c_T), \quad (3)$$

where we have set $H_T = K_T$ for ease of generalization to the multiperiod problem later on and have also set $h_T = k_T \equiv K_T a_T$ and $c_T = a_T' K_T a_T$. Using the model (1) for y_T and taking expectations, we minimize

$$V_T = (A_T y_{T-1} + C_T x_T + b_T)' H_T (A_T y_{T-1} + C_T x_T + b_T)$$
$$- 2(A_T y_{T-1} + C_T x_T + b_T)' h_T + E_{T-1}(u_T' H_T u_T) + c_T, \quad (4)$$

where use is made of the assumption that u_T is independent of y_{T-1}. Differentiation of (4) with respect to the vector x_T gives

$$\frac{\partial V_T}{\partial x_T} = 2C_T' H_T (A_T y_{T-1} + C_T x_T + b_T) - 2C_T' h_T = 0. \quad (5)$$

The solution of (5) yields the optimal policy for the last period

$$\hat{x}_T = G_T y_{T-1} + g_T, \tag{6}$$

where

$$G_T = -(C_T' H_T C_T)^{-1}(C_T' H_T A_T), \tag{7}$$

$$g_T = -(C_T' H_T C_T)^{-1} C_T'(H_T b_T - h_T). \tag{8}$$

Equation 6 is the optimal feedback control equation, which shows that the optimal policy for period T is a linear function of all the variables y_{T-1} that will affect the outcome y_T through system (1). It also shows the multiperiod nature of our solution. The value of $\hat{x}_T$ cannot be set in period 1, for it depends on y_{T-1} which is not yet observed, but the optimal $\hat{x}_T$ can be specified as a function of y_{T-1} and is needed to determine the optimal policies for earlier periods.

To obtain the minimum expected welfare cost for the last period, conditional on the data y_{T-1}, we substitute (6) for x_T in (4):

$$\hat{V}_T = y_{T-1}'(A_T + C_T G_T)' H_T(A_T + C_T G_T)y_{T-1}$$

$$+ 2y_{T-1}'(A_T + C_T G_T)'(H_T b_T - h_T)$$

$$+ (b_T + C_T g_T)' H_T(b_T + C_T g_T) - 2(b_T + C_T g_T)' h_T$$

$$+ c_T + E_{T-1} u_T' H_T u_T. \tag{9}$$

Observe that $\hat{V}_T$ is a quadratic function of y_{T-1}.

Now consider the problem for one more period $T-1$. This two-period problem involves the choice of two policies x_T and x_{T-1}, but we have already found an optimal policy $\hat{x}_T$ for the last period as a function y_{T-1}. The only remaining problem is to choose x_{T-1} optimally. The choice of x_{T-1} will affect y_{T-1}, but, whatever y_{T-1} may be, our previous solution guarantees that x_T will be chosen optimally. This logic carries over to many periods. At the beginning of any period t we are concerned only with the choice of x_t, provided that we have obtained the optimal solution for the remaining periods from $t+1$ on. That optimal solution will depend on y_t which the current decision x_t will influence, but whatever the outcome y_t may be we know that all future decisions $x_{t+1}, \ldots, x_T$ will be optimally chosen. By the principle of optimality in dynamic programming we minimize only with respect to x_{T-1}

$$V_{T-1} = E_{T-2}\left(y_{T-1}' K_{T-1} y_{T-1} - 2y_{T-1}' k_{T-1} + a_{T-1}' K_{T-1} a_{T-1} + \hat{V}_T\right), \tag{10}$$

where the welfare cost includes the contribution from the current period $T-1$ plus the minimum cost $\hat{V}_T$ from period T on. Substituting (9) for $\hat{V}_T$ into (10) will yield

$$V_{T-1} = E_{T-2}(y'_{T-1}H_{T-1}y_{T-1} - 2y'_{T-1}h_{T-1} + c_{T-1}), \tag{11}$$

where

$$H_{T-1} = K_{T-1} + (A_T + C_T G_T)' H_T (A_T + C_T G_T), \tag{12}$$

$$h_{T-1} = k_{T-1} + (A_T + C_T G_T)' (h_T - H_T b_T), \tag{13}$$

$$c_{T-1} = a'_{T-1} K_{T-1} a_{T-1} + (b_T + C_T g_T)' H_T (b_T + C_T g_T)$$
$$- 2(b_T + C_T g_T)' h_T + c_T + E_{T-1} u'_T H_T u_T. \tag{14}$$

The solution for the T-period problem is complete by observing that the expression (11) to be minimized with respect to x_{T-1} has the same form as expression (3). We can thus repeat the process from (3) to (14) and replace the subscript T with $T-1$ and so on. In summary, the optimal control solution consists of choosing the instrument x_t as a linear function $G_t y_{t-1} + g_t$ of the variables y_{t-1} of the preceding period, as in (6). The matrices of coefficients G_t are determined, together with H_t, by solving (7) and (12), alternately, backward in time from $t = T$, and with initial condition $H_T = K_T$; G_t and H_t having been obtained, the vectors h_t and g_t are determined by solving (13) and (8), respectively, backward in time from $t = T$, and with initial condition $h_T = k_T \equiv K_T a_T$. This solution is identical to that in Section 7.7. (See Problem 1.)

Of course, in the beginning of period 1 we need to act only on x_1, but the above derivation shows that the optimal $\hat{x}_1$ depends on G_1 and H_1, which, in turn, depend on the future H_t and G_t, $t = 2, \ldots, T$. By the multiperiod nature of our problem future policies have to be taken into account in the determination of the optimal policy for the first period. Our calculations will also yield the minimum expected welfare cost associated with the optimal policy from periods 1 to T, given the initial condition y_0. It is given by (9), with T replaced by 1. Equation 9 shows that $\hat{V}_1$ is a quadratic function of y_0. In the calculation of $\hat{V}_1$ we need c_1, which can be obtained by solving (14) backward in time.

In the case of time-invariant A_t, C_t, and K_t the solution for G_t and H_t may reach a steady state for t smaller than a certain value, thus satisfying

$$G = -(C'HC)^{-1} C'HA; \tag{15}$$

$$H = K + (A + CG)' H (A + CG). \tag{16}$$

Because (16) can be written as an infinite series,

$$H = K + (A + CG)' K(A + CG) + (A + CG)'^2 K(A + CG)^2 + \cdots, \quad (17)$$

the steady state will exist if and only if the series converges, that is, if and only if all the characteristic roots of the matrix $(A + CG)$ are smaller than 1 in absolute value. Even when G_t and H_t do reach a steady state, g_t and h_t will not do so if k_{t-1} ($\equiv Ka_{t-1}$) and b_t are changing through time, as we can see from (13). If G_t is constant for $t = 1, 2, 3, \ldots$, the optimal policies $\hat{x}_t$ should respond to the data y_{t-1} in the same way, but the intercepts g_t will change because the targets a_t and the effects b_t of the uncontrolled exogenous variables are changing. Optimal policies should take these changes into account.

8.2 STOCHASTIC VERSUS DETERMINISTIC CONTROL PROBLEMS

Several comments, though probably fairly obvious by this time, should be made concerning the nature of a stochastic control problem distinguished from a deterministic control problem in the context of a linear model and a quadratic welfare function.

First, the principle of multiperiod certainty equivalence as expounded in Section 7.9 applies. This principle becomes apparent in the derivation of Section 8.1. Clearly the formulation of and solution to the optimal control problem applies to the deterministic model obtained from (1) by setting $u_t = 0$. Simply erase the expectation signs and set u_t equal to 0 in all derivations of Section 8.1. The linear feedback control equations $\hat{x}_t = G_t y_{t-1} + g_t$ and the computations of G_t and g_t remain the same as in the stochastic case. Hence multiperiod certainty equivalence applies.

Although the optimal control equations remain exactly the same with or without the random disturbances, there is an important difference in the stochastic case because the time path of the optimal $\hat{x}_t$ and the associated y_{t-1} are stochastic, whereas they are nonstochastic in the deterministic problem. When $u_t = 0$, it is possible to calculate in period 1 the optimal $\hat{x}_t$ for all t. With stochastic disturbances present, future y_t become stochastic and optimal $\hat{x}_t$ can be determined numerically only after y_{t-1} are observed.

As in Chapter 7, it is possible to decompose into two parts the stochastic time series y_t generated by system (1) when x_t is optimally controlled. The first part $\bar{y}_t$ is deterministic; the second, $y_t - \bar{y}_t = y_t^*$ is stochastic. When x_t follows the optimal policy $\hat{x}_t = G_t y_{t-1} + g_t$, system (1) will become, with G_t reaching the steady state G and with $\tilde{b}_t$ denoting $b_t + Cg_t$ under the assumption of time-invariant A_t, C_t, and K_t,

$$y_t = (A + CG) y_{t-1} + \tilde{b}_t + u_t = R y_{t-1} + \tilde{b}_t + u_t. \quad (18)$$

After repeated substitutions of y_{t-1} in (18) by $Ry_{t-2} + \tilde{b}_{t-1} + u_{t-1}$ and y_{t-2} by $Ry_{t-3} + \cdots$, etc., we find that

$$y_t = R^t y_0 + \left(\tilde{b}_t + R\tilde{b}_{t-1} + \cdots + R^{t-1}\tilde{b}_1 \right)$$

$$+ u_t + Ru_{t-1} + \cdots + R^{t-1}u_1. \tag{19}$$

The first line on the right-hand side of (19) is the mean of the process, denoted by $\bar{y}_t$. It is identical to the time path of the deterministic system under control. The second line of (19) is the deviation from mean, or the random part of the process, denoted by y_t^*. It is generated by the process

$$y_t^* = Ry_{t-1}^* + u_t. \tag{20}$$

Methods introduced in the first half of this book can be applied to study the dynamic properties of the mean path $\bar{y}_t$ and the deviations y_t^* from mean.

In addition to the characterizations in Chapters 2, 3, and 4, the framework of optimal control provides one summary measure of the dynamic performance of a system through the expected welfare loss. For the optimal policy it is evaluated by (9) with T set equal to 1 and (14). Note that in the deterministic case all $Eu_t'H_t u_t$ in the calculation of $\hat{V}_1$ would be 0. Thus the minimum expected welfare loss in the stochastic case equals the loss for the deterministic case, that is, for the time series $\bar{y}_t$, plus the expected loss due to the random disturbances, that is, from the time series y_t^* of (20). The first component is due to deviations of $\bar{y}_t$ from a_t, whereas the second is due to the deviations of y_t from $\bar{y}_t$ or y_t^*.

This decomposition deserves several comments. First, even assuming that the number of instruments equals the number of target variables, we cannot expect the stochastic time series y_t to be on target because of random disturbances, and the expected squared deviations will contribute to the expected welfare loss, as measured by the second component of $\hat{V}_t$. Under this assumption the first component is 0 because the deterministic path $\bar{y}_t$ is exactly on target, as can be shown by the argument in Section 7.8. Second, when the number of target variables exceeds the number of instruments, as it often does in practice, both components of $\hat{V}_1$ will be positive. Third, calculations presented in Chapter 9 based on the simple aggregative macroeconomic model described in Chapter 5 show that for a number of reasonable welfare functions the component of $\hat{V}_1$ due to the stochastic disturbances is much larger than the deterministic component, indicating the importance of the stochastic aspect of the problem in the measurement of welfare cost. Thus, although the introduction of random disturbances does not change the optimal control equations, it materially affects the nature of the time series generated by the controlled system and increases the optimal expected welfare cost.

8.3* A CONTROL PROBLEM WITH OBSERVATION ERRORS

To provide a bridge to the control engineering literature and to present an interesting problem in its own right we set up an elementary control problem similar to a standard problem found in the engineering textbooks. We retain Equations 1 and 2 but assume that y_t can be measured only indirectly through an $m \times 1$ vector s_t which is governed by

$$s_t = M_t y_t + \eta_t, \tag{21}$$

where M_t is a known $m \times p$ matrix and η_t is a random vector with $E\eta_t = 0$, $E\eta_t \eta_t' = Q$, and $E\eta_t \eta_r' = 0$ for $t \neq r$. The covariance matrix Q is assumed known. In the engineering literature the standard symbol for the vector of control variables is u_t. Furthermore, the vector of control variables in (1) is dated $t - 1$ under the assumption that their effects are delayed at least for one period. These minor features are not adopted here to conform to the engineering literature.

To solve the problem in Section 8.1, modified by (21) and using dynamic programming, consider first the last period T. Equation 3 remains valid, but (4) will have to be changed because y_{T-1} can no longer be treated as given. Instead $s_t(t = 1, \ldots, T - 1)$ are given in this problem. Using the initial conditions $H_T = K_T$, $h_T = k_T = K_T a_T$ and $c_T = a_T' K_T a_T$, (4) is replaced by

$$
\begin{aligned}
V_T &= E_{T-1}(y_T' H_T y_T - 2y_T' h_T + c_T) \\
&= E_{T-1}[(A_T y_{T-1} + C_T x_T + b_T + u_T)' H_T (A_T y_{T-1} + C_T x_T + b_T + u_T)] \\
&\quad - 2E_{T-1}(A_T y_{T-1} + C_T x_T + b_T + u_T)' h_T + c_T \\
&= E_{T-1}(y_{T-1}' A_T' H_T A_T y_{T-1}) + 2(x_T' C_T' + b_T') H_T A_T (E_{T-1} y_{T-1}) \\
&\quad + (x_T' C_T' + b_T') H_T (C_T x_T + b_T) - 2(E_{T-1} y_{T-1}') A_T' h_T \\
&\quad - 2(x_T' C_T' + b_T') h_T + c_T + E_{T-1} u_T' H_T u_T, \tag{22}
\end{aligned}
$$

where, as in (4), u_T is assumed to be statistically independent of y_{T-1}. Differentiating (22) with respect to x_T gives

$$\frac{\partial V_T}{\partial x_T} = 2C_T' H_T [A(E_{T-1} y_{T-1}) + C_T x_T + b_T] - 2C_T' h_T = 0. \tag{23}$$

The solution of (23) yields the optimal policy for the last period,

$$\hat{x}_T = G_T(E_{T-1}y_{T-1}) + g_T, \tag{24}$$

where G_T and g_T are the same as given in (7) *and* (8). Thus the same feedback control equation applies here as in Section 8.1, except that the variable on the right-hand side is $E_{T-1}y_{T-1}$ rather than y_{T-1} itself. Presumably, using (21), we can evaluate $E_{T-1}y_{T-1}$. This problem is reserved for Section 8.4. Meanwhile, we take $E_{T-1}y_{T-1}$ as given and complete the derivation of optimal control by using dynamic programming.

The minimum expected welfare cost in the last period is obtained by substituting (24) for x_T in (22). To simplify the resulting expression we denote $E_{T-1}y_{T-1}$ by $\bar{y}_{T-1}$, write $\bar{y}_{T-1} = y_{T-1} + (\bar{y}_{T-1} - y_{T-1})$, and use the identity $(A_T + C_T G_T)' H_T C_T = 0$:

$$\hat{V}_T = E_{T-1}[(A_T y_{T-1} + C_T G_T \bar{y}_{T-1} + b_T + C_T g_T)'$$

$$\times H_T(A_T y_{T-1} + C_T G_T \bar{y}_{T-1} + b_T + C_T g_T)]$$

$$- 2(A_T \bar{y}_{T-1} + C_T G_T \bar{y}_{T-1} + b_T + C_T g_T)' h_T + c_T + E_{T-1}u_T' H_T u_T$$

$$= E_{T-1}[y_{T-1}'(A_T + C_T G_T)' H_T(A_T + C_T G_T)y_{T-1}]$$

$$+ E_{T-1}[(\bar{y}_{T-1} - y_{T-1})' G_T' C_T' H_T C_T G_T(\bar{y}_{T-1} - y_{T-1})]$$

$$+ 2\bar{y}_{T-1}'(A_T + C_T G_T)' H_T(b_T + C_T g_T) + (b_T + C_T g_T)' H_T(b_T + C_T g_T)$$

$$- 2\bar{y}_{T-1}'(A_T + C_T G_T)' h_T - 2(b_T + C_T g_T)' h_T + c_T + E_{T-1}u_T' H_T u_T$$

$$= E_{T-1}[y_{T-1}'(A_T + C_T G_T)' H_T(A_T + C_T G_T)y_{T-1}]$$

$$+ 2\bar{y}_{T-1}'(A_T + C_T G_T)'(H_T b_T - h_T)$$

$$+ (b_T + C_T g_T)' H_T(b_T + C_T g_T) - 2(b_T + C_T g_T)' h_T$$

$$+ c_T + E_{T-1}u_T' H_T u_T$$

$$+ E_{T-1}[(\bar{y}_{T-1} - y_{T-1})' G_T' C_T' H_T C_T G_T(\bar{y}_{T-1} - y_{T-1})]. \tag{25}$$

Comparing (25) with (9), in which the vector y_{T-1} is accurately measured, we note that (25) reduces to (9) if $\bar{y}_{T-1} = y_{T-1}$ and that the term E_{T-1} $[(\bar{y}_{T-1} - y_{T-1})' G_T' C_T' H_T C_T G_T (\bar{y}_{T-1} - y_{T-1})]$ reflects the minimum expected cost of measurement error in (25).

Now consider the problem for the last two periods. By the principle of optimality in dynamic programming, we minimize with respect to x_T the expression

$$V_{T-1} = E_{T-2}\left(y_{T-1}' K_{T-1} y_{T-1} - 2y_{T-1}' k_{T-1} + a_{T-1}' K_{T-1} a_{T-1} + \hat{V}_T\right). \quad (26)$$

To incorporate $\hat{V}_T$ from (25) into (26) we use the identity

$$E_{T-2}[E_{T-1} f(y)] = E_{T-2}[f(y)] \quad (27)$$

or

$$E\{E[f(y)|s_1]|s_2\} = E[f(y)|s_2].$$

Here y is a vector of random variables whose distribution is a function of the parameters s_1 (available in period $t-1$) which include as a subset the parameters s_2 (available in period $t-2$); f is a given function.

The identity (27) can be proved as follows. The symbol p denotes probability density function that is assumed to exist. Multiple integration is indicated by a single integral sign with dy standing for $dy_1 dy_2 \cdots$.

$$
\begin{aligned}
E[f(y)|s_2] &= \int_{-\infty}^{\infty} f(y) p(y|s_2)\, dy \\
&= \int f(y) \cdot \int p(y, s_1|s_2)\, ds_1 \cdot dy \\
&= \int f(y) \cdot \int p(y|s_1, s_2) p(s_1|s_2)\, ds_1 \cdot dy \\
&= \int \left[\int f(y) p(y|s_1)\, dy \right] p(s_1|s_2)\, ds_1 \\
&= \int E[f(y)|s_1] p(s_1|s_2)\, ds_1 \\
&= E\{E[f(y)|s_1]|s_2\}. \quad (28)
\end{aligned}
$$

When (25) is substituted for $\hat{V}_T$ in (26), the identity (27) implies that the expectation sign E_{T-1} operating on any function of y_{T-1} can be dropped.

Therefore (26) becomes

$$V_{T-1} = E_{T-2}\{ y'_{T-1}K_{T-1}y_{T-1} + y'_{T-1}(A_T + C_T G_T)' H_T (A_T + C_T G_T)y_{T-1}$$

$$- 2y'_{T-1}k_{T-1} + 2y'_{T-1}(A_T + C_T G_T)'(H_T b_T - h_T)$$

$$+ a'_{T-1}K_{T-1}a_{T-1} + (b_T + C_T g_T)' H_T (b_T + C_T g_T)$$

$$- 2(b_T + C_T g_T)' h_T + c_T + E_{T-1}u'_T H_T u_T$$

$$+ E_{T-1}[(\bar{y}_{T-1} - y_{T-1})' G'_T C'_T H_T C_T G_T (\bar{y}_{T-1} - y_{T-1})]\}$$

$$= E_{T-1}(y'_{T-1}H_{T-1}y_{T-1} - 2y'_{T-1}h_{T-1} + c_{T-1}), \tag{29}$$

where

$$H_{T-1} = K_{T-1} + (A_T + C_T G_T)' H_T (A_T + C_T G_T), \tag{30}$$

$$h_{T-1} = k_{T-1} + (A_T + C_T G_T)'(h_T - H_T b_T), \tag{31}$$

$$c_{T-1} = a'_{T-1}K_{T-1}a_{T-1} + (b_T + C_T g_T)' H_T (b_T + C_T g_T)$$

$$- 2(b_T + C_T g_T)' h_T + c_T + E_{T-1}u'_T H_T u_T$$

$$+ E_{T-1}[(\bar{y}_{T-1} - y_{T-1})' G'_T C'_T H_T C_T G_T (\bar{y}_{T-1} - y_{T-1})]. \tag{32}$$

We did not drop the expectation sign E_{T-1} before the quadratic function of $(\bar{y}_{T-1} - y_{T-1})$ because its expectation depends on the covariance matrix $E_{T-1}(\bar{y}_{T-1} - y_{T-1})(\bar{y}_{T-1} - y_{T-1})'$ of forecasting errors. This matrix can be computed by the method in Section 8.4 and, as shown by (48) and (51), is not influenced by the policy x_{T-1}, as other functions of y_{T-1} are.

Because (29) is identical to the first line of (22), with T replaced by $T-1$, the steps from (22) to (29) can also be repeated with T replaced by $T-1$, and so on. The optimal control problem is now solved. The optimal feedback equations take the form (24), with t replacing T. Each is a linear function of $E_{t-1}y_{t-1} = \bar{y}_{t-1}$. The coefficients G_t are found by solving (7) and (30) backward in time for G_t and H_t, beginning with $t = T$. The intercepts g_t are found by solving (8) and (31) backward in time for g_t and h_t. Note that (30) and (31) are identical to (12) and (13), respectively. Thus *the optimal control equations, when y_t is observed with errors, are the same as for correctly measured y_t except that the variable y_{t-1} is replaced by $E_{t-1}y_{t-1}$ $= \bar{y}_{t-1}$.*

The result just stated, valid for linear systems with known parameters and quadratic welfare functions, is termed a *separation theorem*, which states that when the dependent variables are measured with errors the optimal control solution can be separated into two parts. The first part is a set of optimal control equations determining x_t as a linear function of the conditional expectation of y_{t-1}. These equations are identical to the optimal control equations applied to y_{t-1} itself when it is known for certain. The second part is a solution to the problem of evaluating $E_{t-1}y_{t-1}$. Observe that the principle of certainty equivalence applies to the first part of the solution. When y_{t-1} is uncertain, the optimal control equations derived by assuming that y_{t-1} is a constant vector remain valid except that y_{t-1} is replaced by $E_{t-1}y_{t-1}$.

As in the problem in Section 8.1, where y_t are correctly measured, $\hat{V}_1$, as computed by (25) with 1 replacing T, gives the minimum expected loss for the policy in all periods from 1 to T. Equation 32 is used to compute c_{t-1} backward in time to obtain c_1 for use in $\hat{V}_1$. Note that in both (25) and (32) we assume that $E_{t-1}(\bar{y}_{t-1} - y_{t-1})(\bar{y}_{t-1} - y_{t-1})'$ is a known matrix because the term

$$E_{t-1}[(\bar{y}_{t-1} - y_{t-1})' G_t' C_t' H_t C_t G_t (\bar{y}_{t-1} - y_{t-1})]$$

$$= \text{tr}[G_t' C_t' H_t C_t G_t \cdot E_{t-1}(\bar{y}_{t-1} - y_{t-1})(\bar{y}_{t-1} - y_{t-1})']$$

8.4* THE KALMAN FILTER

When errors of measurement were introduced in Section 8.3, it was found that the optimal control equations in Section 8.1 are applicable, provided that $E_{t-1}y_{t-1}$ replaces y_{t-1} in $x_t = G_t y_t + g_t$. Furthermore, an equation (25) similar to (9) can be used to evaluate the minimum expected welfare loss if the covariance matrix $E_{t-1}(y_{t-1} - \bar{y}_{t-1})(y_{t-1} - \bar{y}_{t-1})'$ of measurement errors is available. The conditional means and covariance matrix are the topics of this section. The solution is due to Kalman (1960).

Given a model

$$y_t = A_t y_{t-1} + C_t x_t + b_t + u_t \tag{33}$$

and an observation equation

$$s_t = M_t y_t + \eta_t, \tag{34}$$

where $Eu_t = 0$, $Eu_t u_t' = V$, $Eu_t u_r' = 0$ for $t \neq r$, $E\eta_t = 0$, $E\eta_t \eta_t' = Q$, $E\eta_t \eta_r' = 0$ for $t \neq r$, and M_t is assumed to be a known matrix, the problem is to find the mean vector

$$E(y_t | s_t) = \bar{y}_t \tag{35}$$

and the covariance matrix

$$\mathrm{cov}(y_t|s_t) = \Sigma_t. \tag{36}$$

It is assumed that $\bar{y}_0$ and Σ_0 are known. The solution takes the form of a set of updating equations to calculate $\bar{y}_t$ and Σ_t from $\bar{y}_{t-1}$ and Σ_{t-1} by using the current observation s_t.

Because the problem is to predict y_t from s_t, it is natural to consider the regression of y_t on s_t. If the solution is to be a revision of the estimate at time $t-1$, it is appropriate to consider the regression of y_t on s_t, given s_{t-1}. The linear regression can be written as

$$E(y_t|s_t) = E(y_t|s_{t-1}) + D_t[s_t - E(s_t|s_{t-1})]. \tag{37}$$

By definition, a regression of y_t on s_t is a conditional mean of y_t, given s_t. This conditional mean is a linear function of s_t if y_t and s_t satisfy a multivariate normal distribution. For our problem assume that u_t and η_t in (33) and (34) are both normal. By repeated elimination of y_{t-1}, y_{t-2}, and so on, y_t in (33) can be expressed as a linear combination of u_t, u_{t-1}, and so on, and is therefore normal. Using (34) also, we find that both y_t and s_t are linear combinations of $u_t, u_{t-1}, \ldots$, and η_t and therefore they are jointly normal. Hence the regression of y_t on s_t is linear. In the linear regression function (37) D_t is the matrix of regression coefficients. The ith row of D_t is a set of coefficients in the regression of the ith element of y_t on the elements of s_t. By regression theory, if the explanatory variables s_t are measured from their mean vector $E(s_t|s_{t-1})$, as in (37), the intercept of the multivariate regression equation is the mean $E(y_t|s_{t-1})$ of the vector of dependent variables. We take s_{t-1} as given in this analysis so that the means of both s_t and y_t are conditional on s_{t-1}.

To simplify the writing of conditional means and conditional covariance matrix, given s_{t-1}, we introduce the notation

$$E(y_t|s_{t-1}) = y_{t|t-1},$$

$$E(s_t|s_{t-1}) = s_{t|t-1},$$

and

$$\Sigma_{t|t-1} = E(y_t - y_{t|t-1})(y_t - y_{t|t-1})'.$$

In this notation $\bar{y}_t$, defined (35), can be written as $y_{t|t}$.

Equation 37 is already an updating equation for $E(y_t|s_t) = y_{t|t}$. The following derivations amount to a simplification of the details in (37). First eliminate $E(s_t|s_{t-1})$ by using

$$s_{t|t-1} = M_t y_{t|t-1}, \tag{38}$$

a result derived from taking conditional expectations on both sides of (34). Using (38) and the newly acquired notations, we write (37) as

$$y_{t|t} = (I - D_t M_t)y_{t|t-1} + D_t s_t. \tag{39}$$

Equation 39 expresses the required estimate $y_{t|t} = \bar{y}_t$ as a linear combination of the forecast $y_{t|t-1}$ from the preceding period and the observation s_t of the current period. Both D_t and $y_{t|t-1}$ have to be evaluated.

The ith row of the matrix D_t, by regression theory, is equal to the row vector of covariances between y_{it} and the row vector s_t', that is, the ith row of the cross-covariance matrix

$$E(y_t - y_{t|t-1})(s_t - s_{t|t-1})' \tag{40}$$

postmultiplied by the inverse of the covariance matrix of the explanatory variables, or

$$E(s_t - s_{t|t-1})(s_t - s_{t|t-1})'. \tag{41}$$

To evaluate (40) we first use (34) and (38) to obtain the equation

$$s_t - s_{t|t-1} = M_t y_t - M_t y_{t|t-1} + \eta_t. \tag{42}$$

Premultiplying the transpose of (42) by $(y_t - y_{t|t-1})$ and taking expectations will give

$$E(y_t - y_{t|t-1})(s_t - s_{t|t-1})' = E(y_t - y_{t|t-1})(y_t - y_{t|t-1})' \cdot M_t'$$

$$= \Sigma_{t|t-1} M_t'. \tag{43}$$

To evaluate (41) we premultiply the transpose of (42) by $(s_t - s_{t|t-1})$ and take expectations,

$$E(s_t - s_{t|t-1})(s_t - s_{t|t-1})' = E(s_t - s_{t|t-1})(y_t - y_{t|t-1})' M_t' + Q$$

$$= M_t \Sigma_{t|t-1} M_t' + Q. \tag{44}$$

By (43) and (44) the matrix of regression coefficients is

$$D_t = [E(y_t - y_{t|t-1})(s_t - s_{t|t-1})'][E(s_t - s_{t|t-1})(s_t - s_{t|t-1})']^{-1}$$

$$= \Sigma_{t|t-1} M_t' [M_t \Sigma_{t|t-1} M_t' + Q]^{-1} \tag{45}$$

This completes the evaluation of the regression coefficients D_t. Recall that each row of D_t is a set of coefficients in the regression of one element of y_t on the vector s_t.

To apply the key result (39), using (45) for D_t, we need to compute $y_{t|t-1}$ and $\Sigma_{t|t-1}$. The first is easily obtained by taking conditional expectation of (33),

$$y_{t|t-1} = E(y_t|s_{t-1}) = A_t y_{t-1|t-1} + C_t x_t + b_t. \tag{46}$$

The second is obtained by rewriting the difference $y_t - y_{t|t-1}$ by using (33) and (46) as

$$y_t - y_{t|t-1} = A_t(y_{t-1} - y_{t-1|t-1}) + u_t \tag{47}$$

and forming the covariance matrix of (47),

$$\Sigma_{t|t-1} = E(y_t - y_{t|t-1})(y_t - y_{t|t-1})'$$

$$= E[A_t(y_{t-1} - y_{t-1|t-1})(y_{t-1} - y_{t-1|t-1})'A_t'] + Eu_t u_t'$$

$$= A_t \Sigma_{t-1|t-1} A_t' + V. \tag{48}$$

To update the covariance matrix $\Sigma_{t|t} = E(y_t - y_{t|t})(y_t - y_{t|t})'$ we refer back to the regression of y_t on s_t, given s_{t-1}. This is the covariance matrix of the residuals of the regression (37) of y_t on s_t. Using (37), we obtain the vector of residuals:

$$y_t - y_{t|t} = y_t - y_{t|t-1} - D_t(s_t - s_{t|t-1}). \tag{49}$$

Taking the expectation of the product of the right-hand side of (49) and its transpose, we have

$$E(y_t - y_{t|t})(y_t - y_{t|t})' = E(y_t - y_{t|t-1})(y_t - y_{t|t-1})'$$

$$+ D_t E(s_t - s_{t|t-1})(s_t - s_{t|t-1})' D_t'$$

$$- E(y_t - y_{t|t-1})(s_t - s_{t|t-1})' D_t'$$

$$- D_t E(s_t - s_{t|t-1})(y_t - y_{t|t-1})'. \tag{50}$$

By (43), (44), and (45) the covariance matrix (50) becomes

$$\Sigma_{t|t} = \Sigma_{t|t-1} - \Sigma_{t|t-1} M_t' (M_t \Sigma_{t|t-1} M_t' + Q)^{-1} M_t \Sigma_{t|t-1}. \tag{51}$$

This completes the solution of the problem in this section.

Two sets of equations to update $\Sigma_{t-1|t-1}$ and $y_{t-1|t-1}$ can now be summarized. The first set consists of (48) and (51); (48) is used to compute $\Sigma_{t|t-1}$ from $\Sigma_{t-1|t-1}$ and (51) is used to compute $\Sigma_{t|t}$ from $\Sigma_{t|t-1}$. The required parameters are the coefficients A_t and M_t in (33) and (34) and the covariance matrices are V and Q in these equations. Note that the $\Sigma_{t-1|t-1}$ is updated without utilizing any observation on s_t. Given the model (33) and (34), we can calculate all $\Sigma_{t|t}$ for $t = 1, 2, \ldots, T$, provided that $\Sigma_{0|0}$ is known.

The second set of equations consists of (39), (45), and (46); (46) is used to compute $y_{t|t-1}$ from $y_{t-1|t-1}$ and (39) is used to compute $y_{t|t}$ as a linear combination of $y_{t|t-1}$ and the observation s_t. The coefficients $(I - D_t M_t)$ and D_t in the linear combination depend on the matrix D_t which is given by (45). These equations are called the *Kalman filter*. The word filter is used because it is a device that cleanses the observations s_t which contain errors in order to form an estimate of the true y_t.

A filtering equation combining (39) and (46) can be written as

$$y_{t|t} = (I - D_t M_t)(A_t y_{t-1|t-1} + C_t x_t + b_t) + D_t s_t$$

$$= (A_t y_{t-1|t-1} + C_t x_t + b_t) + D_t[s_t - M_t(A_t y_{t-1|t-1} + C_t x_t + b_t)]; \quad (52)$$

(52) shows how $y_{t-1|t-1}$ and s_t are combined to form $y_{t|t}$. First $y_{t-1|t-1}$ is applied to the model (33) to form

$$y_{t|t-1} = A_t y_{t-1|t-1} + C_t x_t + b_t.$$

The result is combined with s_t as indicated on the right-hand side of the first line of (52). Alternatively, by the second line of (52) the result $y_{t|t-1}$ is applied to the observation equation to form an estimate

$$s_{t|t-1} = M_t y_{t|t-1}.$$

The error $(s_t - s_{t|t-1})$ of predicting s_t by this estimate is then combined with the forecast $y_{t|t-1}$ from the model (33).

Equation 52 captures the main result of the Kalman filter. Its execution requires knowledge of the coefficients D_t. According to (45), the computation of D_t in turn requires updating the covariance matrix $\Sigma_{t|t}$. Thus the set of equations (48) and (51) are necessary for calculating not only the covariance matrix of estimation errors $(y_t - \bar{y}_t)$ but also the estimate $\bar{y}_t = y_{t|t}$ itself. Recall that in the control problem of the last section $\Sigma_{t|t}$ is required for computing the minimum expected loss by equation (25).

The Kalman filter was derived in this section as a conditional expectation of y_t, given s_t. As such, it is an *optimal estimator* of y_t in the sense that it minimizes the expectation of any positive semidefinite quadratic form

$(y_t - \bar{y}_t)' K(y_t - \bar{y}_t)$ of the estimation errors $y_t - \bar{y}_t$. This optimal property can be shown by considering any vector w and evaluating the conditional expectation of the quadratic form of $y_t - w$, given s_t:

$$E_t(y_t - w)' K(y_t - w) = E_t[(y_t - \bar{y}_t) + (\bar{y}_t - w)]' K[(y_t - \bar{y}_t) + (\bar{y}_t - w)]$$

$$= E_t(y_t - \bar{y}_t)' K(y_t - \bar{y}_t) + E_t(\bar{y}_t - w)' K(\bar{y}_t - w) + 2E_t(\bar{y}_t - w)' K(y_t - \bar{y}_t)$$

$$= E_t(y_t - \bar{y}_t)' K(y_t - \bar{y}_t) + (\bar{y}_t - w)' K(\bar{y}_t - w) + 2(\bar{y}_t - w)' K(E_t y_t - \bar{y}_t)$$

$$\geqslant E_t(y_t - \bar{y}_t)' K(y_t - \bar{y}_t). \tag{53}$$

In (53) the quadratic form $(\bar{y}_t - w)' K(\bar{y}_t - w)$ is known to be positive semidefinite and $E_t y_t - \bar{y}_t$ is 0 by definition. The optimal property of the conditional expectation as an estimator of y_t is proved. When K is set equal to a matrix of 0s, except for one diagonal element, this property implies that the expectation of the squared error in estimating any element of y_t is minimum. Because $E_t(y_t - \bar{y}_t) = 0$, the Kalman filter is a minimum-variance unbiased estimator of y_t.

In this discussion the term estimation has been used differently from the usage in classical statistics. In classical statistics the object to be estimated is a constant (scalar or vector) parameter and the estimator is a random variable; for example, the sample mean is used as an estimator for the population mean. In this section the object to be estimated is a random variable. The estimator is a parameter of a certain distribution. The random vector y_t is to be estimated by the mean $\bar{y}_t$ of the conditional distribution of y_t, given s_t. In the terminology of classical statistics the word "prediction" is used if the object to be estimated is a random variable. We have not restricted ourselves to the fairly specialized concept of estimation adopted in the classical statistics literature. In Bayesian statistics the object to be estimated is also a random variable and the estimator is also a parameter of a certain distribution. It is not surprising that the Kalman filter can be derived in the framework of Bayesian statistics, but the Bayesian interpretation is omitted here. (See Problem 13.)

8.5* SOME ECONOMIC APPLICATIONS OF THE KALMAN FILTER

Possible applications of the Kalman filter to dynamic economics and econometrics are numerous. This section merely provides several illustrative examples and indicates some areas of useful application.

One example is the prediction of interindustry demand relations by Vishwakarma, de Boer, and Palm (1970). Let s_t be an $n \times 1$ vector of outputs from n industries in period t, A_t, an $n \times n$ matrix of input-output coefficients, the $i - j$ element being the amount of output from industry i required to produce one unit of output in industry j, and f_t, an $n \times 1$ vector of final demands for the products of the n industries. The standard input-output relationship in period t is given by the linear equations

$$s_t = A_t s_t + f_t,$$

the solution of which is

$$s_t = (I - A_t)^{-1} f_t = B_t f_t, \tag{54}$$

where $B_t = (I - A_t)^{-1}$. The problem is to estimate the coefficients B_t in the linear relation (54) between industry outputs and final demands under the assumption that the relation contains a vector η_t of random errors; that is,

$$s_t = B_t f_t + \eta_t. \tag{55}$$

In (55) the production s_t and the final demand f_t are assumed to be directly observable. The elements of B_t are to be estimated and B_t is assumed to be time-invariant. The vector η_t is assumed to have mean 0 and covariance matrix Q and to be serially uncorrelated. To cast this problem in the form of equations (33) and (34) for the application of the Kalman filter, let $b'_{i,t}$ denote the ith row of B_t and let the column vector b_t denote the transpose of $(b'_{1,t} b'_{2,t} \cdots b'_{n,t})$. The time-invariance of B_t implies

$$b_t = b_{t-1}. \tag{56}$$

Equation 56 corresponds to (33), with b_t taking the place of y_t, $A_t = I$, and the rest of (33) being 0. The observation equation (34) is formed by rewriting (55) as

$$s_t = \begin{bmatrix} f'_t & 0 & \cdots & 0 \\ 0 & f'_t & \cdots & 0 \\ \multicolumn{4}{c}{\dotfill} \\ 0 & 0 & \cdots & f'_t \end{bmatrix} \begin{bmatrix} b_{1,t} \\ b_{2,t} \\ \vdots \\ b_{n,t} \end{bmatrix} + \eta_t = M_t b_t + \eta_t. \tag{57}$$

The matrix M_t in (34) takes a special form in (57). It is an $n \times n^2$ matrix relating n measurements to n^2 unobservable quantities. By using some estimates of $Q = E\eta_t \eta_t'$, $\bar{b}_0$ and $\Sigma_0 = \text{cov}\, b_0$ we can apply the Kalman filter to estimate b_t in the system (56) and (57). (See Problem 10.)

A second example is the use of the Kalman filter in econometric forecasting by Mariano and Schleicher (1972). Let the reduced form of a system of linear econometric equations be given by (33). In addition, assume that outside information on a subset of y_t is available one period ahead; for example, if (33) contains an equation for investment expenditures, the outside information may consist of data on investment expectations. If one variable in (33) is GNP or one of its components, the additional information may be a preliminary estimate of GNP or its component. The set of outside estimates constitutes the elements of s_t in (34). The matrix M_t takes the special form $M_t = [I \quad O]$ if (33) is rearranged to place the subset of variables with outside measurements before the remaining variables in y_t. By applying the Kalman filtering equation (52) we would combine the outside measurements s_t with the estimate $y_{t-1|t-1}$ of the last period to form an estimate $y_{t|t}$ for the current period. Under the assumptions of the model (33) and (34), y_t cannot be measured accurately, and $y_{t-1|t-1}$ represents only an estimate of y_{t-1} to be applied to the filtering equation (52). In the setup of Mariano and Schleicher (1972) y_{t-1} can be accurately measured. The authors suggest using y_{t-1} in place of $y_{t-1|t-1}$ in (52) for the calculation of the forecast $y_{t|t}$.

A third application of the Kalman filter to economics is the explanation of the formation of expectations by economic agents. This should perhaps be called an area of application rather than an example. Several economists have written on the subject, including John F. Muth (1960 and 1961), Martin Bailey (1962), Lance Taylor (1970), and Marc Nerlove (1972), among others. Muth's papers laid much of the theoretical foundation for subsequent studies in economics and contained results on optimal forecasting independent of Kalman's work. Bailey's discussion of the subject in Section 9.4 of his book (1962) and in Section 11.6 of its second edition (1971) refers to Muth's work and to another paper of Bailey's own (1965) but does not mention the Kalman filter. Taylor (1970) and Nerlove (1972) recognize the Kalman filter explicitly. This is not the place to dwell on the intellectual history of the subject but to present an important application of the Kalman filter to economics. These remarks were made partly to avoid puzzling the reader should he fail to find any mention of the Kalman filter in some of the references cited and partly to point out that similar ideas have been developed independently in economics.

One important idea in dynamic economics is that economic behavior is based on expectations or expected variables. An early application is the work of Philip Cagan (1956) on hyperinflation in which an expected price served as an important variable. Friedman (1957) applied the concept of expected income (also called permanent income) to explain consumption behavior; for example, in Cagan's work on hyperinflation, when formulated in discrete time, expected price y_t^e is assumed to satisfy the difference equation

$$y_t^e - y_{t-1}^e = d(s_t - y_{t-1}^e),\qquad(58)$$

where s_t is the observed price; (58) implies

$$y_t^e = ds_t + (1-d)y_{t-1}^e$$

$$= ds_t + (1-d)ds_{t-1} + (1-d)^2 ds_{t-2} + \cdots.\qquad(59)$$

The problem is to provide a justification of the behavioral assumption (58).

Taylor (1970) offers a justification of the formation of expectations described by (58). He assumes that the "permanent component" y_{1t} of an economic variable satisfies a pth order linear stochastic difference equation. We can convert this equation to a first-order system of p equations.

$$y_t = Ay_{t-1} + u_t\qquad(60)$$

Taylor also assumes that the observed (scalar) series s_t is generated by

$$s_t = My_t + \eta_t,\qquad(61)$$

where $M = [1 \quad 0 \cdots 0]$. Applying the filtering equation (52) to this problem, we have

$$y_{t|t} = (I - D_t M)Ay_{t-1|t-1} + D_t s_t,\qquad(62)$$

where D_t is $p \times 1$ vector of coefficients in the regressions of the p elements of y_t on s_t.

In the special case of $p = 1$, M equals 1 and the prediction $y_{t|t}$ of the scalar time series $y_{1t} = y_t$ is a linear combination of the prediction $y_{t-1|t-1}$ in the last period and the currently observed s_t. If $y_{t|t}$ is identified with the expected variable in (59), (62) is identical to (59), provided that $A = 1$ and that the coefficient D_t reaches a steady state d. Taylor (1970) states conditions from Kalman's theory under which D_t will reach a steady state. Thus a rational explanation is provided for the formation of expectations described by (59). The explanation assumes a linear structure (60) for the true variable, an observation equation (61), and the rational behavior on the part of the economic agent in forming minimum-variance unbiased estimates of the true variable by the Kalman filter.

If the economic variable y_{1t} satisfies a higher order linear stochastic difference equation, the solution (62) to optimal forecast still applies; $y_{1,t|t}$ will still be a linear function of past $s_{t-k}(k=0,1,2,...)$ but the coefficients will no longer be geometrically declining with k as in (59). (See Problem 11.) The result will still provide a theory of rational expectations. Bailey (1962) also explains rational expectations in similar terms, and Nerlove (1972) includes other economic examples of the formation of expectations and a list of useful references.

The last area of application of the Kalman filter to be mentioned in this section is the estimation of time-varying coefficients in linear models in econometrics. Consider the random coefficient model

$$\beta_t = \beta_{t-1} + u_t, \tag{63}$$

$$s_t = M_t \beta_t + \eta_t, \tag{64}$$

where s_t is a vector of observations on the dependent variable, M_t is a matrix of observations on the explanatory variables, and (64) is a regression equation with random coefficients β_t that satisfy (63). This model is a special case of (33) and (34), in which β_t takes the place of y_t, $A_t = I$, $x_t = 0$, and $b_t = 0$ in (33). The Kalman filter can be applied to estimate the random coefficient β_t. The subject and its extension to systems of regression equations are interesting but beyond the scope of this book. (See Problems 12 and 13.) Interested readers may refer to the *Special Issue on Time-Varying Parameters, Annals of Economic and Social Measurement*, **2**, (4) (October 1973), and to Athans (1974).

8.6 AN INTERPRETATION OF THE DYNAMIC PROGRAMMING SOLUTION

Sections 8.4 and 8.5 have dealt with a useful estimation technique and its applications to dynamic economics. We now return to the main theme of this chapter—the application of dynamic programming to the solution of multiperiod control problems. This section provides an interpretation of the dynamic programming solutions of Sections 8.1 and 8.3. Section 8.7 describes some other economic applications of the method of dynamic programming.

An important feature of the method of dynamic programming when applied to the control problem in Section 8.1 is that it decomposes a multiperiod problem involving $x_1, x_2, ..., x_T$ to T problems, each involving one x_t. The same remark is valid for the control problem in Section 8.3

when y_t is not accurately measured. Eventually, after the optimal policies for $x_T, x_{T-1}, \ldots, x_2$ have been determined, the problem is reduced to one of choosing the optimal x_1 for the current period. Thus a multiperiod problem is ultimately reduced to a one-period problem because the future strategies $x_2, x_3, \ldots, x_T$ have been successively eliminated.

The reduction of a problem for periods $t, t+1, \ldots, T$ to a problem for period t alone is accomplished by decomposing the objective function from period t onward into two parts. The first part is concerned with the current period:

$$E_{t-1}(y_t' K_t y_t - 2y_t' k_t + a_t' K_t a_t). \tag{65}$$

The second part is the expectation of the minimum welfare loss from period $t+1$ on:

$$E_{t-1} \hat{V}_{t+1} = E_{t-1}(y_t' Q_t y_t - 2y_t' q_t + r_t), \tag{66}$$

where the matrix Q_t and the vectors q_t and r_t are defined by (9) for the problem in Section 8.1 and by (25) for the problem in Section 8.3; in both equations $t+1$ replaces T. Thus the minimum future loss from $t+1$ on is also a quadratic function of y_t. The two parts of the control problem for period t are combined to yield an objective function

$$E_{t-1}(y_t' H_t y_t - 2y_t' h_t + c_t), \tag{67}$$

where $H_t = K_t + Q_t$, $h_t = k_t + q_t$, and $c_t = a_t' K_t a_t + r_t$. Note that the form of (67) for a multiperiod problem is identical to the form of (65) for a one-period problem. This is how the method of dynamic programming converts a multiperiod problem to many one-period problems and eventually to the problem of determining x_1 alone.

An essential function that permits the conversion of a multiperiod problem to a one-period problem in each period t is the minimum welfare loss $\hat{V}_{t+1}$ from period $t+1$ on. The function expresses the result of optimal planning for the entire future in terms of the current variables y_t. Thus in choosing the current x_t we need to be concerned only with its effect on y_t alone, the indirect effects on y_{t+1}, y_{t+2}, and so on, having been accounted for implicitly. By (65) and (66) the impacts of x_t on y_t are decomposed into two parts, the first being the immediate impact on the welfare loss from the current period and the second being the impact on y_t as it affects the outcome of the best possible strategies to be applied in the future; (65) and (66) show the relative importance of y_t in its effects on current welfare loss and on minimum future welfare loss. This interpretation of the dynamic programming solution is based on a welfare loss function that is *additive*, being the sum of quadratic functions of the variables y_t in different periods T. This interpretation is extended to welfare functions that include cross-product terms $y_t' K_{t,s} y_s$ in Chapter 11.

8.7* APPLICATIONS OF DYNAMIC PROGRAMMING TO MICROECONOMICS

As a technique of optimization for a multiperiod problem under uncertainty dynamic programming has numerous applications in microeconomic theory. Once an objective function involving variables in many periods and a set of dynamic equations governing those variables are given, the method of dynamic programming can be applied to solve the problem of maximizing or minimizing the expectation of the objective function with respect to the variables the economic agent can control. This is how the techniques introduced in this book are relevant to dynamic microeconomic theory under uncertainty, as hinted in the introduction to Chapter 3 on stochastic dynamic economics. The applications of dynamic programming to microeconomics are numerous enough to form a separate volume; therefore they are beyond the scope of this book. No discussion of the method is complete, however, without indicating how it is applied to microeconomics by way of an example.

Samuelson's paper (1969) illustrates an important microeconomic problem to be solved by the method of dynamic programming. At the beginning of each period t the consumer-investor chooses to consume an amount C_t and invest a fraction w_t of his remaining wealth $W_t - C_t$ in a risky asset. There is also the alternative of investing in a safe asset that provides $R = (1 + r)$ dollars per dollar invested one period earlier. By contrast, the risky asset produces a random amount of Z_t dollars at the end of the period. To simplify the present discussion the random variables Z_t, $t = 1, \ldots T$, are assumed to be independently and identically distributed with density function $p(Z_t)$. Each Z_t is assumed to take only nonnegative values. The random amount resulting from a dollar invested at the beginning of period t is $(1 - w_t)R + w_t Z_t$. Therefore total wealth at the beginning of the next period is

$$W_{t+1} = (W_t - C_t)[(1 - w_t)R + w_t Z_t]. \tag{68}$$

Equation 68 is a difference equation in W_t. It is nonlinear in the control variables C_t and w_t. The problem is to maximize

$$E_0 \sum_{t=0}^{T} (1 + \rho)^{-t} U(C_t) \tag{69}$$

(ρ being a discount factor) with respect to C_t and w_t ($t=0,...,T-1$), subject to the constraint (68). The welfare function in (69) is a sum of discounted utilities of the amounts consumed in different periods. Assume that the final bequest is 0, so that $W_T = C_T$ and $W_{T+1} = 0$.

Applying the method of dynamic programming, we consider the choice of C_{T-1} and w_{T-1} in the last period $T-1$. It is to maximize

$$V_{T-1} = E_{T-1}\left[U(C_{T-1}) + (1+\rho)^{-1}U(W_T) \right]$$

$$= U(C_{T-1}) + (1+\rho)^{-1}E_{T-1}U\{(W_{T-1} - C_{T-1})$$

$$\times [(1-w_{T-1})R + w_{T-1}Z_{T-1}]\},$$

(70)

where (68) is used for W_T. Differentiation of (70) with respect to C_{T-1} and w_{T-1} yields

$$U'(C_{T-1}) - (1+\rho)^{-1}EU'(W_T)[(1-w_{T-1})R + w_{T-1}Z_{T-1}] = 0 \quad (71)$$

and

$$EU'(W_T)(W_{T-1} - C_{T-1})(Z_{T-1} - R)$$

$$= 0 = \int_0^\infty U'\{(W_{T-1} - C_{T-1})[(1-w_{T-1})R + w_{T-1}Z]\}$$

$$\times (W_{T-1} - C_{T-1})(Z - R)p(Z)dZ,$$

(72)

where the subscript $T-1$ for the operator E and for the variable Z is dropped because the only random variable Z_{T-1} involved has a density function that is independent of time. Equation 71 and 72 can be solved for the optimal decisions $\hat{C}_{T-1}$ and $\hat{w}_{T-1}$ as functions of W_{T-1}. These are the consumption and investment functions. They are the optimal control equations in the terminology of control theory. When $\hat{C}_{T-1}(W_{T-1})$ and $\hat{w}_{T-1}(W_{T-1})$ are substituted for C_{T-1} and w_{T-1} in (70), we obtain the maximum expected welfare $\hat{V}_{T-1}(W_{T-1})$ from period $T-1$ on.

Having found $\hat{V}_{T-1}(W_{T-1})$, consider the problem for an additional period $T-2$. It is to maximize

$$V_{T-2} = U(C_{T-2}) + (1+\rho)^{-1}E\hat{V}_{T-1}(W_{T-1})$$

$$= U(C_{T-2}) + (1+\rho)^{-1}E\hat{V}_{T-1}\{(W_{T-2} - C_{T-2})$$

$$\times [(1-w_{T-2})R + w_{T-2}Z_{T-2}]\}. \quad (73)$$

Differentiation of (73) with respect to C_{T-2} and w_{T-2}, respectively, gives

$$U'(C_{T-2}) - (1+\rho)^{-1}E\hat{V}'_{T-1}(W_{T-1})[(1-w_{T-2})R + w_{T-2}Z_{T-2}] = 0 \quad (74)$$

and

$$E\hat{V}'_{T-1}(W_{T-1})(W_{T-2} - C_{T-2})(Z_{T-2} - R)$$

$$= 0 = \int_0^\infty \hat{V}'_{T-1}\{(W_{T-2} - C_{T-2})[(1-w_{T-1})R + w_{T-2}Z]\}$$

$$\times (W_{T-2} - C_{T-2})(Z - R)p(Z)dZ. \quad (75)$$

Equations 74 and 75 can be used to find the optimal $\hat{C}_{T-2}$ and $\hat{w}_{T-2}$ and thus the maximum $\hat{V}_{T-2}$. This procedure can be repeated for other periods and the multiperiod optimization problem is solved.

If a steady-state solution exists when the time horizon T is large, we have the time-invariant consumption and investment functions, say $\hat{C}_t = f(W_t)$ and $\hat{w}_t = g(W_t)$. These functions satisfy the two functional equations obtained by deleting the time subscripts and substituting $f(W)$ and $g(W)$, respectively, for C and w in (74) and (75).

To obtain an explicit solution let the utility function be

$$U(C) = \gamma^{-1}C^\gamma, \quad (0 < \gamma < 1). \quad (76)$$

Equation 70 becomes

$$V_{T-1} = \gamma^{-1}C_{T-1}^\gamma + (1+\rho)^{-1}E(\gamma^{-1}W_T^\gamma)$$

$$= \gamma^{-1}C_{T-1}^\gamma + (1+\rho)^{-1}\gamma^{-1}(W_{T-1} - C_{T-1})^\gamma$$

$$\times \int_0^\infty [(1-w_{T-1})R + w_{T-1}Z]^\gamma p(Z)dZ. \quad (77)$$

Its derivatives are

$$\frac{\partial V_{T-1}}{\partial C_{T-1}} = C_{T-1}^{\gamma-1} - (1+\rho)^{-1}(W_{T-1} - C_{T-1})^{\gamma-1}$$

$$\times \int_0^\infty [(1-w_{T-1})R + w_{T-1}Z]^\gamma p(Z)dZ = 0 \quad (78)$$

and

$$\frac{\partial V_{T-1}}{\partial w_{T-1}} = (1+\rho)^{-1}\gamma^{-1}(W_{T-1} - C_{T-1})^\gamma$$

$$\times \int_0^\infty [(1-w_{T-1})R + w_{T-1}Z]^{\gamma-1}(Z - R)p(Z)dZ = 0. \quad (79)$$

Equation 79 can be used to obtain the optimal investment decision $\hat{w}_{T-1}$; $\hat{w}_{T-1}$ is the value of w_{T-1} that satisfies

$$\int_0^\infty [(1-w_{T-1})R+w_{T-1}Z]^{\gamma-1}(Z-R)p(Z)dZ=0. \tag{80}$$

We observe that this optimal portfolio policy is independent of the consumption decision. Furthermore, as long as the function V_t to be maximized in a future step of the dynamic programming solution has the same form as (77), the optimal $\hat{w}_t$ is obtained by solving the same equation (80), with t replacing $T-1$, and is thus independent of t. We show in (86) that V_t has the form (77). Meanwhile, denote the solution simply by $\hat{w}$. To simplify future derivations we let

$$k=\int_0^\infty [(1-\hat{w})R+\hat{w}Z]^\gamma p(Z)dZ. \tag{81}$$

Equations 78 and 81 are used to obtain the consumption function

$$\hat{C}_{T-1}=\frac{a_1}{1+a_1}W_{T-1} \tag{82}$$

where

$$a_1=\left[(1+\rho)^{-1}k\right]^{1/(\gamma-1)}. \tag{83}$$

Thus by maximizing expected utility over time we deduce a consumption function that is linear homogeneous in wealth W. When (81) and (82) are substituted into (77), the maximum expected utility is

$$\hat{V}_{T-1}=\gamma^{-1}\left(\frac{a_1}{1+a_1}\right)^\gamma W_{T-1}^\gamma+(1+\rho)^{-1}\gamma^{-1}\left(\frac{1}{1+a_1}\right)^\gamma W_{T-1}^\gamma\cdot k$$

$$=\gamma^{-1}b_1 W_{T-1}^\gamma, \tag{84}$$

where

$$b_1=\left(\frac{a_1}{1+a_1}\right)^\gamma+(1+\rho)^{-1}k(1+a_1)^{-\gamma}. \tag{85}$$

Given (84) for $\hat{V}_{T-1}$, the problem for one additional period $T-2$ is to maximize

$$V_{T-2}=\gamma^{-1}C_{T-2}^\gamma+(1+\rho)^{-1}E(\gamma^{-1}b_1 W_{T-1}^\gamma)$$

$$=\gamma^{-1}C_{T-2}^\gamma+(1+\rho)^{-1}\gamma^{-1}b_1(W_{T-2}-C_{T-2})^\gamma$$

$$\times\int_0^\infty \{(1-w_{T-2})R+w_{T-2}Z\}^\gamma p(Z)dZ; \tag{86}$$

V_{T-2} does have the same form as V_{T-1} in (77). Hence differentiation of (86) with respect to w_{T-2} will yield an equation identical to (79), except for a multiplicative constant b_1. The solution of this equation for $\hat{w}_{T-2}$ is identical to (80), making the optimal fraction to be invested in the risky asset independent of the consumption decision and time. The consumption function is similarly derived by differentiating (86) with respect to C_{T-2}, yielding

$$C_{T-2}^{\gamma-1} - (1+\rho)^{-1} b_1 (W_{T-2} - C_{T-2})^{\gamma-1} k = 0. \tag{87}$$

The solution of (87) is a consumption function linear in the wealth variable, as before. Repeated applications of the same procedure solve the entire problem. (See Problems 14 and 15.)

This description illustrates the application of dynamic programming to a multiperiod decision problem under uncertainty in microeconomics. For further discussion of the economic issue involved the reader is referred to Samuelson (1969). For related papers on portfolio selection and consumption decisions over time see Phelps (1962), Mossin (1968), Hakansson (1970), Fama (1970), and other references listed in these papers.

Besides portfolio selection and consumption decisions, many other topics of dynamic microeconomics been treated by the method of dynamic programming. Insofar as microeconomic theory is based on maximizing behavior and as dynamic programming is a tool to perform multiperiod maximization, the applications are unlimited. Several examples could be mentioned to indicate their scope. Eppen and Fama (1969) and Chitre (1972) use multiperiod cash balance models subject to uncertainty to study the demand for money. Becker and Stigler (1974) deal with optimal compensations by the firm to discourage its employees from malfeasance. Rothschild (1974) studies optimal search behavior of an individual when the probability distribution of the price of a commodity is unknown. One important area for potential and actual application of dynamic programming and optimal control theory in general is investment and factor-demand decisions of the firm. Some relevant studies are cited in Section 4 of Nerlove's survey article (1972).

In Chapter 9 we return to some optimal control problems in macroeconomics.

PROBLEMS

1. Show that (13) is identical to (28) in Chapter 7.
2. Does the principle of certainty equivalence apply if, in addition to u_t, the matrix A in the

model $y_t = Ay_{t-1} + Cx_t + u_t$ is random? Answer this question by considering a one-period control problem in which A and C are scalars.

3. In the control problem in Section 8.1 assume that there is a delay of one period between the setting of the policy variables and their realization. Let the model remain the same and let the setting of the policy variables at time t be z_t so that the realization $x_t = z_{t-1}$. Solve the control problem by using z_t as the control variables by the method of dynamic programming.

4. How can the welfare loss due to the delay in the realization of the policy variables as specified in Problem 3 be measured? Show the formulas used for the measurement.

5. Provide a solution to Problem 3 by using the method of Lagrange multipliers. Formulate the deterministic part of the problem first and then the stochastic part as in Chapter 7. [See Chow (1970).]

6. Decompose the control problem in Section 8.1 into deterministic and stochastic parts as in Chapter 7. Solve the two parts separately by dynamic programming.

7. Define the deterministic and stochastic parts of the expected welfare loss in the control problem in Section 8.1. What determines the size of each part? List all the important factors.

8. In the control problem in Section 8.3 let the matrix M_t be an identity matrix and the variables y_t be measured with random errors having a covariance matrix Q. How does the solution differ from the solution in Section 8.1 in which there is no measurement error? In particular, how are the feedback control equations affected? How do you measure the difference in expected welfare loss?

9. In a Kalman filtering problem let the model be $y_t = y_{t-1} + u_t$ and the observation be $s_t = y_t + \eta_t$, where y_t is a vector of two elements. Further, let $V = I$, $Q = I$, $\bar{y}_0 = 0$, and $\Sigma_0 = I$. Compute $\Sigma_{t|t}$ and D_t for $t = 1, 2, 3$. How would you estimate y_t for $t = 1, 2, 3$? Express your estimate of y_3 as a function of s_1, s_2, and s_3.

10. In the input-output problem of (56) and (57) how would you provide estimates of $\bar{b}_0$, Σ_0, and $Q = \mathrm{cov}\,\eta_t$ for the purpose of applying the Kalman filter?

11. Let the matrix A in (60) be 2×2. Write out the prediction equation for $y_{t|t}$ as specified by (62) and all associated equations required for its computation. What is the relation between $y_{t|t}$ and the past $s_{t-k}(k = 0, 1, 2, \ldots)$?

12. In the random coefficient model in (63) and (64) let $u_t = 0$. How is the estimate of β_t (say, for $t = 20$) obtained by the Kalman filter related to the estimate by the method of ordinary least squares?

13. In the random coefficient model in (63) and (64) let $u_t = 0$. How is the estimate of β_t (say, for $t = 20$) obtained by the Kalman filter related to any Bayesian estimate that you know?

14. In the consumption and portfolio selection problem in (68) and (69) let $U(C) = \log(C)$ and solve the problem.

15. Write down V_{T-3} in the dynamic programming solution of the problem following (87). Provide a set of recursion equations for the coefficients c_i in the consumption functions $\hat{C}_{T-i} = c_i W_{T-i}$.

16. In the consumption and portfolio selection problem in (68) and (69) let the random variable Z_t take only two possible values, 1.4 and 1.4^{-1}, with equal probabilities, let $R = 1.04$, and let $U(C) = 2C^{\frac{1}{2}}$. Find the optimal fraction of investment in the risky asset and the optimal consumption function for a three-period problem with no bequest.

17. In Problem 11 in Chapter 7 decompose the matrix H_1 (the coefficients in the quadratic loss function of y_1 to be minimized in the first period) into K_1 and Q_1 for the current period and the future, respectively. Explain your result.

CHAPTER 9

Some Problems of Macroeconomic Policy by Optimal Control

In this chapter we shall study several problems of macroeconomic policy, using the techniques of optimal control developed in the last two chapters. These problems include the setting of policy objectives, the determination of optimal policies, the calculation of minimum expected welfare loss, the measurement of welfare tradeoffs, the comparison of optimal policies with other policies like maintaining constant growth rates for the instruments, the study of the relative effectiveness of monetary versus fiscal policies, and the evaluation of historical policies. The numerical results presented in this chapter are based mainly on the simple macroeconomic model described in Chapter 5. It is convenient therefore to begin by setting up the policy objectives in terms of that model.

9.1 SETTING UP THE POLICY OBJECTIVES AND THE ECONOMETRIC MODEL

Any discussion of quantitative macroeconomic policy should begin with a statement of objectives. This amounts to specifying a welfare function of the economic variables of interest. Because quantitative policy analysis depends on an econometric model that describes the relations among these variables, the specification of a welfare function has to be compatible with the characteristics of the econometric model to be used for policy analysis. Depending on the particular set of variables that is considered important from the viewpoint of the policy makers, an econometric model has to be specified accordingly. In the illustrative calculations reported in this chapter, however, only the simple macroeconometric model described in Chapter 5 is employed. This limits the range of policy issues for which we can provide substantive answers, but it does not restrict our discussion of the methodological problems involved.

The first step in setting policy objectives is the selection of the variables to be included in the welfare function. In the context of macroeconomic policy these variables usually include certain measures of aggregate expenditures such as gross national product and its major components, the price level or its rate of change, the level of employment or unemployment of the labor force and its components, and perhaps certain financial variables such as various rates of interest. In the simple multiplier-accelerator model to be used only aggregate expenditures and the yield of 20-year corporate bonds are included. The numerical calculations will therefore be confined to these variables. Five welfare functions are specified. The first two exclude the long-term interest rate $R = y_4$ from the welfare function, whereas the last three include it. Recall from Chapter 5 that the expenditure variables in the model consist of

$C = y_1 =$ total personal consumption expenditures, in millions of current dollars,

$I_1 = y_2 =$ gross private domestic investment in producers' durable equipment plus change in business inventories, millions,

$I_2 = y_3 =$ new construction, millions,

$Y_1 = y_5 = C + I_1 + I_2.$

There are two policy variables:

$M = x_1 =$ currency and demand deposits adjusted in the middle of the year, millions,

$G = x_2 =$ government purchases of goods and services, millions.

In addition, a variable Y is defined as $(1 - g)(C + I_1 + I_2 + G)$, where the marginal tax rate g is set equal to .21. The first welfare function includes Y_1 and G as the only variables. The second includes C, I_1, I_2, Y, and G. The third includes R and Y. The fourth and the fifth welfare functions include, respectively, the sets R, Y_1, and G and R, C, I_1, I_2, Y, and G. Thus, besides being distinguished by the presence or absence of the long-term interest rate R, these functions can be ordered by the increasing numbers of components included.

After the variables in the welfare function are selected the next step specifies the target paths for these variables. If we wish to penalize period-to-period changes, the target paths can be specified for the first differences of the variables. If period-to-period changes are considered unimportant, the targets can be specified in terms of the levels of the variables themselves. It is also possible to steer both levels and first differences to targets, but such calculations are not presented. Two sets of calculations follow for each of the five sets of variables included in the welfare function, as listed in the last paragraph. The first aims at steering the first differences, the second, at steering the expenditure or interest-rate variables themselves to given target paths. The target paths of all expenditure variables (in first differences or levels) are set to grow by 5 per cent per year, starting from their historical values y_0 in 1964. The target path for the first difference of the long-term interest rate R is set equal to 0 for all periods. The target path for the interest rate variable itself is set equal to 43,300, which equals the 4.33 per cent per year figure in 1964 times 10^4, as this variable is defined in the model.

Within the framework of a quadratic welfare function the step following the specification of target paths is the assignment of relative penalties given to the squared deviations of the chosen variables from their targets. These are the diagonal elements of the matrix K_t in the welfare function. For simplicity we let K_t be a time-invariant diagonal matrix, with its diagonal elements corresponding to the selected target variables set equal to 1 and with all other elements set equal to 0. This means that deviations of an equal amount from targets are penalized equally, regardless of expenditure categories. Because the weight given to the squared deviation of the rate of interest, when this variable is present in the welfare function, is the same as the weight given to the squared deviation of any expenditure variable, a deviation of the interest rate by 1 percentage point (or 10,000 in our units) is regarded as costing as much as a deviation of an expenditure variable by 10,000 million dollars. The time horizon T is chosen to be 10 years.

Having specified the welfare loss functions, we set forth the econometric models to be employed for the optimal control calculations. Besides the five structural equations for the endogenous variables ΔC, ΔI_1, ΔI_2, ΔR, and ΔY_1, four more structural equations in the form of identities are needed to perform the calculations with first differences. An equation for ΔY is needed because this variable appears in some of the welfare function. Furthermore, we need equations for I_1, I_2, and Y because they all appear as lagged endogenous variables in the model. These variables have to be determined in each period in order that the required values for the other endogenous variables may be calculated in the following period. These nine structural equations are presented in Table 9.1. The reduced

Table 9.1 Structural Equations

	ΔC	ΔI_1	ΔI_2	ΔR	ΔY_1	I_1	I_2	ΔY	Y	ΔC_{-1}	$I_{1,-1}$	$I_{2,-1}$	Y_{-1}	ΔM	ΔG	ΔM_{-1}	P_{-1}	Intercept
(1)	−1	0	0	0	.3083	0	0	0	0	.1938	0	0	0	.4079	.0783	0	87.7	−4797
(2)	0	−1	0	0	.2806	0	0	0	0	0	−.6625	0	.0090	0	.1683	0	159	−6125
(3)	0	0	−1	−.2198	.1046	0	0	0	0	0	0	−.5099	.0410	0	0	0	92.8	−5996
(4)	0	0	0	−1	.1109	0	0	0	0	0	0	0	0	−.7389	.1872	.3178	0	−973
(5)	1	1	1	0	−1	0	0	0	0	0	0	0	0	0	0	0	0	0
(6)	0	1	0	0	0	−1	0	0	0	0	1	0	0	0	0	0	0	0
(7)	0	0	1	0	0	0	−1	0	0	0	0	1	0	0	0	0	0	0
(8)	0	0	0	0	.79	0	0	−1	0	0	0	0	0	0	.79	0	0	0
(9)	0	0	0	0	0	0	0	1	−1	0	0	0	1	0	0	0	0	0

Source. Table 1 in Chow (1967), p. 9. See also Equations 9 to 14 in Chapter 5.

form equations, obtained by solving the structural equations for the endogenous variables, are presented in Table 9.2. For the calculations that use the levels of the variables in the welfare function three more structural equations (identities) for C, R, and Y_1 must be introduced; the control variables are M and G rather than ΔM and ΔG. These are straightforward modifications and are not reported in table form.

Table 9.2 Reduced Form Equations

Dependent Variable	Coefficient of	ΔC_{-1}	$I_{1,-1}$	$I_{2,-1}$	Y_{-1}	ΔM_{-1}	ΔM	ΔG
(1) ΔC		0.3744	−0.6173	−0.4751	0.0466	−0.0651	0.9393	0.2697
(2) ΔI_1		0.1644	−1.2243	−0.4324	0.0514	−0.0592	0.4837	0.3425
(3) ΔI_2		0.0470	−0.1606	−0.6335	0.0531	−0.0868	0.3007	0.0087
(4) ΔR		0.0650	−0.2220	−0.1709	0.0168	0.2943	−0.5477	0.2561
(5) ΔY_1		0.5857	−2.0023	−1.5411	0.1511	−0.2111	1.7236	0.6209
(6) I_1		0.1644	−0.2243	−0.4324	0.0514	−0.0592	0.4837	0.3425
(7) I_2		0.0470	−0.1606	0.3665	0.0531	−0.0868	0.3007	0.0087
(8) ΔY		0.4627	−1.5818	−1.2174	0.1194	−0.1668	1.3617	1.2805
(9) Y		0.4627	−1.5818	−1.2174	1.1194	−0.1668	1.3617	1.2805

Source. Solution of the structural equations in Table 9.1. See also Table 5.1 in Chapter 5.

To perform optimal control calculations with the reduced-form equations in Table 9.2 in the setup of (6) in Chapter 7 we augment the vector of endogenous variables by the two control variables ΔM_t and ΔG_t. The vector y_t in the notation in (6) in Chapter 7 is therefore 11×1, the matrix A is 11×11, and the matrix C is 11×2. Because the model does not explain the price level P and lagged price P_{-1} appears in the first three structural equations, we assume that this exogenous variable grows by 2 percent per year from 120.7, its value in 1964. This is absorbed in the intercept b_t. The data on y_0, the initial values of the variables in 1964, are the following:

ΔC	ΔI_1	ΔI_2	ΔR	ΔY_1	C	I_1	I_2
24,300	3,400	2,300	2,300	30,000	399,260	38,800	48,900

R	Y_1	ΔY	Y	M	M_{-1}	G	G_{-1}
43,300	476,960	30,000	487,000	153,331	147,144	128,559	122,559

9.2 NATURE OF THE OPTIMAL POLICIES

Before presenting the optimal control equations

$$x_t = G_t y_t + g_t \tag{1}$$

it may be useful to recall how the coefficients G_t and the intercepts g_t are calculated. Referring to Chapter 7, we see that the matrices G_t and H_t are computed by (22) and (27), with t replacing T, backward in time for $t = 10$ to $t = 1$. The vectors g_t and h_t are computed by (23) and (28), with t replacing T, also backward in time for $t = 10$ to $t = 1$. The specified targets a_t affect only the intercepts (28), but not the coefficients G_t. The role of the feedback coefficients G_t is to reduce the variances of the deviations y_t^* of the variables y_t from their means $\bar{y}_t$. The targets a_t affect the mean paths $\bar{y}_t$ of the system under control by their effects on g_t, but it is up to the feedback reactions of x_t to y_{t-1} to ensure that the deviations from the mean paths remain small.

For the first run in which only ΔY_1 and ΔG in the welfare function are used, the number of target variables equals the number of instruments. By the theorem in Section 7.7 the targets could be met exactly in each period by using the optimal control policy if the vector u_t of random disturbances were absent from the model

$$y_t = A y_{t-1} + C x_t + b_t + u_t. \tag{2}$$

The optimal coefficients in the feedback control equations for all periods would be

$$G_t = -(C'KC)^{-1} C'KA, \qquad (t = 1, \dots, T), \tag{3}$$

and the matrices H_t would all be equal to K.

For the second run, in which ΔC, ΔI_1, ΔI_2, ΔY, and ΔG in the welfare function are used, there are five target variables and only two instruments. The target variables in the solution $\bar{y}_t$ of the deterministic model, ignoring the random disturbance u_t, will deviate from their specified target values in a_t, and the matrices G_t and H_t will change as t goes from T to 1. However, G_t reaches a steady-state matrix rather quickly, as shown in Table 9.3, being virtually identical from period 6 on back to period 1. Calculations of G_t for runs 4 and 5, using first differences and levels, show the similarly rapid convergence of run (2). Run 3 has only two target variables, ΔR and ΔY or R and Y, and G_t is again constant over t, as given by Equation 3. The optimal G_1 for the first period appears in Table 9.4 for each of the five welfare functions using first differences and levels of the variables.

Table 9.3 Coefficients G_t of Optimal Control Equations as Functions of Time—Run 2 Using First Differences

Instruments	Variable ΔC_{-1}	$I_{1,-1}$	$I_{2,-1}$	Y_{-1}	ΔM_{-1}
ΔM_{10}	$-.3571$	.9793	.7800	$-.0760$	.1069
ΔG_{10}	.0081	.1739	.0536	$-.0067$	.0073
ΔM_8	$-.3460$	.9076	.7044	$-.0690$	.1118
ΔG_8	.0003	.2046	.1120	$-.0118$	.0040
ΔM_6	$-.3459$	.9073	.7028	$-.0688$	.1121
ΔG_6	.0002	.2047	.1135	$-.0119$	.0038

From the results in Table 9.4 it can be seen that monetary policy based on ΔM or M is much more active, as measured by the absolute values of the feedback coefficients, than fiscal policy based on ΔG or G except for run 3. In all runs except run 3 government expenditures appear explicitly in the welfare loss function, thus inhibiting the feedback reactions of this policy variable. Once the government expenditures variable is given a small weight, as in run 3, it becomes active. The inactive role played by money supply in run 3 is due not only to the low cost of using government expenditures, but to the high cost assigned to variations in the rate of interest that money supply affects through the liquidity preference relation in the model. Thus we have seen that the values of the feedback control coefficieints are influenced by the costs assigned to the instrument in question, to the variables it directly and significantly affects, and to a substitute instrument. The first two discourage its active use, the last encourages it.

With respect to the signs of the feedback coefficients, if our formulation of the multiplier and the acceleration relations are correct, both money supply and government expenditures should respond negatively to lagged consumption expenditures but positively to lagged investment expenditures. The reason for this result can be traced to the forms of the consumption and investment functions. The consumption function given in Table 9.1 has the form

$$\Delta C_t = .3083\Delta Y_{1,t} + .1938\Delta C_{t-1} + \cdots.$$

The investment function, as derived from the accelerations principle by a stock-flow transformation, as explained in Chapter 5, has the form

$$I_{1,t} = .2806\Delta Y_{1,t} + .3375 I_{1,t-1} + \cdots,$$

Table 9.4 Optimal Feedback Control Matrix G_1

Run		ΔC_{-1}	$I_{1,-1}$	$I_{2,-1}$	Y_{-1}	M_1	M_{-2}	C_{-1}	R_{-1}	$Y_{1,-1}$	G_{-1}
					Using First Differences						
(1)	ΔM	-.3398	1.1616	.8941	-.0877	.1225	-.1225				
	ΔG	.0000	.0000	.00000	.0000	.0000	.0000				
(2)	ΔM	-.3459	.9073	.7028	-.0688	.1121	-.1121				
	ΔG	.0002	.2047	.1136	-.0119	.0038	.0038				
(3)	ΔM	-.0336	.1149	.0885	-.0087	.3997	-.3997				
	ΔG	-.3256	1.1130	.8567	-.0840	-.2947	.2947				
(4)	ΔM	-.2727	.9337	.7198	-.0706	.1821	-.1821				
	ΔG	-.0888	.2929	.2167	-.0214	-.0719	.0719				
(5)	ΔM	-.2547	.6713	.5206	-.0510	.1909	-.1909				
	ΔG	-.0879	.4319	.2881	-.0290	-.0712	.0712				
					Using Levels of Variables						
(1)	M	-.3398	1.1616	.8941	-.0877	1.1225	-.1225			-.5802	.3602
	G	.0000	.0000	.0000	.0000	.0000	.0000			.0000	.0000
(2)	M	-.3612	.7339	.5206	-.2291	1.096	-.096	-.6526			.4463
	G	.0116	.2228	.2179	-.3170	.0156	-.0156	.3757			.2791
(3)	M	-.0336	.1149	.0885	-.2525	1.3997	-.3997		1.2195		.0000
	G	-.3256	1.1130	.8567	-.6056	-.2947	.2947		-1.2967		1.0000
(4)	M	-.2762	.9460	.7298	-.0715	1.1785	-.1785		.2515	-.5007	.2444
	G	-.0675	.2206	.1615	-.0160	-.0530	.0530		-.2602	-.0796	.1286
(5)	M	-.2822	.5693	.4109	-.2130	1.1624	-.1624	-.5172	.2865		.3169
	G	-.0634	.3786	.3217	-.3321	-.0469	.0469	.2471	-.2719		.4024

or

$$\Delta I_{1,\,t} = .2806 \Delta Y_{1,\,t} - .6625 I_{1,\,t-1} + \cdots .$$

A compensatory policy is to react negatively to ΔC_{t-1} because its coefficients in the consumption function and in the reduced-form equations (Table 9.2) for the expenditure variables are positive. It is to react positively to $I_{1,\,t-1}$ and $I_{2,\,t-1}$ because their coefficients in the investment functions and in the reduced-form equations for the expenditure variables are negative. Thus the difference between the formulations of the consumption and investment functions requires different policy responses by the fiscal and monetary instruments to lagged consumption and investment expenditures.

We have just illustrated how the specifications of the welfare function and the characteristics of the econometric model employed affect the nature of the optimal control equations. It might be asked, if the optimal policy depends so heavily on the welfare function and the econometric model used, how policy makers can rely on the results of optimal control calculations. The answer is that, precisely because of the crucial dependence of the optimal policy on these two factors, optimal control calculations of the kind we have described are extremely important to the making of macroeconomic policy. Decisions are being made, for example, by the open market committee of the Board of Governors of the Federal Reserve System. Policy objectives are discussed and economic projections based on alternative policies are presented at the Federal Open Market Committee meetings. Thus it is recognized that decisions ought to be based on the aims to be achieved and on the estimated reactions of the economy to the decisions. Exactly how should a policy be set according to these two factors? Optimal control calculations provide an answer to this question. From these calculations the government authorities would know what their policies are supposed to achieve and what dynamic responses of the economy are being postulated.

One position sometimes articulated is that existing econometric models are unreliable, and it would be dangerous to base policy decisions on these models. First of all, decision makers in the government do, in fact, consider the reactions of the economy in making decisions, however imperfect their knowledge of these reactions may be. The use of a mathematical model merely forces a decision maker to state explicitly, and in quantitative form, the dynamic reactions of the economy he is assuming in making decisions. It is more dangerous for decision makers to hide their assumptions concerning the economy on which policies are based than to make the assumptions explicitly in quantitative form, that is, in the form of a mathematical model. If they postulate only a few simple relations between the major variables, let them be written down as a simple model for the optimal control calculations. Second, as we show in Chapter 10, uncertainty in the model parameters can be incorporated in the optimal control calculations. Most decisions in practice are made in the face of uncertainty, but uncertainty does not justify discarding a rational quantitative approach. In essence, we suggest that such an approach be considered seriously in the determination of macroeconomic policies.

9.3 CALCULATION OF EXPECTED WELFARE LOSS

How would the economy behave under optimal policies? This question is interesting for several reasons. It is certainly important to know how well the economy will behave under the best policies. Carl Christ (1973) in a review of the 1973 *Report of the Council of Economic Advisers* commented that it is dangerous for the government to promise too much by way of reducing inflation and unemployment while the promises are impossible to fulfill. This point is related to the setting of realistic targets in the optimal control calculations. If some important variables, as judged by their mean path $\bar{y}_t$ and the variances around the mean values, behaved rather poorly under the optimal policies whereas others behaved satisfactorily, we might consider changing the relative penalties (the matrix K_t) and/or the targets a_t to find out whether the first variables could be made to behave better by accepting tolerable deterioration in the behavior of the second. This is the question of welfare tradeoffs discussed in Section 9.4. The examination of the characteristics of the system under one optimal policy precedes the calculations of tradeoffs. Third, it is interesting to compare the behavior of the system under certain optimal and other alternative policies. This topic is also discussed in later sections. In short, comparative stochastic dynamic analysis requires the examination of the dynamic characteristics of the system under each set of specific conditions.

As we pointed out in Chapter 7, once the optimal control equations are obtained, the tools of the preceding chapters can be used to study the dynamic characteristics of the system under control. Within the framework of optimal control, it is useful to plot the mean path $\bar{y}_t$ for the important variables along with the target path a_t. The deviations $(\bar{y}_t - a_t)$ show how well the means of these variables behave according to the objectives set for them. The variances around their means can be calculated and summed over the different periods $t = 1, \ldots, T$. These are the diagonal elements in the covariancre matrices $\Gamma_{\cdot t}$ given by (48) in Chapter 7. If the deviations $(\bar{y}_t - a_t)$ are squared and weighted by the diagonal elements of K_t and the results are summed over different variables and t, we obtain the deterministic component $W_1 = \sum_{t=1}^{T}(\bar{y}_t - a_t)' K_t(\bar{y}_t - a_t)$ of the expected welfare loss. If the variances of $(y_t - \bar{y}_t)$ are weighted by the diagonal elements of K_t and the results are summed over different variables and t, we have the stochastic component $EW_2 = E\sum_{t=1}^{T}(y_t - \bar{y}_t)' K_t(y_t - \bar{y}_t)$ of the expected welfare loss. Thus W_1 and EW_2 provide summary measures of the performance of the system under control.

For the five welfare functions specified, using first differences and levels of the variables separately, W_1, EW_2, and $W_1 + EW_2 = EW$ computed from the optimal policies are presented in Table 9.5. The rows that correspond to the suboptimal policies in this table are explained in Section 9.5. As we have pointed out, the deterministic component W_1 is 0 for runs 1 and 3 because the number of target variables equals the number of instruments. For run 1, using first differences, $(1/T)EW_2$ is 388.4. The target variables are ΔY_1 and ΔG in this run. From the discussion surrounding (56) in Chapter 7 EW_2 equals $\sum_{t=1}^{T} Eu_t' K_t u_t$, but ΔG is set by the optimal policy to follow its target values exactly, leaving ΔM to control the remaining target variable ΔY_1; EW_2 thus becomes the sum of the variances of ΔY alone over the 10 periods. The variance for each period is 388.4(10^6), or the standard deviation of ΔY_1 is approximately 19.6 billions of dollars because the variable itself is measured in millions of dollars. For run 2 the sum of the variances of ΔC, ΔI_1, ΔI_2, ΔY, and ΔG equals 451.8 billions squared, if an average is taken over the 10 periods. Because the number of target variables exceeds the number of instruments, H_t no longer equals K_t for $t < T$ and G_t also varies with time, making the variances of the variables dependent on time also. Similar interpretations apply to the figures for $(1/T)W_1$ and $(1/T)EW_2$ given in Table 9.5 for the optimal policies of the remaining runs.

It is worth noting in Table 9.5 that the stochastic part of the expected welfare cost is much larger than the deterministic part in each run. This illustrates the importance of accounting for the stochastic disturbances in the calculation of the minimum expected welfare loss.

9.4 MEASUREMENT OF WELFARE TRADEOFFS

In an optimal control calculation the specified values set for the target variables will not be achieved perfectly. Even if there were no random disturbances in the model, the solution to the deterministic system under control will be exactly on target only when the number of target variables does not exceed the number of instruments. The existence of random disturbances will cause the variables to deviate from their target values.

Table 9.5 Measuring Welfare Costs for Optimal and Suboptimal Policies (10^6 millions)

	Welfare Weights on First Differences or Levels of								Using First Differences			Using Levels		
Run	C	I_1	I_2	R	Y_1	Y	G		$\frac{1}{T}W_1$	$\frac{1}{T}EW_2$	$\frac{1}{T}(W_1+EW_2)$	$\frac{1}{T}W_1$	$\frac{1}{T}EW_2$	$\frac{1}{T}(W_1+EW_2)$
(1)				1			1	Optimal	0.0	388.4	388.4	0.0	388.4	388.4
								suboptimal	24.1	547.6	571.7	2.1	639.2	641.3
								ratio		1.41			1.64	
(2)	1	1	1			1	1	Optimal	12.5	451.8	464.3	59.7	417.1	476.8
								suboptimal	43.5	586.8	630.3	69.9	778.8	848.7
								ratio		1.30			1.87	
(3)			1		1			Optimal	0.0	257.5	257.5	0.0	257.5	257.5
								suboptimal	21.6	358.9	380.4	1.0	460.7	461.7
								ratio		1.39			1.79	
(4)			1	1			1	Optimal	12.7	426.3	439.0	7.3	487.0	494.4
								suboptimal	33.4	564.7	598.1	9.6	701.0	710.6
								ratio		1.32			1.44	
(5)	1	1	1	1	1		1	Optimal	21.9	477.4	499.2	64.1	520.3	584.4
								suboptimal	50.8	603.9	654.7	73.3	840.6	913.9
								ratio		1.26			1.62	

The exceptional case occurs when the variable is not governed by a stochastic equation. An example is the variable ΔG or G in run 1, as already commented on. This variable is one of two target variables but there are two instruments. The optimal solution is to set ΔG or G equal to its target values, leaving the remaining instrument ΔM or M to steer the remaining target variable ΔY_1 or Y_1 to its specified values as closely as possible; ΔG or G is not governed by any feedback control equation, as seen by the 0 coefficients in Table 9.4. Otherwise, it would become stochastic because the variables in the feedback control equation are stochastic. Hence, except for an instrument that is set equal to its target values when the total number of instruments is as large as the number of target variables, we expect the target variables to deviate from their target values because of random disturbances.

In general, there are tradeoffs between the expected squared deviations of different variables from their targets. In computing an optimal control solution, we set the terms of trade between the expected squared deviations by the elements in the diagonal matrix K_t. After the optimal control solution is obtained we should examine the expected squared deviation of each variable from its targets, averaged over the time periods 1 to T, to determine whether the averages for different target variables bear a satisfactory relation to one another. If the average squared deviation for one variable is large compared with the average for another, we can revise the welfare function by assigning more weight to the element of K_t which corresponds to the first variable, or by setting a more ambitious target for that variable. The resulting optimal control calculations will yield a smaller average squared deviation for the first variable or an average squared deviation from a more ambitious target. By varying the welfare function we can trace the tradeoff possibilities in terms of the expected squared deviation of different variables from their targets.

We provide some illustrative calculations of the tradeoffs between the expected squared deviations of three variables, Y_1, G, and R, in our model by using the levels of these variables in the welfare function. The results are obtained by assigning the weights 1, 1, and 0 in K_t, compared with 1, 1, and 1, to these three variables. As presented in Table 9.6, they are the expected squared deviations of these variables from the targets specified in Section 9.1, averaged over 10 periods, and decomposed into the squared deviations of the means from targets (the deterministic part) and the variances of the variables from their means (the stochastic part). The welfare functions are those in runs 1 and 4 in Table 9.5, using the levels of the variables.

As far as the average squared deviations of $\bar{y}_{it}$ from a_{it} are concerned, Table 9.6 shows that giving a positive weight to the interest rate variable in the welfare function in run 4 reduces the term from 9.1 to 5.9 for R but increases the terms for Y_1 and G from 0 to .7 and .8, respectively. The total for the three variables is reduced from 9.1 to 7.4. The variance of R, averaged over the 10 periods, is reduced from 118.8 to 82.4 by assigning a positive weight to the interest rate variable in the welfare function, whereas the variances of Y_1 and G increase, respectively, from 388.4 and 0.0 to 396.2 and 8.4. The last two columns of Table 9.6 show the expected squared deviation of each of the three variables, averaged over the 10 periods, when R is absent or present in the welfare function. The reduction of the expected welfare loss attributable to R and the increases attributable to Y_1 and G due to the change in the penalty for R can be clearly discerned. This kind of calculations would be useful to measure the tradeoffs between inflation and unemployment if the two variables were included in the model employed for the optimal control calculations.

Table 9.6 Illustrative Calculations of Welfare Tradeoffs (10^6 millions)

Variable i	$\frac{1}{T}\sum_{t=1}^{T}(\bar{y}_{it}-a_{it})^2$		$\frac{1}{T}\sum_{t=1}^{T}E(y_{it}-\bar{y}_{it})^2$		$\frac{1}{T}\sum_{t=1}^{T}E(y_{it}-a_{it})^2$	
	Run 1	Run 4	Run 1	Run 4	Run 1	Run 4
Y_1	0.0	0.7	388.4	396.2	388.4	396.9
G	0.0	0.8	0.0	8.4	0.0	9.2
R	9.1	5.9	118.8	82.4	127.9	88.3
Total	9.1	7.4	507.2	487.0	516.3	494.4

9.5 GAIN FROM THE OPTIMAL CONTROL POLICIES

When a policy is proposed, it is necessary to examine the dynamic performance of the economy subject to this policy in order to evaluate it. For this evaluation it would also be useful to compare the economic performance under the proposed policy with the performance under an optimal policy. Some illustrative calculations for a comparison are given in this section. The main purpose of these calculations is to measure the possible gain, in terms of the expected squared deviations of the target variables from their specified paths, from applying the optimal control policies compared with the rule of maintaining a constant growth rate for each control variable.

Recall the discussion in Chapter 7 in which an optimal control policy is decomposed into two parts. The first is a deterministic control policy to steer the solution $\bar{y}_t$ of the deterministic model under control (in which the random disturbances are ignored) to specified targets. The policy consists of a feedback control equation $\bar{x}_t = G_t\bar{y}_{t-1} + g_t$. Because $\bar{y}_{t-1}$ is deterministic, $\bar{x}_t$ for all future time periods can be determined in period 1 without considering future events in the economy. The second is a stochastic control policy to minimize a weighted sum of the variances of the target variables from their means. The deviations y_t^* of the variables y_t from their means $\bar{y}_t$ are governed by the system

$$y_t^* = (A + CG_t)y_{t-1}^* + u_t, \tag{4}$$

when they are subject to the optimal stochastic control policy $x_t^* = G_t y_{t-1}^*$. The values of x_t^* cannot be specified in period 1; they depend on the future deviations y_{t-1}^*. Accordingly, we have also decomposed the expected welfare loss into two parts. The deterministic part W_1 is due to the deviations of $\bar{y}_t$ from $\bar{a}_t$, the stochastic part EW_2, to the variances of y_t (from $\bar{y}_t$).

Because the rule of maintaining a constant growth rate for each control variable is a deterministic control policy, the gain of the optimal policy compared with this policy can be decomposed into two parts. The first is the gain of the optimal deterministic control $\bar{x}_t = G_t\bar{y}_{t-1} + g_t$ over this deterministic policy. It is measured by the difference between W_1 and the value of $\sum_{t=1}^{T}(\bar{y}_t - a_t)' K_t(\bar{y} - a_t)$ computed under the latter policy. The second is the gain of the optimal stochastic control $x_t^* = G_t y_{t-1}^*$ over any deterministic policy that does not utilize any future observations on y_{t-1}^*. It is measured by the difference between EW_2, which is a weighted sum of the variances of y_t^* generated by (4), and $\sum_{t=1}^{T} Ey_t^{*'} K_t y_t^*$, where y_t^* is not subject to any feedback policy, that is,

$$y_t^* = Ay_{t-1}^* + u_t. \tag{5}$$

These two parts are derived below.

The reader will have noticed that the proposal to maintain a constant growth rate for just one control variable such as money supply is by itself not a sufficiently well specified proposition for the purpose of a rigorous analysis. To complete the specification of a meaningful proposition, we add that all control variables should grow at constant rates and that a quadratic welfare function be used to measure the performance of the economy. Furthermore, for the benefit of the proponents of such a proposition the particular constant growth rates will be found by minimizing the expected loss determined by the given welfare function.

For the first part of the gain W_1 for the optimal deterministic policy has been explained and exhibited in Table 9.5. For the proposed suboptimal policy the control equation is constrained to be, with D denoting a diagonal matrix,

$$\bar{x}_t = D\bar{x}_{t-1} \tag{6}$$

or, alternatively,

$$\bar{x}_t = G\bar{y}_{t-1} = (0 \cdots 0 \quad D \quad 0 \cdots 0), \tag{7}$$

where the matrix G has 0 elements except for the submatrix corresponding to the vector $\bar{x}_{t-1}$, which is imbedded in the vector $\bar{y}_{t-1}$. To obtain the growth rates most suitable to the given welfare function we minimize W_1 with respect to the diagonal elements $(d_1, d_2, \ldots, d_q) = d$ of the matrix D. There are various ways of performing this minimization. The method that we have used for the calculations below is the gradient method described in Goldfeld, Quandt, and Trotter (1966). Given any guess of the unknown vector d, the value of the function $W_1(d)$ can be calculated by using the definition $\Sigma_t(\bar{y}_t - a_t)' K_t(\bar{y}_t - a_t)$, the deterministic model for $\bar{y}_t$, and the control rule (7). So can the gradient of $W_1(d)$ at that point, either numerically or analytically. By using a quadratic approximation to the function to be minimized near that point, the Goldfeld-Quandt-Trotter method ensures that the matrix of second derivatives used to calculate the unknown for the next iteration is positive definite, even if the function itself is not convex. This method can be used to obtain the suboptimal d and the associated welfare cost W_1 can be calculated.

For the second part of the gain we have explained the calculation of EW_2 under the optimal policy that utilizes (48) in Chapter 7 for the covariance matrices of y_t^* generated by (4). Because the suboptimal policy ignores any information on y_t^*, it amounts to setting $G_t = 0$ in (4). The stochastic part EW_2 of the welfare cost is then computed by using the covariance matrix $Ey_t^* y_t^{*'}$ given by

$$Ey_t^* y_t^{*'} = V + A_t(Ey_{t-1}^* y_{t-1}^{*'})A_t', \qquad t = 1, \ldots, T. \tag{8}$$

The difference between these two parts of the expected welfare costs measures the gain from using optimal stochastic control over the best deterministic control policy.

The results of the calculations are given in Table 9.5, with the suboptimal policies denoting the best deterministic policy that specifies a constant growth rate for each of the instruments described. As far as the five runs with first differences are concerned, the gain of the optimal solution in the stochastic part of welfare cost varies between 30 to 40 percent and is much more important than the deterministic part. Hence, if the economic model contains stochastic disturbances, we can hardly afford to ignore them in the study of optimal policy. As far as the five runs with the levels of the variables are concerned, the gain of the optimal solution in the stochastic part of the welfare cost varies from 40 to 80 percent and again dominates the deterministic part. These percentage gains are substantial.

We might wish to ask why the relative gains of the optimal policies are greater for the calculations that utilize levels of the variable than the corresponding gains that utilize first differences. To answer this question let us re-examine the calculation of the stochastic part of the welfare cost. For the optimal policy we choose a linear feedback control equation $y_t^* = G_t y_{t-1}^*$ in a way that the system under control

$$y_t^* = A y_{t-1}^* + C x_t^* + u_t = (A + C G_t) y_{t-1}^* + u_t$$

will have a small weighted sum of variances. More precisely, we choose the matrix $G_t = -(C' H_t C)^{-1} C' H_t A$ to make the matrix $(A + C G_t)$ small, in that it will have a minimum $\text{tr}(A + C G_t)' H_t (A + C G_t)$, as pointed out in Chapter 7. The gain from the optimal policy is the gain (in reducing variances) produced by using a smaller matrix $(A + C G_t)$ in the above stochastic system rather than the matrix A itself. If the lag structure of the system, as reflected in the matrix A, becomes so complicated, with reference to a given matrix C, that the ratio of $\text{tr} A' H_t A$ to $\text{tr}(A + C G_t)' \times H_t (A + C G_t)$ becomes larger for the optimal G_t, the gain from optimal control will be greater. Intuitively speaking, the sparser the A, $A = 0$ being the extreme case, the smaller the effects of the lagged variables on the current state, and thus the smaller the gain from optimal feedback control. If this point is valid, we might expect larger gains from optimal control if we employ a quarterly instead of an annual model because there will be more lagged variables and the matrix A will be of higher order and less sparse.

It is recognized that the percentage welfare gains presented above are dependent on the econometric model used, as are measurements in econometric studies in general. Furthermore, the gains of the optimal policies over the suboptimal policies are dependent on the welfare functions chosen. To determine how the gains would alter when judged by a different welfare function, I have calculated the ratios of the suboptimal stochastic welfare costs to the optimal for runs 1, 2, 4 and 5 by using the welfare weights of run 3 which are unity for R and Y only. These ratios are, respectively, 1.24, 1.32, 1.30, and 1.34 for the calculations in first differences; they are 1.27, 1.24, 1.42, and 1.39. respectively, for the calculations in levels. They are not very different from the ratios shown in Table 9.5. The optimal policies are thus seen to be fairly robust against different welfare functions; recall that the optimal policies of runs 1 and 2 were derived without including the rate of interest in their welfare functions.

9.6 RELATIVE EFFECTIVENESS OF MONETARY VERSUS FISCAL INSTRUMENTS

The framework of optimal control can be applied to evaluate other suboptimal policies than the one described in the last section. A suboptimal policy may consist of holding a subset of the instruments at constant growth rates while letting the remaining instruments behave optimally. If the subset of instruments to be restricted consists of monetary policy instruments, we can evaluate how crucial discretionary monetary policy is by comparing the expected welfare losses of the optimal policy (freeing these instruments) and of the suboptimal policy. An evaluation of the importance of discretionary fiscal policy can be similarly performed. The suboptimal policies derived by restricting the monetary and fiscal instruments can be compared not only with the optimal policy but with one another in regard to the expected welfare loss.

The question of the relative effectiveness of monetary and fiscal policies has been approached by considering the relative magnitudes of the various multipliers of these variables in the determination of selected major economic aggregates. These multipliers are derived from the matrices A and C in reduced form, as explained in Section 5.5. Comparison of the various multipliers, however, does not provide a precise answer to the question at hand. The interactions of the current and delayed effects of the instruments through the elements of the matrices A and C have to be properly incorporated, the dependent variables affected by them have to be properly weighted, and the instruments must be given an opportunity to perform their best in terms of the stated objectives of policy. All these considerations are systematically taken into account in the optimal control calculations for the comparison of relative effectiveness of subsets of policy instruments.

To state a method applicable to such a comparison let the control variables be divided into two subsets x_{1t} and x_{2t}, the latter being the subset to be restricted. In the spirit of Section 9.5 we choose to examine a nondiscretionary policy for x_{2t} in the form

$$x_{2t} = Dx_{2,t-1}, \tag{9}$$

where D is a diagonal matrix. The policy (9) should be compatible with the welfare function employed. Given a matrix D whose diagonal elements are denoted by the vector d, the remaining instruments x_{1t} are assumed to follow an optimal policy that minimizes the expected welfare loss by using a specified econometric model. The result is denoted by $W(d)$. We then minimize $W(d)$ with respect to d. The minimum gives the lowest possible expected welfare cost by using the two sets of instruments x_{1t} and x_{2t} when the latter set is restricted to the form of (9). This procedure solves the minimization problem with respect to the two sets of variables in the two stages. First, given any admissible policy for x_{2t} in the form of (9), we minimize the objective function with respect to x_{1t}. The result is a function of the strategy d for x_{2t}. Second, we minimize the result with respect to d.

To carry out the minimization in the first stage we perform an optimal control calculation with respect to x_{1t}, given a strategy (9) for x_{2t}. This calculation is straightforward because the given time path for x_{2t} can be absorbed in the intercept b_t of the reduced form equation to be denoted, say, by b_t^*.

$$y_t = Ay_{t-1} + C^* x_{1t} + b_t^* + u_t. \tag{10}$$

The minimum expected welfare loss associated with the optimal policy for x_{1t} can be calculated by methods previously set forth. If some elements of x_{2t} are included in the welfare function, the weighted sum of squared deviations of these variables from their targets should be added to the minimum expected loss. Thus, given any strategy d for x_{2t}, the function $W(d)$ can be evaluated. Now using the method of Goldfeld, Quandt, and Trotter (1966), as described in Section 9.5, $W(d)$ can be minimized with respect to d. The result can be compared with the minimum expected loss for the optimal policy by using x_{1t} and x_{2t} freely and with the minimum expected loss for the policy of using restricted x_{1t} and optimal x_{2t} as we have suggested. (See Problems 8 and 9, and Section 11.6.)

9.7 EVALUATION OF HISTORICAL POLICIES

The approach to a comparison of optimal with alternative policies, as described in the last two sections, can be applied to the evaluation of historical policies. One method assumes that historical policies can be summarized by a set of feedback control equations. These equations are termed reaction functions by some economists who study the behavior of government authorities in making policy decisions. If government behavior is consistent enough to be described by a set of reaction functions or feedback control equations, the dynamic performance of the econometric system incorporating these equations can be derived by analytic means or stochastic simulations, and the government policy described by these equations can be evaluated accordingly. If government behavior contains more irregularities than can be described by control equations with fixed coefficients, perhaps stochastic equations with random coefficients may be more appropriate. Using these equations, we can study the dynamic performance of the econometric system. In Chapter 10 we shall study optimal control of linear systems with random coefficients. The method presented is applicable to government behavioral equations with random coefficients. In this section we consider briefly the fixed time-invariant coefficients for the government reaction functions and present some illustrative calculations by using the simple model in Table 9.2.

As a crude method of evaluating the actual government policies during the period 1948 to 1963 (the post-World War II sample period of our model), the following two regressions of ΔM_t and ΔG_t on ΔC_{t-1}, $(I_1 + I_2)_{t-1}$, ΔM_{t-1} and linear trend t ($t = 1948, \ldots, 1963$) are reported; the ratio of each regression coefficient to its standard error is placed in parentheses.

Observed Feedback Control Regressions
1958–1963

	ΔC_{-1}	$(I_1 + I_2)_{-1}$	ΔM_{-1}	t	Intercept	s	R^2
ΔM	−.3183	.0957	.0898	−135.8	266,539.	2,673	.23
	(−1.66)	(.58)	(.30)	(−.31)	(.31)		
ΔG	−.2115	.7270	−.1343	−2004.	3,886,846.	6,401	.29
	(−.46)	(1.85)	(.19)	(−1.89)	(1.90)		

In these regressions the two investment variables are combined to avoid too much multicollinearity; the trend and the intercept are used to represent g_t in the feedback control equation. Allowing for their standard errors, we find that the coefficients of ΔC_{-1} and $(I_1 + I_2)_{-1}$ tend to be negative and positive, respectively, as prescribed for the optimal policy in Section 9.2, and that the first three coefficients (or four if the coefficient of investment expenditures counts as two) are not so different from the optimal coefficients in Table 9.4; discount the row of zeros in the optimal equations for ΔG in run 1 because ΔG is given too much weight and discount the row for ΔM in run 3 because ΔR is given too much weight.

If uncertainties were ignored and the above regression coefficients were used for the matrix G_t to calculate the stochastic part of welfare, we would obtain the following results for EW_2:

Run	1	2	3	4	5
Historical regression	566.1	536.2	288.8	583.0	553.1
Suboptimal policy ($G_t = 0$)	547.6	586.8	358.9	564.7	603.9
Ratio	.97	1.09	1.24	.97	1.09

Thus a set of (nonstochastic) feedback control equations based on historical observations would not compare unfavorably with the suboptimal policy of $G_t = 0$, which the rule of maintaining a constant growth rate for each instrument would imply. Note that these estimates of welfare gains are biased in favor of the observed regression policy because the standard errors of the regressions and of the regression coefficients, which have been ignored in these calculations, would increase the variances of the system. The result from this very crude analysis, however, does not support the contention that monetary and fiscal policies in the period 1948 to 1963 were destabilizing. The brief treatment in this section is intended only to be suggestive of an approach to an important problem. Much further research is required. (See Problem 10.)

9.8 REFERENCE TO OTHER APPLICATIONS

In this chapter we have presented some illustrative applications of the methods in Chapters 7 and 8 to problems of macroeconomic policy. Part of the material was drawn from Chow (October 1972) and Chow (December 1973). The applications are meant to be suggestive and are far from exhaustive. It is expected that the dynamic stochastic method of multiperiod control will be integrated with the main body of policy analysis in macroeconomics. The important macroeconomic policy problems that can be fruitfully studied by the methods discussed here appear to be unlimited. The literature is growing. Interested readers should consult the three issues of the *Annual of Economic and Social Measurement* (October 1972, January 1974 and April 1975) devoted to recent developments in the theory and application of optimal stochastic control methods and the references cited in these issues. Other useful references include Fox, Sengupta, and Thorbecke (1966), Paryani (1971), Tintner and Sengupta (1972), B. M. Friedman (1973), R. S. Pindyck (1973), and K. P. Vishwakarma (1974).

PROBLEMS

1. If the weights in the welfare function for run 3 are changed from 1 and 1 for ΔR and ΔY (or R and Y) to 0.5 and 1 respectively, how would you expect the feedback control coefficients G_t in Table 9.4 to change?

2. Using 0.5 and 1.0, respectively, as weights in the welfare function for the only two target variables ΔR and ΔY in the model in Table 9.2, find the coefficients in the optimal feedback control equations for ΔM and ΔG. Why are these coefficients time invariant?

3. Find the stochastic part of the minimum expected welfare cost associated with the optimal policy in Problem 2, assuming a time horizon of 10 years.

4. Using 0.1, 1.0, and 0.5, respectively, for the only nonzero diagonal elements in K_t corresponding to ΔR, ΔY_1, and ΔG of the model in Table 9.2, find the coefficients G_t in the feedback control equations for periods 2 and 1, assuming a time horizon of 2.

5. From the dynamic programming derivation of Section 8.1, how can the formula for $\hat{V}_t$ be utilized to calculate the stochastic part EW_2 of the minimum expected welfare cost associated with the optimal policy?

6. Using the method in Problem 5, find the stochastic part EW_2 of the minimum expected welfare cost associated with the optimal policy of Problem 4.

7. Explain why the coefficient of ΔM_{-1} in the optimal control equation for ΔG is positive for run 2 in Table 9.4. Why are the similar coefficients for runs 3, 4, and 5 negative? Use the reduced-form equations in Table 9.2 to answer these questions.

8. Refer to the method described in Section 9.6 for evaluating the minimum expected welfare loss $W(d)$ associated with the optimal policy when a subset x_{2t} of instruments is restricted to the form of (9). How can the gain of the truly optimal policy, with x_{2t} free to function optimally, over the above suboptimal policy be divided into deterministic and stochastic parts?

9. Reformulate the control problem in Section 9.6 in which a subset x_{2t} of instruments is restricted to the form of (9) by redefining the matrices A and C in the reduced-form equations, thus incorporating x_{2t} in the original welfare function W directly (and not using an additional term to be appended in EW, as described in Section 9.6).

10. Present a critical appraisal of the policy analysis of one of the following studies. Formulate an alternative method of analysis if you can:
  a. E. C. Brown (1955)
  b. M. Bronfenbrenner (1961)
  c. F. Modigliani (1964)
  d. R. S. Holbrook (1972)
  e. J. Kmenta and P. E. Smith (1973)

11. In the light of the approach presented in Chapters 7, 8, and 9, present a review of the classic papers by A. W. Phillips (1954 and 1957) on stabilization policy.

12. How can the method of optimal control be applied to the analysis of macroeconomic policy during the 1930s in the United States by Bogaard and Theil (1959)?

CHAPTER 10

Control of Unknown
Linear Systems
Without Learning

In this chapter we allow for uncertainty in the parameters of a linear econometric system for the purpose of deriving optimal control policies with a quadratic welfare function. Uncertainty is expressed in terms of the variances and covariances of the parameters. Optimal control equations and the associated minimum expected welfare loss are derived. They are found to depend on the mean vector and the covariance matrix of these parameters. Methods for estimating the latter are discussed. Comparison with the solution obtained under the assumption of known parameters is provided, and the effect of uncertainty on the optimal control policy is examined.

10.1 AN INTRODUCTORY EXAMPLE INVOLVING SCALAR VARIABLES

To understand how uncertainty in the parameters of the econometric model affects the optimal control policy we consider the simple case of a scalar dependent variable y_t to be controlled by a scalar instrument x_t through the dynamic equation

$$y_t = a_t y_{t-1} + c_t x_t + u_t. \tag{1}$$

The parameters a_t and c_t are unknown but they are assumed to be random variables with a known mean vector and covariance matrix. It is assumed

226

that the random residual u_t has 0 mean and is independent of u_s for $t \neq s$ and of the random coefficients a_t and b_t. The problem is to minimize Ey_1^2, given y_0.

Using (1), we minimize

$$Ey_1^2 = E(a_1 y_0 + c_1 x_1 + u_1)^2$$
$$= (Ea_1^2)y_0^2 + 2(Ea_1 c_1)y_0 x_1 + (Ec_1^2)x_1^2 + Eu_1^2 \qquad (2)$$

with respect to x_1 by differentiation, which yields the optimal control equation

$$x_1 = -(Ec_1^2)^{-1}(Ea_1 c_1)y_0. \qquad (3)$$

If the parameters a_1 and c_1 were known for certain and were equal to $\bar{a}_1$ and $\bar{c}_1$, say, the optimal control equation would be

$$\bar{x}_1 = -(\bar{c}_1^2)^{-1}(\bar{a}_1\bar{c}_1)y_0 = -\bar{c}_1^{-1}\bar{a}_1 y_0. \qquad (4)$$

Several characteristics of this special case should be stressed. First, when uncertainty exists, the optimal feedback control equation remains a linear function of the lagged dependent variable, as shown by (3). Second, the coefficient of the linear feedback control equation is a function of the second moments of the unknown parameters in the model. Third, certainty equivalence no longer holds; that is, if the random variables a_1 and c_1 in the model were replaced by their means, to be denoted by $\bar{a}_1$ and $\bar{c}_1$, respectively, and the resulting model were used to obtain an optimal control policy, the policy would not be optimal for the original problem with randomness in the parameters a_1 and c_1. The certainty-equivalent solution, obtained by replacing a_1 and c_1 in the model by $\bar{a}_1$ and $\bar{c}_1$, respectively, is given by (4). It is different from the optimal solution of (3). Fourth, with uncertainty present, the optimal policy tends to be more conservative in that it has a smaller feedback coefficient, *provided that* a_1 and c_1 are uncorrelated. To show this we rewrite (3) as

$$x_1 = -(\bar{c}_1^2 + \operatorname{var} c_1)^{-1}[\bar{a}_1\bar{c}_1 + \operatorname{cov}(a_1, c_1)]y_0. \qquad (5)$$

If the covariance of a_1 and c_1 is 0, the coefficient of y_0 in (5) will be smaller in absolute value than the coefficient in (4) because $\operatorname{var} c_1$ in the denominator is positive. The right-hand side of (5) shows exactly how the variance of c_1 affects the reaction coefficient. If $\operatorname{cov}(a_1, c_1)$ is positive, however, assuming that the product $\bar{a}_1 \bar{c}_1$ is positive, we find that the numerator of the reaction coefficient in (5) is larger than in (4). This may offset the larger denominator due to $\mathrm{var}\, c_1$. The relative magnitudes of the coefficients in (4) and (5) thus remain indeterminate.

These four characteristics of the solution to the optimal control problem with uncertain parameters are re-examined in the general case when both x_t and y_t are vectors and the time horizon is T periods. In Section 10.2 derive the optimal control equation and the associated expected loss for a multivariate system. The solution will depend on the second moments of the unknown parameters, as in the special case described. Bayesian and classical statistical methods for evaluating these moments are given, and comparison with the certainty case is then made.

10.2 OPTIMAL CONTROL OF LINEAR SYSTEMS WITH RANDOM PARAMETERS

As in Chapters 7 and 8, the econometric model employed is assumed to be linear:

$$y_t = A_t y_{t-1} + C_t x_t + b_t + u_t. \tag{6}$$

The parameters A_t, C_t, and b_t, however, are now treated as random. There are two interpretations of these random coefficients. First, the true econometric model has constant but unknown coefficients. The policy analyst forms posterior probability density functions for them from data made available by Bayesian statistical methods. The definition and derivation of the posterior density function is given in Section 10.4. Second, the coefficients in (6) are random in some objective sense, and we are dealing with a random-coefficient model. The general formulas provided in this section for optimal control calculations are valid for both interpretations under the crucial assumption stated after Equation 14. When, however, the general formulas are implemented in later sections by specifying the distribution of the unknown coefficients, we take the viewpoint of a system (6) with constant and time-invariant coefficients, thus accepting the first interpretation.

As before, (6) may have been converted from a higher order system to begin with, and the control variables x_t may be imbedded in the vector y_t by the use of identities. The random residuals u_t are assumed to be statistically independent of $u_s (t \neq s)$ and of A_t, C_t, and b_t. We assume that joint distributions of A_t, C_t, and b_t exist without being concerned with how they are derived until later sections. The welfare cost is again assumed to be quadratic:

$$W = \sum_{t=1}^{T} (y_t - a_t)' K_t (y_t - a_t) = \sum_{t=1}^{T} (y_t' K_t y_t - 2y_t' K_t a_t + a_t' K_t a_t). \quad (7)$$

The control problem is to minimize the expectation of W, given the system (6).

The method of dynamic programming is applied to this problem. The development here parallels that in Section 8.1. By this method we first solve the problem for the last period, given the information up to the end of period $T-1$. Using E_t for conditional expectation, given all information available at the end of period t, we minimize

$$V_T = E_{T-1}(y_T' K_T y_T - 2y_T' K_T a_T + a_T' K_T a_T)$$

$$= E_{T-1}(y_T' H_T y_T - 2y_T' h_T + c_T), \quad (8)$$

where, in anticipation of generalization to the multiperiod problem, we have let

$$H_T = K_T; \quad h_T = K_T a_T; \quad c_T = a_T' K_T a_T. \quad (9)$$

Substituting the system (6) for y_T in (8) and taking expectations, we have

$$V_T = E_{T-1}(A_T y_{T-1} + C_T x_T + b_T)' H_T (A_T y_{T-1} + C_T x_T + b_T) + E_{T-1} u_T' H_T u_T$$

$$- 2E_{T-1}(A_T y_{T-1} + C_T x_T + b_T)' h_T + E_{T-1} c_T$$

$$= E_{T-1}(A_T y_{T-1} + b_T)' H_T (A_T y_{T-1} + b_T) + x_T' E_{T-1}(C_T' H_T C_T) x_T$$

$$+ 2x_T' E_{T-1} C_T' H_T (A_T y_{T-1} + b_T) + E_{T-1} u_T' H_T u_T$$

$$- 2E_{T-1}(A_T y_{T-1} + b_T)' h_T - 2x_T'(E_{T-1} C_T') h_T + E_{T-1} c_T. \quad (10)$$

Minimization of (10) with respect to x_T by differentiation,

$$\frac{\partial V_T}{\partial x_T} = 2E_{T-1}(C_T' H_T C_T) x_T + 2E_{T-1}(C_T' H_T A_T) y_{T-1}$$

$$+ 2E_{T-1}(C_T' H_T b_T) - 2(E_{T-1} C_T') h_T$$

$$= 0, \quad (11)$$

yields the optimal control policy for the last period

$$\hat{x}_T = G_T y_{T-1} + g_T,\tag{12}$$

where

$$G_T = -(E_{T-1}C_T'H_TC_T)^{-1}(E_{T-1}C_T'H_TA_T),\tag{13}$$

$$g_T = -(E_{T-1}C_T'H_TC_T)^{-1}[(E_{T-1}C_T'H_Tb_T)-(E_{T-1}C_T')h_T].\tag{14}$$

The optimal feedback control equation (12) may appear to be a linear function of y_{T-1}, but this in general is not the case. The coefficients G_T and g_T, insofar as they depend on the conditional expectations at the end of period $T-1$, are functions of the observations $y_{T-1}, y_{T-2},\ldots,y_1$, and $x_{T-1}, x_{T-2},\ldots,x_1$. If we are willing, however, to *approximate the joint density of A_T, C_T, and b_T as being conceived at the end of $T-1$ by their density as of the end of period 0*, thus ignoring possible revisions of the density by observations on y_t and x_t from period 1 on, the feedback control equation (12) can be treated as linear. This is the approximation taken in this chapter. The optimal policy thus derived can be improved on by a better approximation, to be presented in Chapter 11. Nevertheless, the result of this chapter provides a lower bound to the value of control or an upper bound to the loss arising from uncertainty in the parameters. Furthermore, if the information on the coefficients available at time 0 is large, compared with future information from period 1 to period $T-1$, as it is likely to be when an econometric model is employed for policy analysis, the solution of this chapter will be close to optimal. Illustrative calculations presented in Chapter 11 substantiate this claim.

The minimum expected welfare cost for the last period is obtained by substituting (12) for x_T in (10),

$$\begin{aligned}
\hat{V}_T = {}& E_{T-1}[(A_T+C_TG_T)y_{T-1}+b_T+C_Tg_T]'\\
& H_T[(A_T+C_TG_T)y_{T-1}+b_T+C_Tg_T]\\
& + E_{T-1}u_T'H_Tu_T - 2E_{T-1}[(A_T+C_TG_T)y_{T-1}+b_T+C_Tg_T]'h_T + E_{T-1}c_T\\
= {}& y_{T-1}'E_{T-1}(A_T+C_TG_T)'H_T(A_T+C_TG_T)y_{T-1}\\
& + 2y_{T-1}'E_{T-1}(A_T+C_TG_T)'(H_Tb_T-h_T)\\
& + E_{T-1}(b_T+C_Tg_T)'H_T(b_T+C_Tg_T)\\
& + E_{T-1}u_T'H_Tu_T - 2E_{T-1}(b_T+c_Tg_T)'h_T + E_{T-1}c_T.
\end{aligned}\tag{15}$$

Here $\hat{V}_T$ can be treated as a quadratic function of y_{T-1} if the conditional density of A_T, C_T, and b_T as of $T-1$ is assumed to be independent of $y_{T-1},\ldots,y_1$ and $x_{T-1},\ldots,x_1$.

Now, consider including the period $T-1$ in our optimization problem. By the principle of optimality this two-period problem involving x_T and x_{T-1} is reduced to one having only one unknown x_{T-1}. It is to minimize, with respect to x_{T-1}, the expression

$$V_{T-1} = E_{T-2}\big(y'_{T-1}K_{T-1}y_{T-1} - 2y'_{T-1}K_{T-1}a_{T-1} + a'_{T-1}K_{T-1}a_{T-1} + \hat{V}_T\big).$$
(16)

When (15) is used for $\hat{V}_T$, (16) becomes

$$V_{T-1} = E_{T-2}(y'_{T-1}H_{T-1}y_{T-1} - 2y'_{T-1}h_{T-1} + c_{T-1}),$$
(17)

where

$$H_{T-1} = K_{T-1} + E_{T-1}(A_T + C_T G_T)'H_T(A_T + C_T G_T)$$

$$= K_{T-1} + E_{T-1}(A'_T H_T A_T) + G'_T(E_{T-1}C'_T H_T A_T),$$
(18)

$$h_{T-1} = K_{T-1}a_{T-1} + E_{T-1}(A_T + C_T G_T)'(h_T - H_T b_T)$$

$$= K_{T-1}a_{T-1} + E_{T-1}(A_T + C_T G_T)'h_T$$

$$- E_{T-1}(A'_T H_T b_T) - G'_T(E_{T-1}C'_T H_T b_T),$$
(19)

$$c_{T-1} = E_{T-1}(b_T + C_T g_T)'H_T(b_T + C_T g_T) - 2E_{T-1}(b_T + C_T g_T)'h_T$$

$$+ a'_{T-1}K_{T-1}a_{T-1} + E_{T-1}u'_T H_T u_T + E_{T-1}c_T.$$
(20)

The problem of minimizing (17) is seen to be identical to that of minimizing (8), with $T-1$ replacing T. The solution will therefore take the form of (12), with supplementary equations (13) and (14), and with $T-1$ replacing T in these equations. Once $\hat{x}_{T-1}$ is found, we can evaluate $\hat{V}_{T-1}$ as in (15) and include the additional period $T-2$ in the three-period optimization problem. That problem is to minimize, with respect to x_{T-2}, the expression

$$V_{T-2} = E_{T-3}\big(y'_{T-2}K_{T-2}y_{T-2} - 2y'_{T-2}K_{T-2}a_{T-2} + a'_{T-2}K_{T-2}a_{T-2} + \hat{V}_{T-1}\big).$$
(21)

This process can be continued until the optimal x_1 is found. The associated $\hat{V}_1$ gives the minimum expected loss for the T-period optimization problem.

To recapitulate, the optimal feedback control equation for each period is linear in y_{t-1}, as given by (12), *if* the conditional expectations required in evaluating the coefficients G_t and g_t in (13) and (14) can be computed independently of $y_{t-1}, y_{t-2}, \ldots, x_{t-1}, x_{t-2}, \cdots$. Under this assumption we compute G_T, H_{T-1}, $G_{T-1}, \ldots$, backward in time, using the pair of equations (13) and (18) and the initial condition (9). Similarly, we compute g_T, $h_{T-1}, g_{T-1}, \ldots$, backward in time, using the pair of equations (14) and (19). In these computations it is essential that the conditional expectations be evaluated. We turn to this subject in the next three sections.

10.3* CONDITIONAL EXPECTATIONS IN TERMS OF REDUCED-FORM PARAMETERS

In the remainder of this chapter we implement the result of Section 10.2 for econometric systems with time-invariant coefficients, that is, systems that take the form of (6) with the time subscript t omitted from A_t and C_t. Computation of the optimal control solution with (13), (14), (18), and (19) requires the evaluation of the expectations $E(C'H_tC)$, $E(C'H_tA)$, $E(C'H_tb_t)$, $E(A'H_tA)$, and $E(A'H_tb_t)$. These are functions of the second moments of the elements of A, C, and b_t. If the original system leading to (6) is of the first order, if lagged control variables do not appear, and if exogenous variables other than the control variables are also absent, the parameters of the reduced-form equations are simply A, C, and b; but because the original reduced-form equations may include y_{t-k} $(k > 1)$, $x_{t-k}(k > 0)$, and r other exogenous variables w_t the parameters in these equations must be related to A, C, and b_t in the notation of (6) before its result can be applied. The purpose of this section is to express the expectations $E(C'H_tC)$, and so on, as expectations of functions of the parameters in the original reduced form equations.

The original reduced-form equations are assumed to be

$$y_t = A_1 y_{t-1} + \cdots + A_m y_{t-m} + C_1 x_{t-1} + \cdots + C_n x_{t-n} + \Pi_2 x_t + \Pi_3 w_t + u_t.$$

$$(22)$$

They are rewritten as

$$
\begin{bmatrix} y_t \\ y_{t-1} \\ \vdots \\ y_{t-m+1} \\ x_t \\ x_{t-1} \\ \vdots \\ x_{t-n+1} \end{bmatrix} = \left[\begin{array}{ccc|c|ccc|c} A_1 \cdots & & A_m & C_1 \cdots & & C_n \\ I \cdots & & 0 & & 0 \cdots & & 0 & 0 \\ \hline 0 \cdots & & I & 0 & 0 \cdots & & 0 & 0 \\ \hline 0 \cdots & & 0 & 0 & 0 \cdots & & 0 & 0 \\ \hline 0 \cdots & & 0 & 0 & I \cdots & & 0 & 0 \\ \hline 0 \cdots & & 0 & 0 & 0 \cdots & & I & 0 \end{array} \right] \begin{bmatrix} y_{t-1} \\ y_{t-2} \\ \vdots \\ y_{t-m} \\ x_{t-1} \\ x_{t-2} \\ \vdots \\ x_{t-n} \end{bmatrix} + \begin{bmatrix} \Pi_2 \\ 0 \\ \vdots \\ 0 \\ I \\ 0 \\ \vdots \\ 0 \end{bmatrix} x_t + \begin{bmatrix} \Pi_3 w_t \\ 0 \\ \vdots \\ 0 \\ 0 \\ 0 \\ \vdots \\ 0 \end{bmatrix} + \begin{bmatrix} u_t \\ 0 \\ \vdots \\ 0 \\ 0 \\ 0 \\ \vdots \\ 0 \end{bmatrix}
$$

$$(23)$$

Equation 23 is put in a more compact form as

$$y_t = A y_{t-1} + C x_t + b_t + u_t \tag{24}$$

by letting

$$\Pi_1 = (A_1 \ldots A_m C_1 \ldots C_n), \tag{25}$$

and, according to the partitions of (23),

$$
A = \left[\begin{array}{cccc} \multicolumn{4}{c}{\Pi_1} \\ \hline 0 & 0 & 0 & 0 \\ 0 & 0 & 0 & 0 \\ 0 & 0 & 0 & 0 \end{array} \right] + \left[\begin{array}{cccc} 0 & 0 & 0 & 0 \\ \hline I & 0 & 0 & 0 \\ 0 & 0 & 0 & 0 \\ 0 & 0 & I & 0 \end{array} \right], \tag{26}
$$

$$
C = \begin{bmatrix} \Pi_2 \\ 0 \\ I \\ 0 \end{bmatrix}, \tag{27}
$$

$$b_t = \begin{bmatrix} \Pi_3 w_t \\ 0 \\ 0 \\ 0 \end{bmatrix}. \tag{28}$$

To express $E(C'H_t C)$, $E(C'H_t A)$, and so on, in terms of Π_1, Π_2, and Π_3 we follow the partitions of (26), (27), and (28), and partition the matrix H (omitting time subscript t) as

$$H = \begin{bmatrix} H_{11} & H_{12} & H_{13} & H_{14} \\ H_{21} & H_{22} & H_{23} & H_{24} \\ H_{31} & H_{32} & H_{33} & H_{34} \\ H_{41} & H_{42} & H_{43} & H_{44} \end{bmatrix}. \tag{29}$$

If the system (22) has p endogenous variables and q control variables, the submatrices H_{11}, H_{22}, H_{33}, and H_{44} will have, respectively, p, $(m-1)p$, q, and $(n-1)q$ rows or columns, which can be verified from (23). Multiplications of these partitioned matrices give

$$C'HC = \Pi_2' H_{11} \Pi_2 + \Pi_2' H_{13} + H_{31} \Pi_2 + H_{33}, \tag{30}$$

$$C'HA = \Pi_2' H_{11} \Pi_1 + \Pi_2'(H_{12} \quad 0 \quad H_{14} \quad 0) + H_{31} \Pi_1 + (H_{32} \quad 0 \quad H_{34} \quad 0), \tag{31}$$

$$C'Hb_t = \Pi_2' H_{11} \Pi_3 w_t + H_{31} \Pi_3 w_t, \tag{32}$$

$$A'HA = \Pi_1' H_{11} \Pi_1 + \Pi_1'(H_{12} \quad 0 \quad H_{14} \quad 0) + (H_{12} \quad 0 \quad H_{14} \quad 0)' \Pi_1,$$

$$+ \begin{bmatrix} H_{22} & 0 & H_{24} & 0 \\ 0 & 0 & 0 & 0 \\ H_{42} & 0 & H_{44} & 0 \\ 0 & 0 & 0 & 0 \end{bmatrix}, \tag{33}$$

$$A'Hb_t = \Pi_1' H_{11} \Pi_3 w_t + \begin{bmatrix} H_{21} \Pi_3 w_t \\ 0 \\ H_{41} \Pi_3 w_t \\ 0 \end{bmatrix}. \tag{34}$$

Our task is to evaluate the expectations of (30) to (34) in terms of the first two moments of the elements of

$$\Pi = (\Pi_1 \quad \Pi_2 \quad \Pi_3). \tag{35}$$

Denote the mean of Π by $\overline{\Pi}$. Letting the $s = (pm + qn) + q + r$ columns of Π be $\pi_1 \cdots \pi_s$, we write the ps elements of Π as a column vector π. Denote the covariance matrix of π by Q so that

$$E\pi\pi' = \overline{\pi}\overline{\pi}' + Q = \begin{bmatrix} \overline{\pi}_1\overline{\pi}_1' \cdots \overline{\pi}_1\overline{\pi}_s' \\ \cdots \cdots \\ \overline{\pi}_s\overline{\pi}_1' \cdots \overline{\pi}_s\overline{\pi}_s' \end{bmatrix} + \begin{bmatrix} Q_{11} \cdots Q_{1s} \\ \cdots \cdots \\ Q_{s1} \cdots Q_{ss} \end{bmatrix}. \tag{36}$$

We evaluate the expectations of the leading terms of (30) to (34) by noting that they are submatrices of $E(\Pi' H_{11} \Pi)$ according to the partition of Π by (35). The $i - j$ element of $E(\Pi' H_{11} \Pi)$ is

$$E(\Pi' H_{11} \Pi)_{ij} = E\pi_i' H_{11} \pi_j = E \operatorname{tr}(H_{11} \pi_j \pi_i')$$

$$= \operatorname{tr} H_{11} E\pi_j \pi_i' = \overline{\pi}_i' H_{11} \overline{\pi}_j + \operatorname{tr} H_{11} Q_{ji}. \tag{37}$$

Thus the required expectations of (30) to (34) can be computed by using the mean vector and the covariance matrix of π.

10.4* MEAN AND COVARIANCE MATRIX OF REDUCED-FORM PARAMETERS BY BAYESIAN METHODS

In this section we evaluate the mean vector and the covariance matrix of π by Bayesian methods. Using the definition of conditional probability distribution, we have

$$p(\pi|y) = \frac{p(\pi, y)}{p(y)}, \tag{38}$$

where p is the probability density function and π and y are values of two random variables. This definition is valid when π and y are vectors or even matrices. It is applied by Bayesian statisticians to form a probability density function of the unknown parameters π, given some data y. To do so both the numerator and the denominator of the right-hand side of (38) are rewritten to yield

$$p(\pi|y) = \frac{p(y|\pi)p(\pi)}{\int_{-\infty}^{\infty} p(\pi, y)\, d\pi} = \frac{p(y|\pi)p(\pi)}{\int_{-\infty}^{\infty} p(y|\pi)p(\pi)\, d\pi}. \tag{39}$$

The denominator is not a function of π. It can be treated as a constant as far as the probability density function of π is concerned. Using the symbol $\propto$ for "proportional to," we have

$$p(\pi|y) \propto p(y|\pi)p(\pi). \tag{40}$$

Let π be a vector of values for s unknown parameters, in which case the integrals in (39) should be interpreted as multiple integrals and $d\pi$ stands for $d\pi_1 \cdot d\pi_2 \cdots d\pi_s$. Let y be a vector of observations. In (40) $p(y|\pi)$ is the *likelihood function* and $p(\pi)$ is the *prior density function* of the unknown parameters before the observations y are obtained.

Our problem is to find the posterior density function of the unknown parameters π of the system (22), given N observations available before the control problem has to be solved. Using the posterior density, we evaluate the mean vector and the covariance matrix of π to be applied to (37). Given (37), the required expectations of (30) to (34) can be computed.

The posterior density function of π is proportional to the product of the likelihood function and the prior density function, according to (40). To obtain the likelihood function we assume that the residual vector u_t in (22) has a p-variate normal density with mean 0 and covariance matrix $V = R^{-1}$,

$$p(u_t|R) \propto |R|^{\frac{1}{2}} \exp(-\tfrac{1}{2} u_t' R u_t). \tag{41}$$

The joint density of N independent observations of u_t, $t = 1, \ldots, N$, is

$$p(U|R) \propto |R|^{\frac{1}{2}N} \exp\left(-\tfrac{1}{2} \sum_{t=1}^{N} u_t' R u_t\right)$$

$$\propto |R|^{\frac{1}{2}N} \exp(-\tfrac{1}{2} \operatorname{tr} URU')$$

$$\propto |R|^{\frac{1}{2}N} \exp(-\tfrac{1}{2} \operatorname{tr} RU'U), \tag{42}$$

where U is an $N \times p$ matrix with u_t' as its tth row. To derive the joint density of $y_1, \ldots, y_N$ from the joint density of $u_1, \ldots, u_N$ we write the N observations of the model (22) as

$$Y = Z\Pi' + U, \tag{43}$$

where Y is an $N \times p$ matrix consisting of columns of observations on p endogenous variables, Z is an $N \times s$ matrix consisting of columns of observations on the s explanatory variables on the right-hand side of (22), and $\Pi = (\Pi_1 \quad \Pi_2 \quad \Pi_3)$ is defined by (25) and (35). Using (43) to change the variables from U to Y, we have the required likelihood function

$$p(Y|\Pi, R) \propto |R|^{\frac{1}{2}N} \exp[-\tfrac{1}{2} \operatorname{tr} R(Y' - \Pi Z')(Y - Z\Pi')]. \tag{44}$$

For the prior density of the parameters Π and R of (44) there are two convenient possibilities. The first is

$$p(\Pi, R) = p(\Pi) \cdot p(R) \propto \text{constant} \cdot |R|^{-\frac{1}{2}(p+1)}, \tag{45}$$

which is known as a *diffuse prior density*. In (45) Π and R are assumed to be independent. The probability density function of Π is assumed to be constant, indicating that the statistician has no prior opinion to assign higher densities to some values of these parameters than other values. To motivate $p(R)$ recall that R is the inverse of the covariance matrix of the conditional distribution of y_t, given all the explanatory variables of the system (22). If y_t is a scalar, R is the inverse of the variance v. For a diffuse prior density of the variance v it would be unreasonable to assume that v itself is uniformly distributed because very large values of v would appear to be less likely than smaller values. Some Bayesian statisticians assume that the density of $\log v$ is uniform. Given that the probability density of $\log v$ is a constant, the density of v itself is proportional to

$$\left| \frac{d \log v}{dv} \right| = v^{-1}, \tag{46}$$

which is the Jacobian for the transformation of variable from $\log v$ to v. Furthermore, if the density of v is proportional to v^{-1}, the density of $R = v^{-1}$ is proportional to

$$R \cdot \left| \frac{dv}{dR} \right| = R \cdot \left| \frac{dR^{-1}}{dR} \right| = R^{-1} \tag{47}$$

as a result of change of variable from v to $R = v^{-1}$, using the Jacobian $|dv/dR|$. This specification of the diffuse prior density for R is a special case of (45) when the number p of variables in y_t equals 1; (45) is a generalization of this diffuse prior density in which the determinant of the matrix R is used when $p > 1$. It is to a large extent motivated by mathematical convenience, as we shall see presently.

Besides using (45) as a diffuse prior, it is possible to use an *informative prior density*. In this case the statistician has a certain opinion concerning which values of the parameters Π and R are more likely than others. Mathematical convenience also weighs heavily in choosing an informative prior density function. Recall that the prior density is to be multiplied by the likelihood to form the posterior density. It is therefore convenient to have a mathematical form for the prior density such that when it is multiplied by the likelihood function the product retains the same form as the prior density. The posterior density obtained after using one set of observations will serve as the prior density for the analysis of a new set of observations, and so on. It will be convenient to preserve the same mathematical form for the density functions in a series of statistical analyses. If the product of a prior density function and the likelihood function has the same mathematical form as the prior density function, these two functions are called *natural conjugate functions* to each other. For our problem we first use the diffuse prior density of (45) to obtain the posterior density of Π and R and then determine whether the result possesses the mathematically convenient property. If it does, we will use the same family of functions for an informative prior density.

Forming the product of (44) and (45), we obtain the posterior density function of Π and R, given the observations Y,

$$p(\Pi, R \mid Y) \propto |R|^{\frac{1}{2}(N-p-1)} \exp\left[-\tfrac{1}{2} \operatorname{tr} R(Y' - \Pi Z')(Y - Z\Pi')\right]. \quad (48)$$

If a new set of N_1 observations is available, as given by an $N_1 \times p$ matrix Y_1 of dependent variables and an $N_1 \times s$ matrix Z_1 of explanatory variables, the likelihood function for the set is the same as (44) with N_1, Y_1, and Z_1 replacing N, Y, and Z, respectively. Using (48) as the prior density function will yield a second posterior density function

$$p(\Pi, R \mid Y_2) \propto |R|^{\frac{1}{2}(N_2-p-1)} \exp\left[-\tfrac{1}{2} \operatorname{tr} R(Y_2' - \Pi Z_2')(Y_2 - Z_2\Pi')\right], \quad (49)$$

where we have let $N_2 = N + N_1$,

$$Y_2 = \begin{bmatrix} Y \\ Y_1 \end{bmatrix} \quad \text{and} \quad Z_2 = \begin{bmatrix} Z \\ Z_1 \end{bmatrix},$$

and have observed

$$\operatorname{tr} R(Y_2' - \Pi Z_2')(Y_2 - Z_2\Pi') = \operatorname{tr}(Y_2 - Z_2\Pi')R(Y_2' - \Pi Z_2')$$

$$= \operatorname{tr}(Y - Z\Pi')R(Y' - \Pi Z')$$

$$+ \operatorname{tr}(Y_1 - Z_1\Pi')R(Y_1' - \Pi Z_1').$$

Because (49) has the same form as (48), either of these functions can serve as a natural conjugate prior density for the multivariate normal likelihood function. Either can be used as an informative prior density function.

It will be useful to rewrite the density function (48) for the purpose of evaluating the means and covariance matrix of the elements of Π as required by the optimal control calculations. To do so we use the identity

$$(Y' - \Pi Z')(Y - Z\Pi') = S + (\hat{\Pi} - \Pi)Z'Z(\hat{\Pi} - \Pi)', \qquad (50)$$

where

$$\hat{\Pi}' = (Z'Z)^{-1}Z'Y \qquad (51)$$

and

$$S = (Y' - \hat{\Pi}Z')(Y - Z\hat{\Pi}'). \qquad (52)$$

(See Problem 1.) Note that each row of $\hat{\Pi}$ is a set of coefficients in the regression of one variable in y_t on the explanatory variables of system (22) obtained by the method of least squares, using the N observations. The matrix S consists of sums of squares and cross products of the residuals in the p regressions of the variables in y_t on the explanatory variables. Using (50), (51), and (52), we rewrite (48) as

$$p(\Pi, R \mid Y) \propto |R|^{\frac{1}{2}s} \exp\left[-\frac{1}{2}\operatorname{tr} R(\Pi - \hat{\Pi})Z'Z(\Pi - \hat{\Pi})' \right]$$

$$\times |R|^{\frac{1}{2}(N-p-1-s)} \exp(-\frac{1}{2}\operatorname{tr} RS)$$

$$\propto p(\Pi \mid R, Y) \times p(R \mid Y). \qquad (53)$$

The joint posterior density of Π and R is thus written as the product of the conditional posterior density $p(\Pi \mid R, Y)$ of Π, given R and the posterior density $p(R \mid Y)$ of R.

The conditional posterior density of Π, given R,

$$p(\Pi \mid R, Y) \propto |R|^{\frac{1}{2}s} \exp\left[-\frac{1}{2}\operatorname{tr} R(\Pi - \hat{\Pi})Z'Z(\Pi - \hat{\Pi})' \right] \qquad (54)$$

is a multivariate normal density function for the column vector π consisting of the s columns $\pi_1, \pi_2, \ldots, \pi_s$ of the matrix Π. To show this we first rewrite the exponent on the right-hand side of (54), denoting the ith column of Z by z_i so that the $i-j$ element of $Z'Z$ is $z_i'z_j$ and using the definition $\operatorname{tr} AB = \sum_i \sum_j a_{ij} b_{ji}$,

$$
\operatorname{tr}\left[R(\Pi - \hat{\Pi})Z'Z(\Pi - \hat{\Pi})' \right] = \operatorname{tr}\left\{ \left[(\Pi - \hat{\Pi})'R(\Pi - \hat{\Pi}) \right](Z'Z) \right\}
$$

$$
= \operatorname{tr}\left\{ \left[(\pi_i - \hat{\pi}_i)'R(\pi_j - \hat{\pi}_j) \right][z_i'z_j] \right\}
$$

$$
= \sum_i \sum_j (\pi_i - \hat{\pi}_i)'R(\pi_j - \hat{\pi}_j) \cdot (z_j'z_i)
$$

$$
= \sum_i \sum_j (\pi_i - \hat{\pi}_i)'\left[(z_i'z_j)R \right](\pi_j - \hat{\pi}_j)
$$

$$
= \left[(\pi_1 - \hat{\pi}_1)' \cdots (\pi_s - \hat{\pi}_s)' \right]
\begin{bmatrix}
(z_1'z_1)R & (z_1'z_2)R \ldots (z_1'z_s)R \\
(z_2'z_1)R & (z_2'z_2)R \ldots (z_2'z_s)R \\
\cdot \cdot \cdot \cdot \cdot \cdot \cdot \cdot \cdot \cdot \cdot \cdot \cdot \cdot \\
(z_s'z_1)R & (z_s'z_2)R \cdots (z_s'z_s)R
\end{bmatrix}
\begin{bmatrix}
\pi_1 - \hat{\pi}_1 \\
\pi_2 - \hat{\pi}_2 \\
\vdots \\
\pi_s - \hat{\pi}_s
\end{bmatrix}.
$$

$$(55)$$

The exponent of (54) is thus seen to be that of a multivariate normal density function of the vector π, with a *mean vector equal to* $\hat{\pi}$ and a *covariance matrix equal to the inverse of*

$$
\begin{bmatrix}
(z_1'z_1)R & (z_1'z_2)R \ldots (z_1'z_s)R \\
(z_2'z_1)R & (z_2'z_2)R \ldots (z_2'z_2)R \\
\cdot \cdot \cdot \cdot \cdot \cdot \cdot \cdot \cdot \cdot \cdot \cdot \cdot \cdot \\
(z_s'z_1)R & (z_s'z_2)R \ldots (z_s'z_s)R
\end{bmatrix}
\equiv (Z'Z) \otimes R, \qquad (56)
$$

where $(Z'Z) \otimes R$ stands for the Kronecker product of $(Z'Z)$ and R, defined by the left-hand side of (56). This multivariate normal density function requires a multiplicative factor equal to the square root of the determinant of (56). Using the identity for an $s \times s$ matrix $(Z'Z)$ and a $p \times p$ matrix R

$$|(Z'Z) \otimes R| = |Z'Z|^p \times |R|^s, \tag{57}$$

which can be found on page 348 of Anderson (1958), for example, and noting that the determinant $|Z'Z|^p$ is not a function of the variables Π and R, we include the multiplicative factor $|R|^{\frac{1}{2}s}$ in the multivariate normal density function given in (54).

The special case $p = 1$ deserves comment. The problem is reduced to a multiple regression of a scalar variable y on the explanatory variables Z, with R^{-1} as the variance of the residual. The regression coefficients obtained by the method of least squares are elements of the $1 \times s$ vector $\hat{\pi}$. By (54) and (56) the unknown regression coefficient vector π has a multivariate normal density function with a mean vector $\hat{\pi}$ and a covariance matrix $(Z'Z)^{-1} \times R^{-1}$, R being a scalar. This result has a striking resemblance to the result of the classical regression problem involving the vector y of observations on a dependent variable and the matrix Z of observations on a set of explanatory variables regarded as fixed, given R^{-1} as the variance of the residual. Classical statisticians speak of the distribution of the least-squares estimates $\hat{\pi}$ as a function of the unknown parameters π rather than the distribution of the unknown parameters π, given the least squares estimate $\hat{\pi}$. The distribution of $\hat{\pi}$, according to classical statistical theory, has a mean vector equal to the unknown π and a covariance matrix equal to $(Z'Z)^{-1} \cdot R^{-1}$.

We now turn to the second component of the posterior density function (53), namely the posterior density of R:

$$p(R|Y) \propto |R|^{\frac{1}{2}(N-s-p-1)} \exp(-\tfrac{1}{2} \operatorname{tr} RS). \tag{58}$$

Equation 58 is a *Wishart density* of the $p \times p$ matrix R, with the matrix S and the scalar $N - s$ as parameters. The reader may refer to Anderson (1958), pp. 154–159, for a description of this density function and its most common use in classical statistics, that is, in specifying the distribution of a sample covariance matrix of data generated by a multivariate normal distribution. If R is a scalar, (58) is reduced to the form

$$p(R) \propto R^\alpha \exp(-\beta R), \tag{59}$$

which is a gamma density function. We need only to use the following fact concerning (58). If a $p \times p$ matrix R is distributed according to (58), the expectation of R^{-1} is

$$E(R^{-1}) = (N - s - p - 1)^{-1} S. \tag{60}$$

The result (60) can be found in Kaufman (1967), p. 14.

The posterior density (53) of Π and R is known as the *normal-Wishart density*, the product of a normal density for Π, given R, and a Wishart density for R. Our control problem requires the evaluation of the expectation and the covariance matrix of the column vector π consisting of the columns of Π. To find the first, we integrate over all real values for the elements of Π and all positive values for the elements of R,

$$E\Pi = \int \int \Pi p(\Pi | R, Y) \cdot p(R | Y) d\Pi \cdot dR$$

$$= \int \left[\int \Pi p(\Pi | R, Y) d\Pi \right] p(R | Y) dR$$

$$= \hat{\Pi} \int p(R | Y) dR = \hat{\Pi}. \tag{61}$$

In (61) we have utilized the result from (54) and (55) that the mean of the conditional distribution of Π, given R, is the least squares estimate $\hat{\Pi}$ and the fact that $\hat{\Pi}$ is not a function of R. To find the covariance matrix of the vector π we write

$$\mathrm{cov}\,\pi = \int \int (\pi - \hat{\pi})(\pi - \hat{\pi})' p(\Pi | R, Y) \cdot p(R | Y) d\Pi \cdot dR$$

$$= \int [(Z'Z) \otimes R]^{-1} p(R | Y) dR$$

$$= \int \left[(Z'Z)^{-1} \otimes R^{-1} \right] p(R | Y) dR$$

$$= (Z'Z)^{-1} \otimes (E R^{-1})$$

$$= (N - s - p - 1)^{-1} (Z'Z)^{-1} \otimes S. \tag{62}$$

In (62) we have applied the inverse of (56) as the covariance matrix of the conditional distribution of π, given R, used the identity

$$[(Z'Z) \otimes R]^{-1} = (Z'Z)^{-1} \otimes R^{-1}, \tag{63}$$

and applied (60) to $E R^{-1}$. [See Problem 2 for the identity (63).] This completes the derivation of $E\pi$ and $\mathrm{cov}\,\pi$ for application in (37).

To simplify the resulting expression let

$$(N-s-p-1)^{-1}(Z'Z)^{-1}=(c_{ij});\qquad(64)$$

(62) then implies that Q_{ij} in (36) equals $c_{ij}S$ and (37) becomes

$$E(\Pi'H_{11}\Pi)_{ij}=\hat{\pi}_i'H_{11}\hat{\pi}_j+(\operatorname{tr}H_{11}S)c_{ji}.\qquad(65)$$

Given (65) as its $i-j$ element, the expectation of the matrix $\Pi'H_{11}\Pi$ is

$$E(\Pi'H_{11}\Pi)=\hat{\Pi}'H_{11}\hat{\Pi}+(\operatorname{tr}H_{11}S)(c_{ji})$$

$$=\hat{\Pi}'H_{11}\hat{\Pi}+(N-s-p-1)^{-1}(\operatorname{tr}H_{11}S)(Z'Z)^{-1}.\qquad(66)$$

Submatrices of (66) can be used to evaluate the expectations of the leading terms of (30) to (34). The other terms simply require the submatrices of $E\Pi=\hat{\Pi}$.

10.5* AN APPROXIMATE SOLUTION

According to the viewpoint of classical statistics, uncertainty in the parameters is expressed by the variances of certain statistics pertinent to the parameters. In the case of the regression coefficients π, as we have pointed out, classical statisticians speak of the covariance matrix of the least squares estimator $\hat{\pi}$ and not of the covariance matrix of the unknown π. If this viewpoint is strictly adhered to, it is not meaningful to write $E\pi$ and $\operatorname{cov}\pi$. It is possible, however, to reinterpret the optimal control problem from the classical statistical point of view.

Consider the example on Section 8.1. Let $\hat{a}_1$ and $\hat{c}_1$ be the least squares estimates of the constant parameters a_1 and c_1 in the model (1). If the purpose of control is to make Ey_1^2 as small as possible, given y_0 and the estimates $\hat{a}_1$ and $\hat{c}_1$, it may be reasonable to minimize the expectation of the square of

$$\hat{y}_1=\hat{a}_1y_0+\hat{c}_1x_1+u_1\qquad(67)$$

with respect to x_1. In (67) the residual u_1 is assumed to be independent of $\hat{a}_1$ and $\hat{c}_1$. If we do not know a_1 and c_1, it seems reasonable to choose a policy x_1 such that the expectation of the square of the random estimate $\hat{y}_1$ of y_1 based on the random estimates $\hat{a}_1$ and $\hat{c}_1$ be as small as possible. The justification is similar to that of choosing a random estimator with minimum expected squared deviation from the unknown parameter. This argument can be generalized to a nonzero target by using the deviation of the random estimate $\hat{y}_1$ from this target. Generalizations to a multivariate system and to a multiperiod control problem are also straightforward. The matrices A_t, C_t, and b_t in Section 8.2 should be replaced by $\hat{A}_t$, $\hat{C}_t$, and $\hat{b}_t$, respectively, which are random estimators of these parameters. Under the assumption that the parameters are time-invariant and using the approximation of the conditional expectation E_t for all t by E_0, we replace $EC'H_tC$, and so on, in Section 8.3 with $E\hat{C}H_t\hat{C}$, and so on. Thus, if we adhere to the classical viewpoint, we can apply $E\hat{\pi}$ and $\mathrm{cov}\,\hat{\pi}$ to the optimal control calculations in which a Bayesian would use $E\pi$ and $\mathrm{cov}\,\pi$.

When a macroeconometric model is constructed, parameters of the structural equations are often estimated and a covariance matrix of these estimates is obtained. It is then necessary to transform these estimates and their covariance matrix to estimates of the reduced-form parameters and the associated covariance matrix. Let N observations on the linear structure, which corresponds to the reduced form (43), be written as

$$YB' + Z\Gamma' = E, \tag{68}$$

where each row of (B Γ) consists of coefficients of one structural equation, Y and Z are matrices defined for (43), and each row of E has a p-variate normal distribution, independent of any other row. The reduced-form coefficients Π are related to the coefficients of the structure by

$$\Pi = -B^{-1}\Gamma. \tag{69}$$

For optimal control calculations using classical statistical methods, it is required to obtain the mean vector and the covariance matrix of the elements of the estimates $\tilde{\pi}$ of π.

To obtain approximations of the required mean vector and covariance matrix based on large-sample distribution theory the following approach can be used. Let $(\tilde{B}\ \tilde{\Gamma})$ be consistent and asymptotically unbiased estimates of (B Γ) and let the $(p+s)$ columns of these estimates have an asymptotic covariance matrix W. By Theorem 1 of Goldberger, Nagar, and Odeh (1961), the columns of the reduced-form estimates $\tilde{\Pi} = -\tilde{B}^{-1}\tilde{\Gamma}$ will have an asymptotic covariance matrix that may be approximated by

$$\tilde{Q} = \left[(\tilde{\Pi}'\ \ I_s)\otimes\tilde{B}^{-1}\right]W\left[(\tilde{\Pi}'\ \ I_s)\otimes\tilde{B}^{-1}\right]', \tag{70}$$

where I_s is an identity matrix of order s. To prove this theorem Goldberger *et al.* use the following:

Lemma. Given a sequence of random vectors $\tilde{\alpha}_N$, $N = 1, 2, 3, \ldots$, which has an asymptotic expectation $\lim_{N \to \infty} E\tilde{\alpha}$, a probability limit $\text{plim}_{N \to \infty} \tilde{\alpha}$, both equal to α, an asymptotic covariance matrix $\lim_{N \to \infty} E(\tilde{a} - \alpha)(\tilde{a} - \alpha)'$ equal to W, and a vector $\tilde{\pi} = f(\tilde{\alpha})$ of differentiable functions of $\tilde{\alpha}$, the asymptotic expectation of $\tilde{\pi}$ and its probability limit are

$$\lim_{N \to \infty} E\tilde{\pi} = \text{plim}_{N \to \infty} \tilde{\pi} = f(\alpha), \tag{71}$$

and the asymptotic covariance matrix of $\tilde{\pi}$ is

$$\lim_{N \to \infty} E[\tilde{\pi} - f(\alpha)][\tilde{\pi} - f(\alpha)]' = DWD', \tag{72}$$

where D is the matrix of first-order partial derivatives of the elements of $\tilde{\pi}$ with respect to the elements of $\tilde{\alpha}$ evaluated at $\tilde{\alpha} = \alpha$.

To apply this lemma to prove the theorem of equation (70), we need only to evaluate the matrix of partial derivatives of $\tilde{\pi}$ with respect to $\tilde{\alpha}$, where the vector $\tilde{\pi}$ is composed of the columns $\tilde{\pi}_1 \cdots \tilde{\pi}_s$ of $\tilde{\Pi} = -\tilde{B}^{-1}\tilde{\Gamma}$ and the vector $\tilde{\alpha}$ is composed of the columns $\tilde{\beta}_1 \cdots \tilde{\beta}_p, \tilde{\gamma}_1 \cdots \tilde{\gamma}_s$ of $(\tilde{B} \quad \tilde{\Gamma})$. The relation between the estimates of the reduced-form coefficients and the estimates of the structural coefficients can be written as

$$\tilde{\pi} = \begin{bmatrix} \tilde{\pi}_1 \\ \vdots \\ \tilde{\pi}_s \end{bmatrix} = - \begin{bmatrix} \tilde{B}^{-1}\tilde{\gamma}_1 \\ \vdots \\ \tilde{B}^{-1}\tilde{\gamma}_s \end{bmatrix}.$$

Omitting the " ~ " sign when understood, we can write the ith element $\tilde{\pi}_{ij}$ of the vector $\tilde{\pi}_j$ as

$$\pi_{ij} = - \sum_{m=1}^{p} \beta^{im}\gamma_{mj}, \tag{74}$$

where β^{im} is the i-m element of B^{-1} and γ_{mj} is the m-j element of Γ. Using (74), we obtain the derivative of π_{ij} with respect to β_{kl}:

$$\frac{\partial \pi_{ij}}{\partial \beta_{kl}} = - \sum_{m=1}^{p} \frac{\partial \beta^{im}}{\partial \beta_{kl}} \cdot \gamma_{mj} = \beta^{ik} \sum_{m=1}^{p} \beta^{lm} \cdot \gamma_{mj} = -\beta^{ik}\pi_{lj}, \tag{75}$$

where we have applied the differentiation rule

$$\frac{\partial \beta^{im}}{\partial \beta_{kl}} = -\beta^{ik}\beta^{lm}. \tag{76}$$

(See Problem 3 for this differentiation rule.) Equation 75 implies that the matrix of derivatives of the elements of π_j with respect to the elements of β_l, with $\partial \pi_{ij}/\partial \beta_{kl}$ as its $i\text{-}k$ element, is

$$\frac{\partial \pi_j}{\partial \beta_l} = -\pi_{lj}B^{-1}. \tag{77}$$

Using (74), we obtain the derivative of π_{ij} with respect to γ_{kl}:

$$\frac{\partial \pi_{ij}}{\partial \gamma_{kl}} = \begin{cases} -\beta^{ik} & \text{for } l=j, \\ 0 & \text{otherwise.} \end{cases} \tag{78}$$

Equation 78 implies that the matrix of derivatives of the elements of π_j with respect to the elements of γ_l, with $\partial \pi_{ij}/\partial \gamma_{kl}$ as its $i\text{-}k$ element, is

$$\frac{\partial \pi_j'}{\partial \gamma_l} = \begin{cases} -B^{-1} & \text{for } l=j, \\ 0 & \text{otherwise.} \end{cases} \tag{79}$$

Equations 77 and 79 together give the $ps \times p(p+s)$ matrix $-[(\Pi' \quad I_s) \otimes B^{-1}]$ of partial derivatives of the ps elements of $\tilde{\pi}$ with respect to the $p(p+s)$ elements of $\tilde{\alpha}$. According to the lemma, these derivatives should be evaluated at the true and unknown values of B and Γ. Because these values are not available, we use the consistent estimates $\tilde{B}$ and $\tilde{\Gamma}$ in the matrix of partial derivatives as approximations. This completes the proof of the theorem of (70). The matrix $\tilde{Q}$ of (70) can be used for Q and $\tilde{\pi}$ can be used for $\bar{\pi}$ in (37) to provide an approximate solution to the evaluation of the expectations of (30) to (34).

The method just described is approximate, based on large sample theory, whereas the method in Section 8.4 is exact. The method in this section, however, has incorporated *a priori* restrictions imposed on the structural parameters B and Γ by the econometrician in the specification of the econometric model. These restrictions often take the form that certain elements of B and Γ are 0 because each equation contains only a subset of the endogenous and the predetermined variables. Other linear restrictions on the structural parameters are also imposed in practice. Because the reduced-form parameters $\Pi = -B^{-1}\Gamma$ are nonlinear functions of the structural parameters, these linear restrictions imply nonlinear restrictions on the parameters of the reduced form. When the estimates of the reduced-form parameters are obtained from the estimates of structural parameters, using the equations $\tilde{\Pi} = -\tilde{B}^{-1}\tilde{\Gamma}$ as proposed in this section, the nonlinear restrictions are properly incorporated. The procedure in Section 10.8 ignores these restrictions and by so doing might yield larger variances for the unknown parameters than if the restrictions were imposed. This subject is discussed in Dhrymes (1973).

10.6 COMPARISON WITH THE CERTAINTY-EQUIVALENT SOLUTION

In this section we attempt to compare the optimal control solution in Section 10.4 for random parameters with the solution that treats the random parameters as constants equal to their means. The latter is a certainty-equivalent solution. A comparison will bring out the consequences of accounting for the uncertainty of the parameters in an econometric model. Two parts of the optimal solution are compared. They are the optimal feedback equation and the optimal welfare cost.

If the random parameters in system (24) are reduced to their means for the certainty case, we simply replace A, C, and b_t in our solution in Section 10.3 with the mean values. The optimal feedback control coefficients G_t and g_t, given by (13) and (14) with t replacing T, can thus be compared with the corresponding coefficients in the certainty case in which A becomes $\bar{A}$, and so on. Similarly, the optimal welfare cost $\hat{V}_1$, given by (15) with 1 replacing T, can also be compared with the corresponding cost in the certainty case. The analytical results in Sections 10.3 and 10.4 can be used to compute the solutions in both the certainty and the uncertainty situations for the purpose of comparison. In the remainder of this section we ask whether qualitative results in such a comparison can be ascertained.

To facilitate comparison we rewrite the optimization problem in Section 10.2 in a slightly simplified form to make the optimal control equation linear homogeneous, eliminating the intercept g_t. First, the target a_t can be set equal to 0 because it can be absorbed in intercept of the reduced-form equation. In the notation of (22) we replace $\Pi_3 w_t$ with $\Pi_{3t} w_t$, where $\Pi_{3t} = (\Pi_3, a_t)$ and the last element of the new vector w_t is set equal to -1 identically. The resulting expression is $y_t - a_t$. Second, w_t can be formally treated as a set of endogenous variables by introducing the equation $w_t = D_t w_{t-1}$, where the elements of the diagonal matrix D_t are simply the ratios of the known elements of w_t to those of w_{t-1}. Combining this equation with (24) and denoting by z_t the column vector consisting of y_t and w_{t-1}, we have

$$z_t = \alpha_t z_{t-1} + \Gamma x_t + v_t, \tag{80}$$

where

$$z_t = \begin{bmatrix} y_t \\ w_{t+1} \end{bmatrix}; \quad \alpha_t = \begin{bmatrix} A & \Pi_{3t} \\ 0 & D_t \end{bmatrix}; \quad \Gamma = \begin{bmatrix} C \\ 0 \end{bmatrix}; \quad \nu_t = \begin{bmatrix} u_t \\ 0 \end{bmatrix}. \quad (81)$$

The welfare function will be

$$W = \sum_{t=1}^{T} z_t' Q_t z_t, \quad (82)$$

where, according to the partition of z_t into y_t and w_{t+1},

$$Q_t = \begin{bmatrix} K_t & 0 \\ 0 & 0 \end{bmatrix}. \quad (83)$$

Following the dynamic programming approach in Section 10.2, we easily find, analogous to (12), (13), (15), (18), and (20) respectively,

$$\hat{x}_t = G_t z_{t-1}, \quad (84)$$

$$G_t = -(E_{t-1}\Gamma'H_t\Gamma)^{-1}(E_{t-1}\Gamma'H_t\alpha_t), \quad (85)$$

$$\hat{V}_t = z_{t-1}'E_{t-1}(\alpha_t + \Gamma G_t)'H_t(\alpha_t + \Gamma G_t)z_{t-1} + c_{t-1}, \quad (86)$$

$$H_{t-1} = Q_{t-1} + E_{t-1}(\alpha_t + \Gamma G_t)'H_t(\alpha_t + \Gamma G_t), \quad (87)$$

$$c_{t-1} = E_{t-1}\nu_t'H_t\nu_t + E_{t-1}c_t, \quad (88)$$

with initial conditions $H_T = Q_T$, and $c_T = 0$.

Having rewritten our solution, we will now try to compare the optimal feedback coefficients G_t and the optimal expected welfare cost $\hat{V}_1$ with the corresponding results for the certainty case. In the certainty case the random coefficients α_t and Γ in (80) are assumed to reduce to their mean values $\bar{\alpha}_t$ and $\bar{\Gamma}$. Let α^* and Γ^* denote the deviations of α and Γ from their means. Some elements of α^* and Γ^* are obviously 0 because the coefficients are known for certain. Consider the problem for the last period T. From (85) for the uncertainty case we have

$$G_T = -(\bar{\Gamma}'H_T\bar{\Gamma} + E\Gamma^{*'}H_T\Gamma^*)^{-1}(\bar{\Gamma}'H_T\bar{\alpha}_T + E\Gamma^{*'}H_T\alpha^*), \quad (89)$$

whereas in the certainty case (89) reduces to

$$\overline{G}_T = -\left(\overline{\Gamma}' H_T \overline{\Gamma}\right)^{-1}\left(\overline{\Gamma}' H_T \overline{\alpha}_T\right). \tag{90}$$

What can be said about the relative magnitudes of $\overline{G}_T$ and G_T?

We can begin by making two elementary observations. First, if the coefficients α_t and Γ are subject to uncertainty, the principle of "certainty equivalence" does not apply. According to this principle, we can derive an optimal policy under uncertainty simply by replacing α_t and Γ with $\overline{\alpha}_t$ and $\overline{\Gamma}$. Clearly, the contrast of (89) and (90) shows that this is not the case. Second, as we pointed out in Section 7.7, if the number of instruments q is greater than the number of target variables (which equals the rank of $H_T = Q_T$), we can select a subset of instruments to achieve an optimal policy in the certainty case. In other words, to solve the equation

$$\left(\overline{\Gamma}' H_T \overline{\Gamma}\right) G_T = -\left(\overline{\Gamma}' H_T \overline{\alpha}_T\right) \tag{91}$$

for G_T we can arbitrarily set $q - \text{rank}(\overline{\Gamma}' H_T \overline{\Gamma})$ rows of G_T equal to 0. In the uncertainty case, however, more and possibly all instruments will be required, even if there are more instruments than targets. This point, already observed in Brainard (1967), can be easily seen by noting that the rank of $\overline{\Gamma}' H_T \overline{\Gamma} + E\Gamma^{*\prime} H_T \Gamma^*$ in (89) will in general be greater than the rank of $\overline{\Gamma}' H_T \overline{\Gamma}$ in (90) when the latter is smaller than q.

After making these two elementary observations, we may ask whether uncertainty will call forth smaller policy responses to recent economic data as manifested by the smaller magnitudes of the elements of G_T than $\overline{G}_T$; for example, can we say that the sum of squares of the elements of each column of $\overline{G}_T$ in the certainty case is necessarily larger than that of the corresponding column of G_T? This assertion would mean more policy response to each observed variable in z_{T-1}. To answer this question we recall in the discussion of Section 7.7 that the model for (90) is mathematically identical with the multivariate regression model

$$\overline{\alpha}_T = \overline{\Gamma}\left(-\overline{G}_T\right) + \overline{R}, \tag{92}$$

where the columns of $\overline{\alpha}_T$ are the dependent variables, the columns of $\overline{\Gamma}$ are explanatory variables, and the columns of $-\overline{G}_T$ are the regression coefficients obtained by generalized least squares; α^* and Γ^* can be regarded as measurement errors that yield α_T and Γ as observed variables and $-G_T$, as regression coefficients, given by (89). In Section 10.1 we have considered the special case of only one dependent variable and one piece of datum for the instrument to respond to (so that α_T becomes a scalar) in which $H_T = 1$. In this case the optimal feedback coefficient $\overline{G}_T$ under certainty will be larger than G_T if the errors Γ^* and α^* are uncorrelated; $\overline{G}_T$ may be smaller in absolute value than G_T if Γ^* and α^* are correlated. Let $\overline{\alpha}_T > 0$, $\overline{\Gamma} > 0$, and thus $\overline{G}_T = -(\overline{\alpha}_T/\overline{\Gamma}) < 0$. A positive covariance between Γ^* and α^* can make G_T bigger in absolute value than $\overline{G}_T$ by (89) and (90).

If the system (90) has p dependent variables, the p elements of each column of $\overline{\alpha}_T$ will be explained by the q columns of $\overline{\Gamma}$, with q regression coefficients given by the corresponding column of $-\overline{G}_T$. In this multiple regression situation, if the explanatory variables $\overline{\Gamma}$ are measured with errors Γ^*, and the dependent variable is also measured with errors (the corresponding column of α^*), we cannot conclude that each column of $-\overline{G}_T$ will have a greater sum of squares than the corresponding column of $-G_T$. Even if we assume that the columns of Γ^* are uncorrelated with those of α^*, that is, $E\Gamma^{*\prime}H_T\alpha^* = 0$, we still cannot deduce a greater sum of squares for the columns of $\overline{G}_T$. If $E\Gamma^{*\prime}H_T\alpha^* = 0$, (89) and (90) imply

$$\overline{G}_T'\overline{G}_T = G_T'\left[I + \left(\overline{\Gamma}'H_T\overline{\Gamma}\right)^{-1}(E\Gamma^{*\prime}H_T\Gamma^*)\right]'\left[I + \left(\overline{\Gamma}'H_T\overline{\Gamma}\right)^{-1}(E\Gamma^{*\prime}H_T\Gamma^*)\right]G_T$$

$$= G_T'G_T + G_T'\left\{\left(\overline{\Gamma}'H_T\overline{\Gamma}\right)^{-1}(E\Gamma^{*\prime}H_T\Gamma^*) + (E\Gamma^{*\prime}H_T\Gamma^*)\left(\overline{\Gamma}'H_T\overline{\Gamma}\right)^{-1}\right.$$

$$\left. + (E\Gamma^{*\prime}H_T\Gamma^*)\left(\overline{\Gamma}'H_T\overline{\Gamma}\right)^{-1}\left(\overline{\Gamma}'H_T\overline{\Gamma}\right)^{-1}(E\Gamma^{*\prime}H_T\Gamma^*)\right\}G_T. \qquad (93)$$

The matrix in curly brackets is not in general positive semidefinite, so that the diagonal elements of $\overline{G}_T'\overline{G}_T$ is not necessarily greater than or equal to the diagonal elements of $G_T'G_T$, although this may often turn out to be the case for specific applications. The deduction would be valid if, in addition to $E\Gamma^{*\prime}H_T\alpha^* = 0$, both $\overline{\Gamma}'H_T\overline{\Gamma}$ and $E\Gamma^{*\prime}H_T\Gamma^*$ were diagonal, but this is a special case indeed. Intuitively, one reason for a possibly larger policy response in the uncertainty situation is that, whereas the variances in Γ^* per se may lead to reduction in the magnitudes of G_T, the covariances between Γ^* and α^* can be exploited in the design of active control policies.

Next to be studied is the optimal welfare cost $\hat{V}_T$, still under the assumption $E\Gamma^{*\prime}H_T\alpha^* = 0$. By (86) and (88) it is a quadratic form in the variables z_{T-1} (the given data of the feedback control equation), plus a constant. Will uncertainty necessarily lead to an increase in the minimum expected loss? The answer is no. If uncertainty is to increase, or at least not to reduce, the minimum expected welfare cost for the last period, given any initial conditions z_{T-1}, the matrix

$$E(\alpha_T + \Gamma_T G_T)' H_T (\alpha_T + \Gamma_T G_T) - (\bar{\alpha}_T + \bar{\Gamma}_T \bar{G}_T)' H_T (\bar{\alpha}_T + \bar{\Gamma}_T \bar{G}_T)$$

$$= E\alpha^{*\prime} H_T \alpha^* + G_T'(\bar{\Gamma}' H_T \bar{\Gamma} + E\Gamma^{*\prime} H_T \Gamma^*) G_T - \bar{G}_T' \bar{\Gamma}' H_T \bar{\Gamma} \bar{G}_T \qquad (94)$$

has to be positive semidefinite. Equation 94 can be identified as the difference between the covariance matrix of the (weighted) multivariate regression residuals when measurement errors exist and the covariance matrix when the errors are absent. Errors in the dependent variables α_T^* alone will make the former matrix bigger by $E\alpha_T^{*\prime} H_T \alpha_T^*$. The remaining two matrices on the right-hand side of (94) are the covariance matrices of the explained parts of the regressions for error and no-error. There is no guarantee that the difference between these two matrices is positive semidefinite. As the study of Cochran (1970) on the effects of measurement errors on the multiple correlation suggests, without special assumptions, it is difficult to ascertain a net increase in the variance of the residuals of a multiple regression as a result of measurement errors, even though errors in the dependent variable alone will tend to cause such an increase. If, however, uncertainty does increase expected welfare cost in period T through the positive semidefiniteness of (94), the effect will tend to accumulate backward to the total expected welfare cost computed in period 1, the process of accumulation being given by (86) and (87).

This chapter has provided an analytical solution for the optimal feedback control equation and the associated expected (quadratic) welfare cost when the parameters of the linear econometric model employed are uncertain. The solution can be used to study the effects of uncertainty by comparison with the result when all parameters are reduced to constants. It seems difficult, however, to ascertain a priori qualitative results concerning such a comparison, although the partial effects of certain factors have been pointed out. By ignoring the possibility of reducing uncertainty through observations during the control process the solution of this chapter exaggerates, and thus sets an upper limit to, the effect of uncertainty on the optimal control policy and the associated welfare cost. Chapter 11 provides a method that takes learning into account in the determination of control policies. It also contains numerical examples that compare the methods of certainty equivalence of this chapter and of the new method that incorporates learning. The material in this chapter is based on Chow (October 1973).

PROBLEMS

1. Prove the identity given in (50).

2. Using the definition of the Kronecker product of two square matrices A and B, given in (56), show that $(A \otimes B)^{-1} = A^{-1} \otimes B^{-1}$, assuming that both A and B are nonsingular.

3. Denoting the i-m element of the inverse of a matrix B by b^{im}, show the differentiation rule (76). *Hint.* Start with the identity $I = BB^{-1}$. Differentiate both sides of this identity with respect to the scalar b_{kl}, using the convention that the derivative of a matrix with respect to a scalar is the matrix of derivatives of the corresponding components. The result is $0 = (\partial B / \partial b_{kl}) B^{-1} + B(\partial B^{-1} / \partial b_{kl})$.

4. The data for y_t (GNP in billions of current dollars) from 1952 to 1972 are, respectively, 345, 365, 398, 419, 441, 447, 484, 504, 520, 560, 591, 632, 685, 750, 794, 864, 930, 977, 1056, and 1155. Data for x_t (government purchases of goods and services in billions of current dollars) from 1953 to 1972 are 81.6, 74.8, 74.2, 78.6, 86.1, 94.2, 97.0, 99.6, 107.6, 117.1, 122.6, 128.7, 137.0, 156.8, 180.1, 199.6, 210.0, 219.5, 234.3, and 255.0. Let the model be $y_t = ay_{t-1} + cx_t + u_t$, where u_t is independently and identically normal with mean 0 and variance $v = R^{-1}$. Find the least-squares estimates of a and c by using the 20 observations provided. Write down the diffuse prior density of a, c, and R. Find the means of the posterior density function of a and c, using this prior density.

5. Find the posterior density function of a, c, and R in Problem 4. Find the conditional posterior density of a and c, given R, and the posterior density of R.

6. From the posterior density function of R in Problem 5 derive the posterior density function of $v = R^{-1}$.

7. Find the expectation of R^{-1} in Problem 5.

8. Find the covariance matrix of the posterior density function of a and c in Problem 4.

9. Using the data and the model in Problem 4, let the objective of the control problem be to steer GNP in 1973 6 percent higher than its value in 1972 and to steer government purchases in 1973 5 percent higher than its value in 1972, with equal weights assigned to the squared deviations of these variables from their targets. Find the optimal feedback control equation of this one-period problem by using the method of certainty equivalence. What is the optimal setting for government expenditures in 1973?

10. Find the minimum expected welfare loss for Problem 9. Interpret this expected loss. Decompose it into a deterministic part and a stochastic part, using the theory in Chapter 7.

11. Find the optimal feedback control equation for Problem 9, allowing for the uncertainty in the coefficients of the model. Compare this equation with the control equation obtained by the method of certainty equivalence. Are the differences to be expected? Explain. Compare also the optimal setting for government expenditures in 1973 in the two cases and comment on the difference.

12. Find the minimum expected welfare loss for Problem 11. Compare the result with that of the certainty equivalence solution and comment on the difference.

13. Change the one-period problem stated in Problem 9 to a two-period problem, using the same growth rates for the targets of GNP and government purchases, that is, 6 and 5 percent annual rates, respectively, from 1972 on. Find the optimal feedback control equations for 1973 and 1974, using the method of certainty equivalence.

14. Find the minimum expected welfare loss for Problem 13.

15. For Problem 13 find the optimal control equations for 1973 and 1974 by using the method in Sections 10.3 and 10.4. Compare them with the equations in Problem 13.

16. Find the minimum expected welfare loss for Problem 15 and compare the result with that of Problem 14.

17. State a set of necessary and sufficient conditions for the parameters G_t and g_t in the optimal feedback control equations in Sections 10.2 to reach steady-state values. Compare these conditions with the corresponding conditions obtained for the method of certainty equivalence.

18. Using a simple simultaneous-equations macroeconomic model of your own choice, show how the covariance matrix of the reduced-form parameters can be derived from the covariance matrix of the structural parameters.

CHAPTER 11

Control of Unknown Linear Systems with Learning

In Chapters 7 and 8 it was assumed that the parameters of the linear econometric model employed for the purpose of control are known for certain. In Chapter 10 this assumption was relaxed and the uncertainty of the parameters based on partial knowledge in the beginning of the planning horizon is taken into account in determining the policy for the first period. One important lesson to be drawn from the solutions to the multiperiod control problems presented so far is that in deciding what to do in the first period we have to look ahead and consider what we will do in future periods under different contingencies. In applying the method of dynamic programming in Chapters 8 and 10 and in calculating the feedback control policies for future periods as intermediate steps in the solution of the optimal policy in the first period, we in effect anticipate future actions in setting the current policy. If this point is carried to its logical conclusion when the parameters are uncertain, we ought to look ahead and consider the optimal feedback policies based on *future knowledge* of the system as a basis of the current policy. In the solution in Chapter 10 it is assumed that the decision maker considers the future optimal policies based only on *current knowledge* of the system, as the discussion following (14) of that chapter reveals. In this chapter we outline an approximate method of optimal control of linear systems that anticipates future learning about the unknown parameters in the design of the current policy.

We can justifiably question the practical importance of the refinement, to be presented below, of the method in Chapter 10 to take future learning into account. If the economy has been observed for many periods, the standard errors of the coefficients in the model are not too large, and if the planning horizon is short compared with the historical sample period how much difference can the incorporation of learning make in the optimal policy for the first period? Note also that passive learning is always possible; at the end of the first period we can and often will revise our estimates of the model parameters before calculating the policy to be applied to the second period. The material in this chapter is not required for an understanding of Chapter 12. The reader may therefore choose to omit the remainder.

Its importance, however, is twofold. First, from the viewpoint of the theory of optimal control, it deals with an important problem. The problem to be treated is that of a multiperiod decision under uncertainty in which the decision maker is allowed to learn from observations. Second, from the practical point of view, how much difference learning will make can be decided only by actually applying a method presented here and comparing the results with those of the methods in Chapters 8 and 10. This chapter throws some light on both the theoretical and practical questions.

We begin in Section 11.1 with a restatement of the method of dynamic programming, which is necessary for the development of the method of this chapter. Section 11.2 describes the approximate method of optimal control that takes learning into account. This method is compared in Section 11.3 with the methods of Chapters 8 and 10, both in conceptual terms and in terms of computations. It will be seen as a generalization of the latter methods. In Section 11.4 we present two simpler and modified versions of the method in this chapter, simpler to compute but still taking learning partly into account. Sections 11.5 and 11.6 provide two applications of this chapter's method to policy problems concerning the United States economy and compare the results with those obtained by the methods in Chapters 8 and 10.

11.1 A RESTATEMENT OF THE METHOD OF DYNAMIC PROGRAMMING

The method of dynamic programming is restated in a more general setting than in Chapters 8 and 10 in order that the more general solution can be derived. Consider a T-period decision problem involving random vectors $y_1, y_2, \ldots, y_T$ and a welfare cost function $W(y_1, y_2, \ldots, y_T)$. Note that each y_t includes both endogenous variables and control variables x_t. It is assumed that x_t will influence y_t through some stochastic model such as

$$y_t = Ay_{t-1} + Cx_t + u_t, \tag{1}$$

where A and C are matrices of unknown parameters and u_t is a random vector with mean 0 and unknown covariance matrix V. For our statement of the method of dynamic programming, however, it is not necessary to assume that the stochastic model is linear. We assume merely that the distribution of y_t depends on x_t. The problem is to choose $x_1,\ldots,x_T$ sequentially to minimize the expected welfare cost in the beginning of period 1.

Because the control variables for later periods need not be chosen until the outcomes of earlier decisions are available, the problem should be solved by first minimizing with respect to the control variables for the later periods, given the outcomes of earlier decisions, and proceeding successively backward in time until the control variable x_1 for the first period is chosen. The logical structure of the solution is as follows. Given a welfare cost function $W = W(y_1,\ldots,y_T)$, we first eliminate x_T by minimizing the conditional expectation of W, given all the data up to $T-1$; we then eliminate x_{T-1} by minimizing the conditional expectation of the minimum, given all the data up to $T-2$, and so on, until we minimize, with respect to x_1, a conditional expectation, given the data at the end of time 0. Thus we first solve the problem of

$$\min_{x_T} E_{T-1} W, \tag{2}$$

and, having obtained (2), we proceed to the solution of the problem of

$$\min_{x_{T-1}} E_{T-2}\left(\min_{x_T} E_{T-1} W \right), \tag{3}$$

and so on. The entire problem can be written as

$$\min_{x_1} E_0\left(\cdots \left\{ \min_{x_{T-2}} E_{T-3}\left[\min_{x_{T-1}} E_{T-2}\left(\min_{x_T} E_{T-1} W \right)\right] \right\} \cdots \right). \tag{4}$$

We show that if the welfare function W is additive, that is,

$$W(y_1,\ldots,y_T) = \sum_{t=1}^{T} W_t(y_t), \tag{5}$$

the solution of the problem stated in (4) will amount to an application of the principle of optimality to determine the optimal control policy x_t for each period t stated in (16) and (21) in Chapter 10. Using the function (5) and the fact that x_t has no effect on y_s for $s < t$, we can rewrite the problem of (2) as

$$\min_{x_T} E_{T-1} W = \min_{x_T} E_{T-1}\left[W_T(y_T) + \sum_{t=1}^{T-1} W_t(y_t) \right]$$

$$= \min_{x_T} E_{T-1} W_T(y_T) + \sum_{t=1}^{T-1} W_t(y_t). \tag{6}$$

Accordingly, (3) becomes

$$\min_{x_{T-1}} E_{T-2}\left(\min_{x_T} E_{T-1} W \right) = \min_{x_{T-1}} E_{T-2}\left[W_{T-1}(y_{T-1}) + \min_{x_T} E_{T-1} W_T(y_T) \right]$$

$$+ \sum_{t=1}^{T-2} W_t(y_t). \tag{7}$$

The first line on the right-hand side of (7) is precisely the problem of minimizing the expression V_{T-1} in (16) in Chapter 10, where $W_{T-1}(y_{T-1})$ is a quadratic function and $\hat{V}_T$ stands for $\min_{x_T} E_{T-1} W_T(y_T)$. As Problem 1, the reader is asked to show that carrying out the minimization for one more period, that is, with respect to the control variables x_{T-2}, will lead to minimizing the expression V_{T-2} in (21) in Chapter 10.

The multiperiod decision problem under uncertainty has now been restated for a more general welfare loss function W. If W is additive, the minimization with respect to the control variables x_t of each period will lead to the application of the principle of optimality as stated in Chapters 8 and 10. In Section 7.2 we commented critically on the validity of the additive welfare function. We now solve a problem with a nonadditive welfare function as a by-product of solving an optimal control problem with unknown parameters.

11.2 DESCRIPTION OF THE METHOD

Let the model be given by (1). There may be exogenous variables z_t in the system that are not subject to control and have unknown coefficients B, but we have omitted Bz_t on the right-hand side of (1) to simplify the algebra in the following. The reader may wish to incorporate Bz_t in the system and modify the solution accordingly. (See Problems 2 and 3.) The welfare cost for the planning period 1 to T is assumed to be quadratic but not additive; that is

$$W = \tfrac{1}{2} \sum_{t=1}^{T} y_t' K_{t,t} y_t + \sum_{t=1}^{T} \sum_{s<t} y_t' K_{t,s} y_s + \sum_{t=1}^{T} y_t' k_t + d, \tag{8}$$

where $K_{t,s} = K_{s,t}'$, k_t, and d are known constants. In preceding chapters it has been assumed that $K_{t,s} = 0$ for $t \neq s$. The more general form (8) is used in this chapter because, as our solution is developed in (13) and (14), we will minimize the expectation of a quadratic function involving the cross products of y_t and y_s for $t \neq s$ even if the original welfare function has no such cross-product terms.

To carry out the first minimization indicated by (6) with respect to x_T we write

$$E_{T-1}W = E_{T-1}\left[\tfrac{1}{2}y_T' H_{T,T}^T y_T + y_T'\left(\sum_{s=1}^{T-1} H_{T,s}^T y_s + h_T^T \right) \right]$$

$$+ \tfrac{1}{2}\sum_{t=1}^{T-1} y_t' H_{t,t}^T y_t + \sum_{t=1}^{T-1}\sum_{s<t} y_t' H_{t,s}^T y_s + \sum_{t=1}^{T-1} y_t' h_t^T + d_T$$

$$= E_{T-1}W_T + W_{NT}. \tag{9}$$

Here the function W has been decomposed into two parts, W_T and W_{NT}. All the terms involving y_T, which can be influenced by x_T, are included in W_T. The terms in W_{NT} are not affected by x_T. We have let $K_{t,s} = H_{t,s}^T$, $k_t = h_t^T$, and $d = d_T$ to facilitate generalization in a step that will follow (14). To minimize $E_{T-1}W_T$, we substitute $Ay_{T-1} + Cx_T + u_T$ for y_T in W_T and take expectations:

$$E_{T-1}W_T = E_{T-1}\left[\tfrac{1}{2}(Ay_{T-1} + Cx_T + u_T)'H_{T,T}^T(Ay_{T-1} + Cx_T + u_T) \right.$$

$$\left. + (Ay_{T-1} + Cx_T + u_T)'\left(\sum_{s=1}^{T-1} H_{T,s}^T y_s + h_T^T \right) \right]$$

$$= \tfrac{1}{2}y_{T-1}'(E_{T-1}A'H_{T,T}^T A + 2E_{T-1}A'H_{T,T-1}^T)y_{T-1}$$

$$+ \tfrac{1}{2}x_T'(E_{T-1}C'H_{T,T}^T C)x_T$$

$$+ x_T'\left[E_{T-1}(C'H_{T,T}^T A + C'H_{T,T-1}^T)y_{T-1} \right.$$

$$\left. + (E_{T-1}C')\left(\sum_{s=1}^{T-2} H_{T,s}^T y_s + h_T^T \right) \right]$$

$$+ y_{T-1}'(E_{T-1}A')\left(\sum_{s=1}^{T-2} H_{T,s}^T y_s + h_T^T \right) + \tfrac{1}{2}E_{T-1}u_T'H_{T,T}^T u_T. \tag{10}$$

When taking expectations, we have adopted the Bayesian view that the matrices A and C have a joint posterior density function at the end of

$T-1$ and have assumed that u_T is distributed independently of them. The main result required for the evaluation of the expectations involving A and C in (10) has been given in (66) in Chapter 10.

Minimization of (10) by differentiation with respect to x_T yields

$$\hat{x}_T = -(E_{T-1}C'H_{T,T}^T C)^{-1}\left[E_{T-1}(C'H_{T,T}^T A + C'H_{T,T-1}^T)y_{T-1}\right.$$
$$\left. +(E_{T-1}C')\left(\sum_{s=1}^{T-2} H_{T,s}^T y_s + h_T^T\right)\right]; \qquad (11)$$

(11) is a feedback control equation, which determines the optimal policy for period T in terms of observations $y_1, y_2, \ldots, y_{T-1}$. These observations affect the posterior density of A and C and thus the expectations involving them, given by (66) in Chapter 10. Substituting the solution (11) for x_T in (10) gives

$$\min_{x_T} E_{T-1}W_T = \tfrac{1}{2}y_{T-1}'(E_{T-1}A'H_{T,T}^T A + 2E_{T-1}A'H_{T,T-1}^T)y_{T-1}$$

$$-\tfrac{1}{2}\left[y_{T-1}'E_{T-1}(A'H_{T,T}^T C + H_{T,T-1}^{T'}C) + \left(\sum_{s=1}^{T-2} y_s'H_{T,s}^{T'} + h_T^{T'}\right)(E_{T-1}C)\right]$$

$$\times[E_{T-1}C'H_{T,T}^T C]^{-1}\left[E_{T-1}(C'H_{T,T}^T A + C'H_{T,T-1}^T)y_{T-1} + (E_{T-1}C')\right.$$

$$\times\left.\left(\sum_{s=1}^{T-2} H_{T,s}^T y_s + h_T^T\right)\right] + y_{T-1}'(E_{T-1}A')\left(\sum_{s=1}^{T-2} H_{T,s}^T y_s + h_T^T\right) + \tfrac{1}{2}E_{T-1}u_T'H_{T,T}^T u_T$$

$$= \tfrac{1}{2}y_{T-1}'\left[E_{T-1}A'H_{T,T}^T A + 2E_{T-1}A'H_{T,T-1}^T - E_{T-1}(A'H_{T,T}^T C + H_{T,T-1}^{T'}C)\right.$$

$$\times(E_{T-1}C'H_{T,T}^T C)^{-1}E_{T-1}(C'H_{T,T}^T A + C'H_{T,T-1}^T)\Big]y_{T-1}$$

$$+y_{T-1}'\left[(E_{T-1}A') - E_{T-1}(A'H_{T,T}^T C + H_{T,T-1}^{T'}C)\right.$$

$$\times(E_{T-1}C'H_{T,T}^T C)^{-1}(E_{T-1}C')\Big]\left(\sum_{s=1}^{T-2} H_{T,s}^T y_s + h_T^T\right)$$

$$-\tfrac{1}{2}\left(\sum_{s=1}^{T-2} y_s'H_{T,s}^{T'} + h_T^{T'}\right)(E_{T-1}C)(E_{T-1}C'H_{T,T}^T C)^{-1}$$

$$\times(E_{T-1}C')\left(\sum_{s=1}^{T-2} H_{T,s}^T y_s + h_T^T\right) + \tfrac{1}{2}E_{T-1}u_T'H_{T,T}^T u_T. \qquad (12)$$

The essense of the method in this chapter is to approximate (12) by a quadratic function in $y_{T-1}, y_{T-2}, \ldots, y_1$. This quadratic function can then be combined with W_{NT} in (9) to yield $\min_{x_T} E_{T-1} W$, which will be quadratic in $y_{T-1}, y_{T-2}, \ldots, y_1$. We then minimize $E_{T-2}(\min_{x_T} E_{T-1} W)$ with respect to x_{T-1}, following the steps from (9) on, with $T-1$ replacing T in all the derivations. To obtain a quadratic approximation of (12) we first choose a tentative path $y_1^0, y_2^0, \ldots, y_{T-1}^0$ that should be reasonably close to the future values of these variables and employ a second-order Taylor expansion of (12), which is a function of $y_1, y_2, \ldots, y_{T-1}$ by virtue of (66) in Chapter 10. The choice of a tentative path is discussed in Section 11.3. Given a tentative path, we can approximate (12) by a second-order Taylor expansion around this path as we did for a nonlinear structural equation in (38) in Chapter 6. This would require evaluating the first and second derivatives of (12) with respect to $y_1, y_2, \ldots, y_{T-1}$. Because (12) is a highly nonlinear function and its derivatives are difficult to express analytically, we evaluate the derivatives numerically by computating the rates of change of the function with respect to small changes in the variables $y_1, y_2, \ldots, y_{T-1}$. Having obtained the first and second derivatives of (12) numerically, we can approximate it by the expression

$$\min_{x_T} E_{T-1} W_T \cong \tfrac{1}{2} \sum_{t=1}^{T-1} y_t' Q_{t,t}^T y_t + \sum_{t=1}^{T-1} \sum_{s<t} y_t' Q_{t,s}^T y_s + \sum_{t=1}^{T-1} y_t' q_t^T + r_T. \quad (13)$$

By combining the right-hand side of (13) with W_{NT} in (9) we get

$$\min_{x_T} E_{T-1} W = \tfrac{1}{2} \sum_{t=1}^{T-1} y_t' H_{t,t}^{T-1} y_t + \sum_{t=1}^{T-1} \sum_{s<t} y_t' H_{t,s}^{T-1} y_s + \sum_{t=1}^{T-1} y_t' h_t^{T-1} + d_{T-1},$$

$$(14)$$

where

$$H_{i,j}^{T-1} = H_{i,j}^T + Q_{i,j}^T, \qquad (i = 1, \ldots, T-1; j \leqslant i),$$

$$h_i^{T-1} = h_i^T + q_i^T,$$

$$d_{T-1} = d_T + r_T. \quad (15)$$

Once (14) is obtained it can be treated in the same way as (9), with $T-1$ replacing T. Thus (14) will be decomposed into two parts, W_{T-1} and $W_{N(T-1)}$, the first involving y_{T-1}, whereas the second does not; $E_{T-2} W_{T-1}$ will then be minimized with respect to x_{T-1} and will yield results analogous to (10) and (12). The analog of (12), namely $\min_{x_{T-1}} E_{T-2} W_{T-1}$, will be approximated by a quadratic function with coefficients $Q_{t,s}^{T-1}$, q_t^{T-1}, and r_{T-1}. This quadratic function will be combined with $W_{N(T-1)}$ to yield $\min_{x_{T-1}} E_{T-2}(\min_{x_t} E_{T-1} W)$ as in (14). The coefficients of the last quadratic function are obtained by recursion formulas (15) with $T-1$ replacing T. Now we are back to minimizing the expectation of a quadratic function in the form of (9). The process continues until we minimize, with respect to x_1, the conditional expectation E_0 of a quadratic function in y_1, namely, $\frac{1}{2} y_1' H_{1,1}^1 y_1 + y_1' h_1^1$.

It may be helpful to describe the method in this chapter diagrammatically for $T=4$. For simplicity the linear terms are omitted in all the quadratic functions used in the following presentation. To begin with, let $K_{t,s}$ in the welfare function be written as $H_{t,s}^4$. The welfare function is decomposed into two parts, W_{N4} and W_4, the latter involving y_4, as given by the left-hand matrix in Figure 11.1; $E_3 W_4$ is then minimized with respect to x_4. The minimum $\min_{x_4} E_3 W_4$ is approximated by a quadratic function with coefficients $Q_{t,s}^4 (t,s=1,2,3)$. These coefficients are added to the coefficients $H_{t,s}^4$ in the original function to form a new matrix with

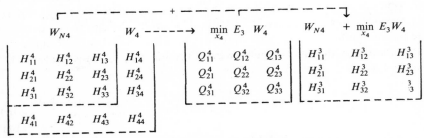

Figure 11.1 Diagramatic presentation of the optimal control method.

coefficients $H_{t,s}^3$, following (15). The resulting quadratic function is then treated in the same way. (See Problem 4.)

11.3 COMPARISON WITH TWO OTHER APPROXIMATE METHODS

In this section the method in Section 11.2 is compared with the two simpler, and less nearly optimal, methods in Chapters 8 and 10. The comparison brings out the logical structure of each method of solution and provides interesting interpretations of the various calculations required in obtaining the solution. Because the methods in Chapters 8 and 10, designated methods I and II, respectively, for convenience, yield linear feedback control equations in the form of $x_t = G_t y_{t-1} + g_t$, either method can be used to provide a tentative path $y_1^0, y_2^0, \ldots, y_{T-1}^0$ for the second-order Taylor expansion in expression (12) required in method III in Section 11.2. A tentative path can be generated by applying either set of feedback control equations to the linear model, given the initial condition y_0, and setting the random disturbances u_t equal to 0. After solving for the optimal control equations by method III in the form of (11), with all expectations in (11) evaluated at the tentative path, we can generate a new path of $y_1, \ldots, y_{T-1}$. The latter can be used as a second tentative path for use in calculating a new solution by method III. We may choose to iterate. The limited experience accumulated so far is that, using method I or II to generate the tentative path for the first iteration, we obtain convergence of the optimal $\hat{x}_1$ for the first period in four or five iterations.

Before making a comparison of the three methods, it may be useful to review the first two as special cases of method III. Recall that method II in Chapter 10 was derived by the procedure given in Section 11.2, *except* that all conditional expectations E_t are treated as E_0. Thus the possibility of learning (revising the posterior density of A and C) is ignored in deriving the optimal policy for x_1. When this approximation is taken, and under the assumption that $K_{t,s}$ in the welfare function (8) is 0 for $t \neq s$, (11) and (12) will be reduced, respectively, to (for $1 \leqslant t \leqslant T$)

$$\hat{x}_t = -(E_0 C' H_t C)^{-1} [(E_0 C' H_t A) y_{t-1} + (E_0 C') h_t], \qquad (16)$$

where $H_t \equiv H_{t,t}^t$ and $h_t \equiv h_t^t$ and

$$\min_{x_t} E_{t-1} W_t = \tfrac{1}{2} y_{t-1}' \Big[E_0 A' H_t A - (E_0 A' H_t C)(E_0 C' H_t C)^{-1} (E_0 C' H_t A) \Big] y_{t-1}$$

$$+ y_{t-1}' \Big[E_0 A' - (E_0 A' H_t C)(E_0 C' H_t C)^{-1} (E_0 C') \Big] h_t$$

$$- \tfrac{1}{2} h_t' (E_0 C)(E_0 C' H_t C)^{-1} (E_0 C') h_t + \tfrac{1}{2} E_0 u_t' H_t u_t. \qquad (17)$$

Equation 17 is truly a quadratic function of y_{t-1} and need not be approximated. Thus the coefficients of the quadratic function (13) will be reduced to

$$Q_{t-1} \equiv Q'_{t-1,t-1} = E_0 A' H_t A - (E_0 A' H_t C)(E_0 C' H_t C)^{-1}(E_0 C' H_t A)$$

$$q_{t-1} \equiv q'_{t-1} = \left[E_0 A' - (E_0 A' H_t C)(E_0 C' H_t C)^{-1}(E_0 C') \right] h_t$$

$$(18)$$

$$r_t = -\tfrac{1}{2} h'_t (E_0 C)(E_0 C' H_t C)^{-1}(E_0 C') h_t + \tfrac{1}{2} E_0 u'_t H_t u_t$$

$$Q'_{i,j} = 0; \quad q'_i = 0, \qquad (i < t-1; j \neq i).$$

Method I is a further simplification of the second method, given by (15), (16), (17), and (18). If the unknown parameters A and C are reduced to $\overline{A}$ and $\overline{C}$, their point estimates (or the means of their posterior density) as of time 0, we simply replace the expectations of functions of A and C by the same functions of $\overline{A}$ and $\overline{C}$. Thus, following (16) to (18), the optimal control for each period is given by the feedback equation

$$\hat{x}_t = -\left(\overline{C}' H_t \overline{C} \right)^{-1} \left[\left(\overline{C}' H_t \overline{A} \right) y_{t-1} + \overline{C}' h_t \right] \tag{19}$$

and the minimum expected welfare cost from period t on is given by

$$\min_{x_t} E_{t-1} W_t = \tfrac{1}{2} y'_{t-1} Q_{t-1} y_{t-1} + y'_{t-1} q_{t-1} + r_t, \tag{20}$$

where

$$Q_{t-1} \equiv Q'_{t-1,t-1} = \overline{A}' H_t \overline{A} - \overline{A}' H_t \overline{C} \left(\overline{C}' H_t \overline{C} \right)^{-1} \left(\overline{C}' H_t \overline{A} \right),$$

$$q_{t-1} \equiv q'_{t-1} = \overline{A}' - \left(\overline{A}' H_t \overline{C} \right) \left(\overline{C}' H_t \overline{C} \right)^{-1} \overline{C}' h_t, \tag{21}$$

$$r_t = -\tfrac{1}{2} h'_t \overline{C} \left(\overline{C}' H_t \overline{C} \right)^{-1} \overline{C}' h_t + \tfrac{1}{2} E_0 u'_t H_t u_t.$$

By (15) H_t, h_t, and d_t are determined by the difference equations

$$H_t \equiv H'_{t,t} = H^{t+1}_{t,t} + Q^{t+1}_{t,t} = K_t + Q_t,$$

$$h_t \equiv h'_{t,t} = h^{t+1}_{t,t} + q^{t+1}_{t,t} = k_t + q_t, \tag{22}$$

$$d_t = d_{t+1} + r_{t+1}.$$

In (22) we have used

$$H_{t,t}^{t+1} = H_{t,t}^{t+2} + Q_{t,t}^{t+2} = H_{t,t}^{t+2} = \cdots = H_{t,t}^{T} = K_{t,t} \equiv K_{t},$$

$$h_{t}^{t+1} = h_{t}^{t+2} + q_{t}^{t+2} = h_{t}^{t+2} = \cdots = h_{t}^{T} = k_{t},$$

(23)

because by (18) $Q_{ij}^{t} = 0$ and $q_{i}^{t} = 0$ for $i < t - 1$.

By method I the optimal policy $\hat{x}_{t}$ for each period t minimizes the expected value of a quadratic function in y_{t}, with H_{t} and h_{t} as coefficients. According to (22), this quadratic function is a sum of two parts:

$$(\tfrac{1}{2}y_{t}'K_{t}y_{t} + y_{t}'k_{t}) + (\tfrac{1}{2}y_{t}'Q_{t}y_{t} + y_{t}'q_{t}).$$

(24)

The first part is the contribution of y_{t} to welfare loss as y_{t} appears directly in the welfare function. The second part is the minimum future welfare cost due to $y_{t+1}, y_{t+2}, \ldots, y_{T}$, assuming that the future $x_{t+1}, \ldots, x_{T}$ is optimally chosen; it is also a quadratic function of the initial condition y_{t} on which future decisions will have to be built. This decomposition shows the relative importance of setting y_{t} for direct contribution to current welfare and for foundation building for the future. The optimal x_{t} is chosen to minimize the sum of the expectations of these two quadratic functions in y_{t}, one for current benefits and the other for future benefits.

The same decomposition and interpretation apply to the second method. The only difference is that in computing the quadratic function for the minimum expected future cost, different coefficients Q_{t} and q_{t} will be used. They are given by (18) rather than (21). When uncertainty in A and C as of time 0 is allowed for, the minimum expected future cost will have to be computed differently from the case in which A and C are treated as given. The difference between these two functions has been explained in detail in Section 10.6. For the present discussion, the comparison of Q_{t} and q_{t} as between (18) and (21) shows how the uncertainty affects the weights given to the preparation for future optimization in the determination of the current policy.

The possibility of learning, as treated in Section 11.2, does not invalidate the present-future decomposition of the quadratic function of y_{t} whose expected value is to be minimized by x_{t}. It does, however, make the future component more complicated. When learning is absent, the only concern for y_{t} from the point of view of the future is that it affects the minimum expected future cost from $t+1$ on; it does not affect minimum expected cost from $t+2$ on because, the model (1) being first-order (an assumption which we are making for method III of this chapter), the latter is a function of y_{t+1} alone and not of y_{t}. When learning is present, y_{t} affects the optimal expected cumulative costs from all future periods on, for (15) implies

$$H_t \equiv H_{t,t}^t = H_{t,t}^{t+1} + Q_{t,t}^{t+1} = K_{t,t} + Q_{t,t}^{t+1} + (Q_{t,t}^{t+2} + \cdots + Q_{t,t}^T) \qquad (25)$$

and similarly for h_t. The terms in parentheses show, respectively, the effects of y_t on the optimal expected future costs from $t+2$ on, from $t+3$ on, and so on. These terms are treated as 0 by the first two methods. They are actually nonzero because, in spite of the first-order system, y_t affects not only y_{t+1} but the planner's conceptions of all future ys through its influence on his posterior densities of A and C in all future periods. Note that $Q_{t,t}^{t+1} \equiv Q_t$ in (25) also differs from the corresponding estimates given by the first two methods as it incorporates the effect of y_t on the posterior density of A and C in $t+1$, whereas the others do not.

Thus by comparing the difference $H_t - K_{t,t}$ in method II with the corresponding quantity for method III, we can measure the effect of learning on the weight given to the future-component of the quadratic function of y_t whose expectation is to be minimized. Presumably, when learning is allowed for, the future-component will receive more weight, as the quantity $H_t - K_{t,t}$ will measure. Therefore, besides providing a solution, the method of this chapter gives an explicit measure of the effect of learning on the quadratic function to be minimized in each period. The comparison of this function for the three methods will provide information about the impact of learning on the optimal control solutions.

11.4 TWO SIMPLIFIED VERSIONS OF METHOD III

In view of the possibly large number of numerical second derivatives that have to be evaluated for the quadratic approximation (13), two simplified versions of method III are suggested. These modifications will save computation, especially when the time horizon T is large. How far they are from being optimal, however, is a difficult question to study analytically. Perhaps numerical results from experiments obtained by applying these methods will throw some light on this question, but we have no such results to report.

The first modification is to omit all the matrices $Q_{i,j}^t$ for $i \neq j$ in the quadratic approximation (13). In the diagrammatic representation of the method given in Figure 11.1 for $T=4$, only the submatrices $H_{i,i}^4$, $Q_{i,i}^4$, $H_{i,i}^3, \ldots, H_{1,1}^1$ along the diagonal will be evaluated. Otherwise, the method is identical with method III as described in Section 11.2. To appreciate the possible loss of accuracy arising from this modification, consider the minimization of $E_{T-1}W_T$ with respect to x_T given by (10), (11), (12), and (13). In (10) all $H_{T,j}^T$ for $j \neq T$ are treated as a null matrix. This affects the optimal feedback control equation (11) for $\hat{x}_T$ and the resulting $\min_{x_T} E_{T-1}W_T$ given by (12). The quadratic approximation of (12) given by (13) is also affected. In (13) not only $Q_{t,s}^T$ for $t \neq s$ are assumed to be 0 but the submatrix $Q_{T-1,T-1}^T$ is also different because of this modification; $Q_{T-1,T-1}^T$ consists of the second derivatives of (12) with respect to y_{T-1}. These derivatives will change if $H_{T,T-1}^T$ in (12) is assumed to be 0. This modified version of method III does take learning partly into account because in the computation of $E_{T-1}W_T$ by the modified version of (10) future observations will be utilized to evaluate $E_{T-1}A'H_{T,T}^T A$, $E_{T-1}C'H_{T,T}^T C$, and $E_{T-1}C'H_{T,T}^T A$.

The second simplification is to apply method III for only M periods ahead, where M is smaller than the time horizon T. In other words, learning from future observations up to period $M-1$ is anticipated in the determination of the policy for the first period, but possible learning from observations to be made after period M will be ignored. This simplification can be implemented by performing the dynamic programming algorithm backward in time from period T to period M, using method II. The control variables $x_{M+1}, x_{M+2}, \ldots, x_T$ can thus be eliminated by the optimization process. From the vantage point at the end of period M the minimum expected welfare loss from period $M+1$ on is a quadratic function of y_M. After obtaining $H_{i,j}^M$ and $h_{i,j}^M$ by method II we then apply method III to an M-period control problem.

Both modified versions of method III incorporate learning to some extent in the determination of the policy for period 1. For measuring the effect of learning on the solution by these methods the decomposition analysis described in Section 11.3 is applicable; that is, for each period, the expectation of a quadratic function of y_t is minimized. As before, $K_{t,t}$ and k_t are the coefficients of the current component of this quadratic function; $H_{t,t}^t - K_{t,t}$ and $h_t^t - k_t$ are the coefficients of the future component affected by learning.

11.5 CONTROL SOLUTIONS FOR A ONE-EQUATION MODEL

In order to illustrate the use of the method in Section 11.2 and to measure the effect of learning on the decomposition between immediate and future welfare and the optimal control policy, we consider first a simple model with a scalar dependent variable y_t^*. The explanatory variables are y_{t-1}^* and a scalar control variable x_t. To include both y_t^* and x_t in a vector y_t of dependent variables we write

$$\begin{bmatrix} y_t^* \\ x_t \end{bmatrix} = \begin{bmatrix} a & 0 \\ 0 & 0 \end{bmatrix} \begin{bmatrix} y_{t-1}^* \\ x_{t-1} \end{bmatrix} + \begin{bmatrix} c \\ 1 \end{bmatrix} x_t + \begin{bmatrix} u_t^* \\ 0 \end{bmatrix} \qquad (26)$$

or, in the notation of (1),

$$y_t = Ay_{t-1} + Cx_t + u_t. \qquad (27)$$

Let n observations on the system (26) be available by the end of time 0; assume that y_t and its explanatory variables have been observed for $t = -n+1, \ldots, 0$. Let the prior density of the parameters a and c be diffuse at the beginning of period $-n+1$. Using the results in Chapter 10, we can evaluate the expectations required in the crucial function (12) as follows. Let $\hat{a}_{T-1}$ and $\hat{c}_{T-1}$ denote, respectively, the least squares estimates of a and c, using data from period $-n+1$ to period $T-1$, and let s_{T-1} denote the sum of squared residuals of y_t from the least-squares fitted regression and from $-n+1$ to $T-1$. Let h_{ij} be the i-j element of the matrix H, omitting the superscript and subscripts of H in (12). The required expectations in (12) are

$$E_{T-1}a = \hat{a}_{T-1}; \qquad E_{T-1}c = \hat{c}_{T-1}, \qquad (28)$$

and

$$E_{T-1}A'HA = \begin{bmatrix} h_{11}E_{T-1}a^2 & 0 \\ 0 & 0 \end{bmatrix};$$

$$(29)$$

$$E_{T-1}C'HA = [h_{11}E_{T-1}ac + h_{12}\hat{a}_{T-1} \quad 0];$$

$$E_{T-1}C'HC = h_{11}E_{T-1}c^2 + 2h_{12}\hat{c}_{T-1} + h_{22},$$

where

$$\begin{bmatrix} E_{T-1}a^2 & E_{T-1}ac \\ E_{T-1}ac & E_{T-1}c^2 \end{bmatrix} = \begin{bmatrix} \hat{a}_{T-1}^2 & \hat{a}_{T-1}\hat{c}_{T-1} \\ \hat{a}_{T-1}\hat{c}_{T-1} & \hat{c}_{T-1}^2 \end{bmatrix}$$

$$+ \frac{s_{T-1}}{(n+T-1)-4} \begin{bmatrix} \sum\limits_{-n+1}^{T-1} y_{i-1}^{*2} & \sum\limits_{-n+1}^{T-1} y_{i-1}^* x_i \\ \sum\limits_{-n+1}^{T-1} y_{i-1}^* x_i & \sum\limits_{-n+1}^{T-1} x_i^2 \end{bmatrix}^{-1} \qquad (30)$$

The covariance matrix of a and c given on the right-hand side of (30) is based on (62) in Chapter 10. Expressions 28 and 29 are substituted into (12) for the computation of numerical derivatives in the quadratic approximation (13). Note that the last term in (12), $E_{T-1} u_T' H_{T,T}^T u_T = \mathrm{tr}\, H_{T,T}^T E_{T-1} u_T u_T'$, though unknown, is not to be influenced by the choice of x_T and can therefore be regarded as a constant for the purpose of deriving the optimal policy. In the computations below the sample covariance matrix of the regression residuals that use the n available observations represents $E_{t-1} u_t u_t'$ $(t = 1, \ldots, T-1)$ in each of the three methods. Although this calculation is only approximate, it will not affect the comparisons between the methods.

In the following example y_t^* of (26) is represented by annual gross national product in billions of current dollars and x_t is represented by annual government purchases of goods and services in billions of current dollars. Annual observations of these variables from 1953 to 1972 constitute the sample of 20 observations available before planning begins. The equation that explains GNP by lagged GNP and government expenditures G can be interpreted as a reduced-form equation from a structure consisting of an identity $GNP = C + I + G$, a consumption function explaining C by GNP and GNP_{-1}, and an investment function explaining I by GNP and GNP_{-1}, although we may not wish to take this structure too seriously. Using the 20 annual observations given in Problem 4 of Chapter 10, we have the regression

$$\hat{y}_t^* = \underset{(.072)}{.890\, y_{t-1}^*} + \underset{(.311)}{.779\, x_t} \qquad \begin{array}{l} R^2 = .998 \\ s^2 = 159.1 \\ DW = 2.109 \end{array} \qquad (31)$$

where the numbers in parentheses are standard errors and DW is the Durbin-Watson statistic.

Assume that at the beginning of period 1 (1973) we used the sample data and the model in (31) to steer GNP and government expenditures toward their target paths by applying one of the three optimal control policies. The target paths specify a 6 percent annual growth rate for GNP from its 1972 figure, and a 5 percent annual growth rate for government expenditures from its 1972 figure. The $K_{t,t}$ matrix is assumed to be a 2×2 identity matrix for all t; equal cost is assigned to the squared deviation of each of the two variables from its target. The planning horizon T is assumed to be 10 years. The regression (31) shows that the coefficient of the control variable x_t is significant at not much better than 5 percent. There seems to be about the right amount of uncertainty in this model to make the example interesting. If the standard errors of the coefficients were much smaller, we might not be able to observe the effects of uncertainty and learning. If they were much larger, the model would probably not be taken seriously for planning purpose. The solution of method II is used to generate a tentative path for use in the quadratic approximation of method III. Table 11.1 presents some results from the application of the three methods of optimal control to this example.

Following the present-future decomposition of the quadratic function of y_t whose expectation is to be minimized at each stage, as discussed in Section 11.3, we observe that for method I (certainty equivalence) the $H'_{t,\,t}$ matrix is reaching a steady-state value as t decreases from 10 to 1. For period 10, $H^{10}_{10,\,10}$ is simply $K^{10}_{10,\,10}$, the 2×2 identity matrix, because there will be no future to speak of after period 10. Some more weight is then added to the square of the y_t^* variable as t decreases, or as the future becomes longer in duration, until the future component becomes .654. The conditions under which the matrix difference equations (21) and (22) for $H_t = H'_{t,\,t}$ will have a steady-state solution were discussed in Section 7.8 and are not repeated here. The present example illustrates a steady-state solution for $H'_{t,\,t}$. There is no weight given to the square of x_t in the future component of the quadratic function because all future ys from $t+1$ on will not be dependent on x_t (y_{t+1}^* is a function of y_t^* and x_{t+1} but not of x_t).

It was pointed out in Section 10.6 that in control under uncertainty without learning the matrix $Q_{t-1} = Q^t_{t-1,\,t-1}$ on (18), namely the matrix of the quadratic function which gives the minimum expected future cost (17) from period t on, can be interpreted as the covariance matrix of residuals in the weighted regression of A on C. Similarly, in the certainty equivalence solution the matrix Q_{t-1} on (21) can be interpreted as the covariance matrix of residuals in the weighted regression of $\bar{A}$ on $\bar{C}$. Insofar as A and C can be viewed as $\bar{A}$ and $\bar{C}$ plus random errors, the former covariance matrix may be expected, under not unusual circumstances, to be not smaller than the latter matrix, in the sense that their difference is a positive semidefinite matrix. The present example illustrates this point; the matrix $Q_{t-1} = H^{t-1}_{t-1,\,t-1} - K_{t-1,\,t-1} = H_{t-1} - I$ has a larger leading term for method II than for method I. To put it in another way, the introduction of uncertainty increases the weight for the future component of the quadratic function to be minimized. When learning is introduced, we find the weight given to the future component further increased, as the $H'_{t,\,t}$ matrices for Method III given in Table 11.1 show. We should care more about the future if we are allowed to learn.

Turning now to the feedback control equations $x_t = G_t y_{t-1} + g_t$ (where y_{t-1} consists of y_{t-1}^* and x_{t-1}), we notice in Table 11.1 that the coefficient of y_{t-1}^* (or lagged GNP) is smaller in absolute value for method II than for

Table 11.1 Comparing Three Control Methods for a One-Equation Model

I. Certainty Equivalence
II. Unknown Parameters without Learning
III. Unknown Parameters with Learning

	I	II	III
$H_{10,10}^{10}$	$\begin{bmatrix} 1 & 0 \\ 0 & 1 \end{bmatrix}$	$\begin{bmatrix} 1 & 0 \\ 0 & 1 \end{bmatrix}$	$\begin{bmatrix} 1 & 0 \\ 0 & 1 \end{bmatrix}$
G_{10}	$[-.4317 \quad 0\ ]$	$[-.3898 \quad 0\ ]$	$[-.4153 \quad 0\ ]$
g_{10}	1261	1181	1230
$\hat{x}_{10}$	418.1	420.2	418.5
$H_{9,9}^{9}$	$\begin{bmatrix} 1.4933 & 0 \\ 0 & 1 \end{bmatrix}$	$\begin{bmatrix} 1.5381 & 0 \\ 0 & 1 \end{bmatrix}$	$\begin{bmatrix} 1.7162 & -.0707 \\ .0707 & .9935 \end{bmatrix}$
G_9	$[\ -.5434 \quad 0\ ]$	$[\ -.4896 \quad 0\ ]$	$[\ -.5549 \quad\quad 0\ ]$
g_9	1400	1302	1421
$\hat{x}_9$	398.0	399.4	398.3
$H_{5,5}^{5}$	$\begin{bmatrix} 1.654 & 0 \\ 0 & 1 \end{bmatrix}$	$\begin{bmatrix} 1.8043 & 0 \\ 0 & 1 \end{bmatrix}$	$\begin{bmatrix} 2.7115 & -.4912 \\ -.4912 & 1.2091 \end{bmatrix}$
G_5	$[-.5726 \quad 0\ ]$	$[-.5266 \quad 0\ ]$	$[\quad .6214 \quad\quad 0\ ]$
g_5	1155	1087	1226
$\hat{x}_5$	318.2	317.7	318.2
$H_{1,1}^{1}$	$\begin{bmatrix} 1.654 & 0 \\ 0 & 1 \end{bmatrix}$	$\begin{bmatrix} 1.8063 & 0 \\ 0 & 0 \end{bmatrix}$	$\begin{bmatrix} 4.7060 & -1.9779 \\ 1.9779 & 2.4456 \end{bmatrix}$
G_1	$[-.5727 \quad 0\ ]$	$[-.5269 \quad 0\ ]$	$[\ -.5074 \quad\quad 0\ ]$
g_1	916	862	841
$\hat{x}_1$	254.3	253.7	254.5

method I. This shows that when uncertainty exists we tend to respond less to changing circumstances. A more thorough discussion of this point was presented in Section 10.6, in terms of the size of the coefficients G_t in the weighted regression of A on C, compared with the coefficients in the regression of $\overline{A}$ on $\overline{C}$ (variables without errors). The coefficient of y_{t-1}^* by method III may be, but is not necessarily, larger in absolute value than by method II, suggesting that if learning is allowed it may pay to pursue a more active policy. An active policy is also indicated by the intercept g_t in the control equation. The role of g_t can be seen by considering the simpler model $y_t^* = cx_t + u_t$. The optimal setting (the intercept) for the one-period problem of minimizing $E_0(y_1^* - z)^2$, z being the target, is $\hat{x}_1 = (E_0 c)z/[\text{var } c + (E_0 c)^2]$. Here more uncertainty measured by a larger var c will tend to reduce the intercept. Insofar as learning and uncertainty may have opposite effects on G_t, the relative magnitudes of these coefficients, as between methods III and I, are indeterminate. It is important to observe that in spite of the noticeable differences in the reaction coefficients the numerical values of the optimal $\hat{x}_t$ by the three methods are remarkably similar. These values are obtained for comparison purposes by applying the different optimal conttrol equations to the same set of values of y_{t-1}^* used in the tentative path for method III. Presumably a smaller (negative) coefficient in the feedback control equation is partly compensated by a larger (positive) intercept g_t, the latter playing the role of steering the variables to targets after the feedback effect of the former coefficient has been allowed for.

For this example we used the solution by method III to provide a tentative path and then iterated. The solution for $\hat{x}_1$ converges to 254.43 (being accurate for all these digits) in three iterations. We also applied the first simplified version of method III described in Section 11.4 and found the results to be similar, as might be expected.

11.6 CONTROL SOLUTIONS FOR A TWO-EQUATION MODEL

The example in this section is based on Abel (1974). The model consists of two stochastic equations that explain consumption expenditures C_t and private investment expenditures I_t, the control variables being government expenditures E_t and money supply M_t; all variables are measured in constant 1958 dollars. To treat M_t in constant dollars as a control variable would require the assumption that for the 40 quarterly observations from 1954-I to 1963-IV used in estimating the model changes in the price level did not respond rapidly enough to changes in the nominal money stock to cancel the effect of the latter. Assuming that consumption depends on

C_{t-1}, I_t, and E_t and investment depends on C_t, C_{t-1}, I_{t-1}, and M_t, we obtain two reduced-form equations with C_{t-1}, I_{t-1}, E_t, and M_t as predetermined variables. The investment function is similar to the one in the model described in Chapter 5, except that the quasi-difference of C_t replaces the quasi-difference of income Y_t and money supply replaces the rate of interest. The estimated reduced-form equations are

$$C_t = 0.9266 C_{t-1} - 0.0203 I_{t-1} + 0.3190 E_t + 0.4206 M_t - 63.2386,$$
$$\quad (0.0534) \quad\quad (0.0916) \quad\quad (0.1389) \quad\quad (0.1863) \quad\quad (25.7719)$$

$$R^2 = 0.9958,$$
$$DW = 1.7084.$$

$$(32)$$

$$I_t = 0.1527 C_{t-1} + 0.3806 I_{t-1} - 0.0735 E_t + 1.538 M_t - 210.8994,$$
$$\quad (0.0781) \quad\quad (0.1339) \quad\quad (0.2031) \quad\quad (0.2724) \quad\quad (37.6899)$$

$$R^2 = 0.8749,$$
$$DW = 1.7582.$$

$$(33)$$

The purpose of Abel's study is to compare the relative effectiveness of the two instruments E_t and M_t along the lines suggested in Section 9.6. In the present example, given a welfare function, three expected welfare losses are computed, respectively, by using the two instruments optimally, by using only the instrument E_t optimally with a passive M_t which is allowed to change at a constant percentage rate, and by using only the instrument M_t optimally with a passive E_t which is allowed to change at a constant percentage rate. Note that the constant rate of change for the passive instrument is determined to minimize expected welfare loss while allowing the active instrument to perform its best according to an optimal feedback control rule. The quadratic welfare cost function is specified by target growth rates of 1.25 percent per quarter for the two target variables C_t and I_t and by a diagonal K_t matrix with 1 corresponding to each of the two variables and 0 elsewhere. The planning horizon T is six quarters. All three control methods have been applied to this problem. The results relevant to the computation of the policy for period 1 are given in Table 11.2.

Consider first the present-future decomposition of the matrix $H_{1,1}^1$ of the quadratic function whose expectation is to be minimized with respect to the policy x_1. The matrix representing the future component is the difference between $H_{1,1}^1$, given in Table 11.2, and the 4×4 matrix K_1 which consists of all 0s except for unity as the two leading diagonal elements. Observe that in method I $H_{1,1}^1 = K_1$ when both instruments E_t and M_t are used actively. (See Problem 16.) For each of the three sets of active instruments $\{E_t, M_t\}$, $\{E_t\}$, and $\{M_t\}$ employed the diagonal elements of the matrix $H_{1,1}^1 - K_1$ increase from method I to method II and from method II to method III. The matrix $H_{1,1}^1 - K_1$ tends to be larger for method II than for method I in that the difference between the matrices tends to be positive semidefinite. The reason for this was explained in Section 10.6. Insofar as learning may make the future more important, the matrix $H_{1,1}^1 - K_1$ for method III will tend to be larger than for method II.

As far as the feedback control equations are concerned, the coefficients G_1 tend to be smaller in absolute value for method II than for method I. Uncertainty tends to make the optimal policy less responsive to changing conditions. As we have observed, it is more difficult to ascertain whether the introduction of learning will make the policy by method III more responsive than by method II; for example, with $\{E_t, M_t\}$ as the active instrument set, the leading coefficient -1.7174 of G_1 by method III is larger in absolute value than the corresponding coefficient -1.6856 for method II, but the $2-2$ element -0.1991 in the G_1 matrix for method III is not larger in absolute value than -0.1997 for method II. With $\{E_t\}$ as the active instrument set, the leading coefficient -1.6065 of G_1 for method III is smaller in absolute value than for method II.

The major conclusions concerning economic policy to be derived from Table 11.2 are that using only one of the two possible instruments E_t and M_t actively the minimum expected welfare loss $\hat{W}_1$ from period 1 on is substantially larger than when both instruments are allowed to be active and that using government expenditures E_t as the sole instrument is slightly superior to using money supply M_t as the sole instrument. These two conclusions are valid no matter which method, I, II, or III, is applied to study the problem. By method I the expected loss is 35.5 when both instruments are used; it increases to 44.2 with E_t as the sole instrument and to 48.1 with M_t as the sole instrument. By method II the expected loss is 51.3 for both instruments, 63.4 for E_t alone, and 64.7 for M_t alone. Note that when uncertainty is accounted for not only the expected losses are greater but the superiority of E_t over M_t becomes more obscure. By method III the expected loss is 48.4 for both instruments, 74.2 for E_t alone, and 77.2 for M_t alone. Observe that the $\hat{W}_1$ figures for method III, unlike those for methods I and II, contain errors of calculation. Each $\hat{W}_1$ figure is calculated by expression (4) by using the quadratic approximations (13) for (12) in all the steps from $t = T$ to $t = 1$. These quadratic approximations can introduce errors. Note, for example, that the $\hat{W}_1$ figure for method III is larger than for method II when either E_t or M_t is the sole instrument. This result is unreasonable because the possibility of learning presumably will decrease expected welfare loss. Hopefully the errors introduced will not

TABLE 11.2 Solutions for a Two-Equation Model Using the Three Control Methods

Instrument Set and Results	Method I	II	III
$\{E_t, M_t\}$ $H^1_{1,1}$	$\begin{bmatrix} 1 & 0 & 0 & 0 \\ 0 & 1 & 0 & 0 \\ 0 & 0 & 0 & 0 \\ 0 & 0 & 0 & 0 \end{bmatrix}$	$\begin{bmatrix} 1.4121 & -0.1265 & 0 & 0 \\ -0.1265 & 1.0789 & 0 & 0 \\ 0 & 0 & 0 & 0 \\ 0 & 0 & 0 & 0 \end{bmatrix}$	$\begin{bmatrix} 1.6397 & -0.1974 & -0.0184 & -0.0125 \\ -0.1974 & 1.1434 & 0.0040 & -0.0331 \\ -0.0184 & 0.0040 & 0.0062 & 0.0016 \\ -0.0125 & -0.0331 & 0.0016 & 0.0563 \end{bmatrix}$
G_1	$\begin{bmatrix} -2.6095 & 0.3666 & 0 & 0 \\ -0.2239 & -0.2298 & 0 & 0 \end{bmatrix}$	$\begin{bmatrix} -1.6856 & 0.0647 & 0 & 0 \\ -0.2497 & -0.1997 & 0 & 0 \end{bmatrix}$	$\begin{bmatrix} -1.7174 & 0.0765 \\ -0.2497 & -0.1991 \end{bmatrix}$
g'_1	$[1013.0050 \quad 243.1293]$	$[709.0372 \quad 249.7009]$	$[719.3068 \quad 249.0192]$
$\hat{x}'_1$	$[111.7262 \quad 142.8996]$	$[111.7831 \quad 142.8544]$	$[111.6835 \quad 142.8729]$
$\hat{W}_1$	35.5479	51.3291	48.3926
$\{E_t\}$ $H^1_{1,1}$	$\begin{bmatrix} 1.1545 & 0.1586 & 0 \\ 0.1586 & 1.1628 & 0 \\ 0 & 0 & 0 \end{bmatrix}$	$\begin{bmatrix} 1.7558 & 0.0296 & 0 \\ 0.0296 & 1.2169 & 0 \\ 0 & 0 & 0 \end{bmatrix}$	$\begin{bmatrix} 2.3791 & 0.2670 & -0.0309 \\ 0.2670 & 1.9123 & 0.0219 \\ -0.0309 & 0.0219 & 0.0116 \end{bmatrix}$
G_1	$[-2.7949 \quad 0.1763 \quad 0]$	$[-1.7012 \quad 0.0018 \quad 0]$	$[-1.6065 \quad -0.0991 \quad 0]$
g_1	1094.6816	719.4268	694.4397
$\hat{x}_1$	110.3943	111.0514	111.1035
$\hat{W}_1$	44.1556	63.4141	74.2185

$$\{M_t\} \quad H^1_{1,1} \quad \begin{bmatrix} 3.0666 & -0.2904 & 0 \\ -0.2904 & 1.0408 & 0 \\ 0 & 0 & 0 \end{bmatrix} \qquad \begin{bmatrix} 3.1216 & -0.2876 & 0 \\ -0.2876 & 1.1196 & 0 \\ 0 & 0 & 0 \end{bmatrix} \qquad \begin{bmatrix} 7.0334 & -1.1361 & -0.0257 \\ -1.1361 & 1.3879 & -0.0329 \\ -0.0257 & -0.0329 & 0.0616 \end{bmatrix}$$

$$G_1 \quad \begin{bmatrix} -0.3827 & -0.2075 & 0 \end{bmatrix} \qquad \begin{bmatrix} -0.3580 & -0.1732 & 0 \end{bmatrix} \qquad \begin{bmatrix} -0.4210 & -0.1451 & 0 \end{bmatrix}$$

$$
\begin{array}{llll}
g_1 & 298.0639 & 286.1260 & 306.2156 \\
\hat{x}_1 & 142.9770 & 142.8952 & 142.8845 \\
\hat{W}_1 & 48.0508 & 64.7409 & 77.2147
\end{array}
$$

affect too seriously the relative magnitudes of the three $\hat{W}_1$ figures computed by method III which are used to compare the values of the three instrument sets $\{E_t, M_t\}$, $\{E_t\}$, and $\{M_t\}$.

The major methodological conclusion to be derived from Table 11.2 is that the optimal first-period policy $\hat{x}_1$ is similar for the three methods; that is, when uncertainty and learning are taken into account, the first-period policy is almost identical in numerical value to the policy obtained by the method of certainty equivalence. This result was also found in Section 11.5 when a one-equation model was used. Two reservations should be made in stating this conclusion. First, strictly speaking, how different the first-period policies are among the three methods cannot be measured by simply inspecting the numerical values; for example, is the figure 111.05 for the optimal E_1 by method II when E_t is the sole instrument close to 110.39 by method I? It would appear so, but the difference between these policies should be measured by the welfare losses resulting from their application. (See Problem 17.) Second, our conclusion is based on examples in which the standard errors of the estimated coefficients in the model based on the sample information are not too large and the sample period is long compared with the planning horizon. If the standard errors are larger, presumably uncertainty should play a more important role. More uncertainty and more opportunity to sample in the future will make learning more important. In spite of these two reservations, the results in Sections 11.5 and 11.6 appear to suggest that the optimal policy calculated by the simplified assumption that the parameter estimates are known constants is probably close to the policies that incorporate uncertainty and/or learning, provided that the uncertainty in the parameters is not much larger than is commonly found in the econometric models used for policy analysis. The methods introduced in Chapter 10 and in this chapter have made it possible for us to reach this tentative conclusion. They can be used to investigate the circumstances under which it is valid.

The material in this chapter, except for Section 11.6, is based on Chow (1973d). In the control-engineering literature numerous proposals can be found that take learning into account in the optimal control solutions; these references are too extensive to cite. For alternative approaches to incorporate learning in an optimal control policy in the economics literature the reader is referred to MacRae (1972) and Tse (1974). Prescott (1972) has dealt with the problem of learning with a very simple model but provides no new·method of solution; his results were computed by enumeration of the possibilities following the method of dynamic programming. Rausser and Freebairn (1974) have provided some calculations by using various suboptimal control solutions applied to the U. S. beef trade policy.

PROBLEMS

1. Continue to solve the multiperiod decision problem in (4) with W specified by (5), following steps (6) and (7), and show that the minimization with respect to x_{T-2} will lead to minimizing the expression V_{T-2} in (21) in Chapter 10.

2. Let Bz_t be added to the right-hand side of (1), with z_t denoting exogenous variables that are not subject to control but are assumed to take given values. How will the solution in Section 11.2 be affected?

3. In Problem 2 let the values of z_t be also random and distributed independently of the remaining parameters. Assume that the first two moments of the distribution of z_t are known. How will the solution in Section 11.2 be affected?

4. Using a diagrammatic presentation similar to Figure 11.1, explain the remaining steps in the solution of a four-period control problem by the method in Section 11.2.

5. Let the model by $y_t = cx_t + u_t$ in which both the dependent variable y_t and the control variable x_t are scalars. Assume that 20 observations are available at the beginning of period 1 for a three-period control problem of steering y_t to certain targets $a_t (t = 1, 2, 3)$. Write out $E_2 W_3$ for this problem as a function of the observations y_t and x_t.

6. Using the assumptions stated in Problem 5, write out the optimal feedback control equation for x_3.

7. Using the assumptions stated in Problem 5, write out $\min_{x_3} E_2 W_3$ as a function of the observations y_t and x_t.

8. Explain how a tentative path can be obtained for use in a quadratic approximation of the function $\min_{x_3} E_2 W_3$ obtained in Problem 7.

9. Using the results in Problems 7 and 8 and the data given in Problem 4 in Chapter 10 for y_t and x_t, provide a set of detailed instructions to a programmer for computing a quadratic approximation of $\min_{x_3} E_2 W_3$. (Actually compute this quadratic approximation if you wish.)

10. Let the model be $y_t = ay_{t-1} + c_1 x_{1t} + c_2 x_{2t} + u_t$, where the dependent variables y_t and the two control variables x_{1t} and x_{2t} are scalars. Assume that 20 observations are available at the beginning of period 1 for a three-period control problem of steering y_t to certain targets a_t $(t = 1, 2, 3)$. Set up the model in the form of (1). What are the orders of the matrices $K_{3,3}$, $H_{3,3}^3$, $Q_{2,2}^3$, and $H_{1,1}^1$?

11. How many second derivatives will have to be evaluated numerically to solve the control problem stated in Problem 10 by the method in Section 11.2?

12. How many second derivatives will have to be evaluated numerically to solve the control problem stated in Problem 10 by the first simplified version of the method described in Section 11.4?

13. How many second derivatives will have to be evaluated numerically to solve the control problem stated in Problem 10 by the second simplified version of the method described in Section 11.4 with $M = 2$?

14. Derive the optimal control solution by method I using the model in (1) and the nonadditive welfare function (8).

15. Derive the optimal control solution by method II using the model in (1) and the nonadditive welfare function (8).

16. Explain why, in Table 11.2, the matrix $H_{1,1}^1$ equals K_1 for method I when the active instrument set $\{E_t, M_t\}$ is employed. *Hint.* Review Chapter 7.

17. Describe a computer simulation experiment by which the expected welfare loss resulting from each of the three control methods I, II, and III can be compared. It should be assumed

CONTROL OF UNKNOWN LINEAR SYSTEMS WITH LEARNING

that the user of any of the three methods can revise his estimates of the unknown parameters before actually carrying out the policy in each future period.

18. It has been pointed out that the computation of $\hat{W}_1$ by method III given in Table 11.2 is inaccurate. Describe a method by computer simulation to obtain a more accurate estimate of the minimum expected loss by using control method III.

CHAPTER 12

Control of
Nonlinear Systems

In the second half of this book we have been concerned with the use of econometric models for the purpose of setting quantitative economic policies. The policies derived can take the form of numerical values for the control variables or of functions of the observed economic variables. The first are *open-loop* policies; the second are *closed-loop* policies. The latter policies are closed-loop feedback control equations. These are particularly useful not only because they are required for the derivation of the current optimal policy when there is uncertainty concerning the econometric system but also because they can be conveniently incorporated into a stochastic model for the purpose of studying its dynamic properties under control. Our discussion so far has been confined to linear stochastic systems. In this chapter the discussion is extended to nonlinear systems. Optimal or approximately optimal policies in both open-loop and closed-loop form are derived.

Methods for the control of nonstochastic systems are studied first. These methods are of interest even if in practice the econometric models employed for policy analysis are nearly always stochastic. The reason is that for the control of a known *linear* stochastic system using a quadratic welfare function the method of certainty equivalence (which ignores the additive random disturbances in the model) provides not only the optimal decision for the first period but also the optimal feedback control equations for future periods, as we pointed out in Chapter 7. If the system is not too far from being linear, we may surmise that the certainty equivalence solution is probably close to being optimal. Hence the interest in the optimal control solution for nonstochastic models. Control methods for nonstochastic systems are discussed in Sections 12.1 to 12.3. We then present methods for controlling nonlinear stochastic systems in the remainder of this chapter.

12.1 SOLUTION OF DETERMINISTIC CONTROL BY LAGRANGE MULTIPLIERS

As we have just remarked, the solution to a control problem can take the form of a set of numerical values for the control variables or a set of feedback control equations. Our treatment of the control of linear systems has emphasized the solution in feedback form. For the control of nonlinear deterministic systems this section presents a solution in feedback form to be derived by the method of Lagrange multipliers. It is a generalization of the method in Section 7.4 to the case of nonlinear econometric systems. In Sections 12.2 and 12.3 solutions take the open-loop form.

Following the notation of Section 6.4, we write the tth observation on the ith nonlinear structural equation as

$$y_{it} - \Phi_i(y_t, y_{t-1}, x_t, z_t) = 0, \qquad (i = 1, \dots, p). \tag{1}$$

As before, y_t denotes a column vector of p endogenous variables, x_t is a column vector of q control variables, and z_t is a vector of noncontrollable exogenous variables whose values are taken as given. The vector x_t is assumed to be imbedded in y_t. Any additive random disturbance has been omitted because we are dealing with a deterministic model. All endogenous variables with higher order lags than the first and all lagged control variables are assumed to have been eliminated by identities. Denoting by Φ the column vector function with Φ_i as its ith element, we can write the system of p structural equations as

$$y_t - \Phi(y_t, y_{t-1}, x_t, z_t) = 0. \tag{2}$$

The welfare loss need not be a quadratic function, but for ease of comparison of the following treatment with Section 7.4 we shall use a quadratic welfare function, leaving a generalization to the case of a nonquadratic function for a later discussion. The problem is to minimize an additive quadratic function in y_t subject to the nonlinear constraint (2) for $t = 1, 2, \dots, T$. Introducing the $p \times 1$ vectors λ_t of Lagrangian multipliers, we differentiate the Lagrangian expression

$$L = \tfrac{1}{2} \sum_{t=1}^{T} (y_t - a_t)' K_t (y_t - a_t) - \sum_{t=1}^{T} \lambda_t'[y_t - \Phi(y_t, y_{t-1}, x_t, z_t)] \tag{3}$$

with respect to y_t, x_t, and λ_t. Note that

$$\lambda_t'\Phi(y_t, y_{t-1}, \cdots) = \sum_{i=1}^{p} \Phi_i(y_t, y_{t-1}, \cdots)\lambda_{it}. \tag{4}$$

Setting the vector of derivatives with respect to y_t equal to 0 gives

$$\frac{\partial L}{\partial y_t} = K_t(y_t - a_t) - \lambda_t + B_{1t}'\lambda_t + B_{2,t+1}'\lambda_{t+1} = 0, \quad (t = 1, \ldots, T; \lambda_{T+1} = 0), \tag{5}$$

where, following the notation in Section 6.4,

$$B_{1t}' = \left(\frac{\partial \Phi_1}{\partial y_t} \cdots \frac{\partial \Phi_p}{\partial y_t}\right); \qquad B_{2t}' = \left(\frac{\partial \Phi_1}{\partial y_{t-1}} \cdots \frac{\partial \Phi_p}{\partial y_{t-1}}\right). \tag{6}$$

The elements of B_{1t} and B_{2t} are functions of y_t, y_{t-1}, x_t, and z_t. Similarly, setting the remaining derivatives to 0 gives

$$\frac{\partial L}{\partial x_t} = B_{3t}'\lambda_t = 0, \qquad (t = 1, \ldots, T), \tag{7}$$

where

$$B_{3t}' = \left(\frac{\partial \Phi_1}{\partial x_t} \cdots \frac{\partial \Phi_p}{\partial x_t}\right); \tag{8}$$

$$\frac{\partial L}{\partial \lambda_t} = -[y_t - \Phi(y_t, y_{t-1}, x_t, z_t)] = 0, \qquad (t = 1, \ldots, T). \tag{9}$$

Using (5), (7), and (9), we wish to solve for the unknowns y_t, x_t, and λ_t, for $t = 1, \ldots, T$. Because these equations are analogous to (14), (15), and (16) in Chapter 7, let us apply the same method of solution, beginning with the last period T. For $t = T$ (5) implies

$$\lambda_T = [I - B_{1T}']^{-1}[K_T(y_T - a_T) + B_{2,T+1}'\lambda_{T+1}]$$

$$= (I - B_{1T}')^{-1}(H_T y_T - h_T) \tag{10}$$

where $\lambda_{T+1} = 0$ by convention and where, anticipating future steps for $t < T$, we have set

$$H_T = K_T \tag{11}$$

and

$$h_T = K_T a_T. \tag{12}$$

Using (7) and (10), we have

$$B_{3T}'\lambda_T = B_{3T}'(I - B_{1T}')^{-1}(H_T y_T - h_T) = 0. \tag{13}$$

In a linear model the next step would be to substitute a linear function $A_T y_{T-1} + C_T x_T + b_T$ for y_T in (13) and solve the resulting equation for x_T as we did in (21) in Chapter 7. This step would permit the expression of y_T as a linear function of y_{T-1}, x_T having been eliminated. By (10) λ_T would become a linear function of y_{T-1}. Using this result and (5) for $t = T-1$, we could express λ_{T-1} as a linear function of y_{T-1}. This function would have the same form as (10) with $T-1$ replacing T. The whole process from (10) on could then be repeated. Now that the model is nonlinear it seems reasonable to use a linear approximation of it and apply the steps just described. We will follow this approach and show afterward that such a method, if applied iteratively and converging, will indeed provide a solution to (5), (7), and (9).

For the purpose of obtaining a linear approximation of the model (1) or (2), we assume that a tentative path x_t^0 $(t = 1, \ldots, T)$ for the control variables is given and that, given this path, the initial condition y_0, and the path of the exogenous variables z_t, not subject to control, the model (2) will yield a unique solution y_t^0 for the endogenous variables. The reader is reminded of Section 6.6 on methods of solving nonlinear structural equations. Thus for each t, given y_{t-1}^0, x_t^0, and z_t, the vector y_t^0 satisfies

$$y_t^0 - \Phi(y_t^0, y_{t-1}^0, x_t^0, z_t) = 0. \tag{14}$$

It is around this tentative path of x_t^0 and y_t^0 that we perform a first-order Taylor expansion of the nonlinear function on the left-hand side of (1):

$$y_{it} - y_{it}^0 - \left(\frac{\partial \Phi_i}{\partial y_t}\right)'(y_t - y_t^0) - \left(\frac{\partial \Phi_i}{\partial y_{t-1}}\right)'(y_{t-1} - y_{t-1}^0) - \left(\frac{\partial \Phi_i}{\partial x_t}\right)'(x_t - x_t^0) \cong 0.$$

$$\tag{15}$$

Collecting the scalar equations (15) for $i = 1, \ldots, p$ and using the definitions (6) and (8), we have

$$(I - B_{1t})(y_t - y_t^0) - B_{2t}(y_{t-1} - y_{t-1}^0) - B_{3t}(x_t - x_t^0) = 0 \tag{16}$$

and, solving for y_t,

$$y_t = A_t y_{t-1} + C_t x_t + b_t, \tag{17}$$

where

$$A_t = (I - B_{1t})^{-1} B_{2t}, \tag{18}$$

$$C_t = (I - B_{1t})^{-1} B_{3t}, \tag{19}$$

$$b_t = y_t^0 - A_t y_{t-1}^0 - C_t x_t^0. \tag{20}$$

Having performed a linear approximation (17) of the model for y_t, we substitute for y_T in (13), also using (19), to obtain

$$C_T' H_T (A_T y_{T-1} + C_T x_T + b_T) - C_T' h_T = 0. \tag{21}$$

Solving (21) for x_T gives

$$x_T = G_T y_{T-1} + g_T, \tag{22}$$

where

$$G_T = -(C_T' H_T C_T)^{-1} (C_T' H_T A_T), \tag{23}$$

$$g_T = -(C_T' H_T C_T)^{-1} C_T' (H_T b_T - h_T). \tag{24}$$

These equations are identical to (21), (22), and (23) in Chapter 7. Using (22) and (17), we have

$$y_T = (A_T + C_T G_T) y_{T-1} + b_T + C_T g_T. \tag{25}$$

Equation 25 can be used to eliminate y_T in (10) to give

$$\lambda_T = (I - B_{1T}')^{-1} [H_T (A_T + C_T G_T) y_{T-1} + H_T (b_T + C_T g_T) - h_T]. \tag{26}$$

In the next step we use (5) again with $t = T - 1$ and substitute the right-hand side of (26) for λ_T:

$$\lambda_{T-1} = (I - B_{1T-1}')^{-1} [K_{T-1} (y_{T-1} - a_{T-1}) + B_{2T} \lambda_T]$$

$$= (I - B_{1, T-1}')^{-1} (H_{T-1} y_{T-1} - h_{T-1}), \tag{27}$$

where

$$H_{T-1} = K_{T-1} + A_T' H_T (A_T + C_T G_T), \tag{28}$$

$$h_{T-1} = K_{T-1} a_{T-1} - A_T' H_T (b_T + C_T g_T) + A_T' h_T. \tag{29}$$

Because (27) has the same form as (10) with $T-1$ replacing T, the same steps from (13) on can be followed repeatedly until x_1 is obtained from an equation analogous to (22) with 1 replacing T.

In summary, the proposed method begins with a tentative path x_t^0 for the control variables and the associated path y_t^0 obtained by solving the nonlinear model (2). Around these tentative paths a linear approximation (16) or (17) of the model is computed for each period. The familiar algorithm described in Section 7.4 and Chapter 8 is applied to obtain the linear feedback control equations for x_t and finally the numerical value of x_1. Using these and the model, we can generate a second tentative path (x_t^0, y_t^0), perform a second linear approximation of the model around it, and continue until the method converges. By convergence we mean the same x_1, the same set of optimal feedback equations, and the same tentative path in two successive iterations.

One possibility of finding a tentative path for the first iteration is to use a linear approximation of the model around a set of reasonable values of the variables and apply the linear feedback control algorithms to this time-invariant linear model. The reasonable set of values referred to may be the historical values (y_0, y_{-1}, x_0) at the beginning of the planning horizon if the model is not too far from being linear. It may be the arithmatic means of y_t, y_{t-1}, and x_t, each averaged over the time interval from 1 to T, obtained by first solving the control problem using (y_0, y_{-1}, x_0) for the linear approximation of the model.

It remains to be shown that if the method converges the solution will satisfy (5), (7), and (9). There is no question concerning (5) and (7) because we have applied these equations faithfully in our solution without any approximation, provided that (9) is satisfied. We did perform an approximation of (9) by substituting (16) or (17) for y_T in (13). No approximation would be involved if (9) had been used to express y_T as an appropriate function of y_{T-1} and x_T. The question is, then, when the method converges, whether y_T, according to the nonlinear model, is indeed equal to the right-hand side of (17). The answer is yes, for when the method converges y_T, y_{T-1}, and x_T in the solution will be equal to y_T^0, y_{T-1}^0, and x_T^0, respectively, and at these solution values the nonlinear function and its linear approximation coincide. In other words, the solution values y_T, y_{T-1}, and x_T satisfy both (9) and (17) exactly. This argument applies to the other steps for $t < T$.

To relax the assumption of a quadratic welfare function, we assume that the welfare function is twice differentiable. Given a tentative path for y_t (which incorporates x_t as a subvector), we can regard the quadratic welfare function in (3) as a second-order Taylor expansion of this function around that path. As the tentative path is changed in each iteration, the quadratic approximation, with parameters K_t and a_t, will have to be recomputed for each t. Otherwise, the method remains the same as before.

The idea of linearizing a nonlinear model and applying the linear-quadratic control algorithm to the resulting time-dependent linear model is not new. One way to implement this idea for nonlinear models in the reduced form is suggested in Athans' (1972) survey paper. The treatment in this section is different because it generalizes the method of Lagrange multipliers in Section 7.4 to nonlinear structural models and derives the computation algorithm directly from the solution to the Lagrangian problem. As another implementation, Garbade (1974b) suggests a somewhat different, but similar, algorithm that requires the computation of the Lagrangian multipliers and shows that its solution satisfies the conditions derived from differentiating a Lagrangian expression, but is not derived directly from them. In addition, Garbade (1974a) has applied his algorithm successfully to a nonlinear econometric model he constructed of the United States which consists of some 43 structural equations.

12.2 SOLUTION TO DETERMINISTIC CONTROL PROBLEM AS A MINIMIZATION PROBLEM

Given a welfare function $W(y_1,...,y_t)$ that is not necessarily quadratic or additive and a deterministic model that can be numerically solved for y_t by using known values of y_{t-1} and x_t (the values of z_t being treated as fixed), we can regard W as a function W^* $(x_1,...,x_T)$ of the control variables. We can then treat the constrained minimization of W subject to the nonlinear model as an unconstrained minimization of W^* with respect to $x_1,...,x_T$. This approach has been taken by Fair (1974) and the material of this section is based on his work.

The subject of minimizing or maximizing a function of many variables is one of the most extensively treated in mathematics, and we cannot begin to deal with it in this section. If, however, we confine our attention to obtaining a numerical solution without going into the mathematical theory, some useful experience reported by Fair can be related. With the continuing advances in computer technology and in numerical methods for minimizing a function, which are available in the form of standard computer programs, the approach stated here will become increasingly useful for the numerical solution to deterministic control problems.

Using the IBM 360-91 computer at Princeton University and three basic algorithms to be mentioned, Fair has solved minimization problems involving as many as 239 variables (four control variables and 60 periods except for one variable having only a lagged effect) in a nonlinear model he constructed. The method requires the evaluation of the function W^* in terms of the control variables to be performed many times. In each functional evaluation we may have to solve the nonlinear structural equations for y_t $(t = 1,\ldots,T)$, given y_{t-1} and x_t, by a method such as the Gauss-Seidel described in Section 6.6. Once the function W^* can be evaluated and the evaluation is programmed in the computer, various algorithms can be applied to minimize (or maximize the negative of) the function. The first basic algorithm used is Powell's (1964) conjugate gradient algorithm which does not require the computation of derivatives. In each iteration this algorithm searches for the maximum successively in each of the n directions, if n variables are involved; the directions need not follow the n Cartesian axes. The second is a gradient algorithm that requires the numerical evaluation of the first derivations of W^*. The third is the quadratic hill-climbing algorithm of Goldfeld, Quandt, and Trotter (1966), already referred to in Section 9.5.

Fair has reported in Table 2 of his article that the time required to solve a control problem with 99 variables (four control variables and 25 periods) in his 19-equation model is 231 seconds by Powell's method and 204 seconds by the gradient method. For the largest problem tried, that of 239 variables (four control variables and 60 periods from 1962III to 1977II), it took 20 minutes to perform 104 iterations, with the solution to the 99-variable problem and some historical or extrapolated values as the starting point. At the end of 104 iterations the value of the objective function was changing in the eighth decimal place and the largest difference between successive values of any unknown was .0007. Fair has pointed out that his model is fairly easy to solve. Because the nonlinear part of the model is recursive, it could be done without the Gauss-Seidel method. If a 100-equation model is used, and five iterations by the Gauss-Seidel method are required to solve the model at each time period, Fair estimates that a 20-period problem with one control variable would take about 2.0 minutes, using the gradient algorithm, that a 20-period problem with two control variables would take about 8.7 minutes, and that a 25-period problem with four control variables would take about 85.2 minutes. These estimates are, of course, dependent on the computer technology and the algorithm used. They do indicate that the method of computing by brute force is applicable to many realistic deterministic control problems of reasonable size.

12.3 GRADIENT METHOD FOR OPEN-LOOP SOLUTION TO DETERMINISTIC CONTROL

The solutions given in Sections 12.1 and 12.2 belong to two extremes. The first is a closed-loop feedback solution, the second, an open-loop solution that gives only the numercial values of $x_1,\dots,x_T$. The first has a detailed structure to the solution, the second has not, at least to the user who merely provides a program for evaluating the function W^* and applies a program to minimize that function. The first divides the control variables according to the time period t and solves many problems for different time periods; the second solves for the unknowns $x_1,\dots,x_T$ for all time periods in each iteration. A method somewhere in between these two extremes provides a linear approximation to the nonlinear model and then applies a gradient method for minimizing the objective function that results. It still belongs to the open-loop category. Unlike the method in Section 12.1, it solves for all the unknown vectors $x_1,\dots,x_T$ simultaneously for the T periods in each iteration, although both methods employ a linearization of the model. The method in this section is due to Holbrook (1974).

Let y denote a column vector consisting of the vectors $y_1,\dots,y_T$ of the dependent variables for T periods, and let x denote a column vector consisting of the vectors $x_1,\dots,x_T$ of control variables for T periods. Given x, y can be calculated by solving the nonlinear model for T periods. Thus we can regard y as some nonlinear vector function of x:

$$y = g(x). \tag{30}$$

Holbrook uses the quadratic welfare function

$$W = (y', x')H\begin{pmatrix} y \\ x \end{pmatrix} \tag{31}$$

by measuring both x and y from the given (possibly nonzero) targets, thus redefining the model (30) accordingly without using new symbols for the deviations of the original x and y from their targets.

Obtain a linear approximation of the model (30) around some tentative path x^0, with y^0 defined as $g(x^0)$,

$$y = y^0 + U(x - x^0) = y^0 + U\Delta x, \tag{32}$$

where U is a matrix whose $i-j$ submatrix is the partial derivative of y_i with respect to x_j, evaluated at x^0, and $x - x^0$ is defined as Δx. Note that the partial derivatives in U cannot in general be expressed analytically but have to be evaluated numerically. Substituting (32) for y in (31) gives

$$W = [(y^0 + U\Delta x)', (x^0 + \Delta x)']H \begin{bmatrix} y^0 + U\Delta x \\ x^0 + \Delta x \end{bmatrix}$$

$$= [(y^{0\prime}, x^{0\prime}) + \Delta x'(U', I)]H \left[\begin{pmatrix} y^0 \\ x^0 \end{pmatrix} + \begin{pmatrix} U \\ I \end{pmatrix}\Delta x \right]. \tag{33}$$

Setting to 0 the derivative of (33) with respect to the vector Δx yields the solution

$$\Delta x = -\left[(U', I)H \begin{pmatrix} U \\ I \end{pmatrix} \right]^{-1} (U', I)H \begin{pmatrix} y^0 \\ x^0 \end{pmatrix}. \tag{34}$$

(See Problem 1.) By using (34) we obtain a new solution for x which in turn can serve as the tentative path x^0 in the next iteration.

Holbrook has applied this method to find optimal control solutions for the Bank of Canada's model of the Canadian economy and provided experimental results with the Michigan model of the United States economy which consists of 61 equations; about 45 of them are a set of simultaneous equations and the remainder is recursive. For these experiments the target variables are unemployment, inflation, and the trade balance; the instruments are federal spending, personal income tax rates, and the reserve base. The same six-quarter period (1968III to 1969IV) was used for these experiments. It took 6, 14, and 10 iterations, respectively, in three experiments to obtain results that are considered reasonably close to being optimal.

Holbrook has also pointed out that the iteration by (34) is a simplification of the Newton-Raphson method described in Section 6.6. To apply the Newton-Raphson method to this problem we substitute the nonlinear function $g(x)$ for y in the quadratic welfare function W and obtain the vector $\partial W/\partial x$ of partial derivatives of W with respect to x (or the gradient of W) and the matrix $(\partial^2 W/\partial x \partial x')$ of second partials of W (the Hessian matrix of W). The iteration formula is

$$\Delta x = -\left(\frac{\partial^2 W}{\partial x \partial x'} \right)^{-1} \left(\frac{\partial W}{\partial x} \right). \tag{35}$$

Equation 35 will be reduced to (34) if the terms in the Hessian matrix $(\partial^2 W/\partial x \partial x')$ which involve the second derivatives of $g(x)$ are ignored. (See Problem 2.) Thus both (34) and (35) are gradient methods that specify the change Δx in each iteration as a linear transformation of the gradient $\partial W/\partial x$. The matrix representing the linear transformation differs in the two cases. If either algorithm converges, however, that is, if $\Delta x = 0$, the gradient must be 0, provided that the matrix representing the linear transformation is nonsingular. Therefore the necessary first-order conditions for a minimum are satisfied by the solution.

12.4 CONTROL OF NONLINEAR STOCHASTIC SYSTEMS BY LINEAR FEEDBACK EQUATIONS

In this section we attempt to apply the method in Section 12.1 to the control of a nonlinear stochastic system. Assume that the model takes the form

$$y_t = \Phi(y_t, y_{t-1}, x_t, z_t) + \epsilon_t, \tag{36}$$

which is a generalization of the deterministic system (2) by including a vector ϵ_t of random disturbances that has mean zero and covariance matrix Σ and is serially uncorrelated. The stochastic nature of this model is represented only by an additive vector of random disturbances. A more general case would be to assume

$$y_t = \Phi(y_t, y_{t-1}, x_t, z_t, \eta_t), \tag{37}$$

where η_t is a vector consisting of random (but not necessarily additive) disturbances as well as a set of unknown parameters in the model. Equation 36 is interesting enough to be considered first. We consider the control of (37) later on in this section.

It may occur to a reader familiar with the material in Section 12.1 that the same feedback control equations derived there could be applied to the control of system (36). When the additive random disturbance ϵ_t is absent from (36), the method is to linearize the model around some tentative path, apply the linear feedback control equations derived from optimization by using the time-dependent linear model, and iterate until the solution path coincides with the tentative path. It was shown that this procedure indeed achieves the necessary conditions for minimizing a quadratic welfare loss function by the method of Lagrange multipliers. It seems reasonable to apply to the stochastic control problem the same set of feedback control equations derived from using the deterministic model with $\epsilon_t = 0$ in (36).

This method of certainty equivalence should be applicable because the model is linear (or at least linearized), the random disturbance is additive, and the welfare function is quadratic. In Chapter 7 we showed that the feedback control equations derived from minimizing a quadratic function subject to a linear deterministic model are also optimal for minimizing the expectation of a quadratic function subject to a linear stochastic model with an additive disturbance.

The certainty equivalence argument just presented is almost correct but not quite. It would be strictly correct if the future values of y_t, y_{t-1}, and x_t (z_t being given as before) satisfied our linear approximation of the function Φ, plus the random disturbance ϵ_t. When $\epsilon_t = 0$, we have argued in Section 12.1 that the solution values of y_t, y_{t-1}, and x_t indeed satisfy the linear approximation of the function Φ. When the random ϵ_t is present, all future y_{t-1} and x_t are random variables whose values are unknown to the decision maker at the beginning of period 1. Therefore, unlike the situation in Section 12.1, there are no deterministic solution values of y_{t-1} and x_t around which the appropriate linearization of Φ can be performed. Accordingly, the resulting calculation of the future optimal feedback control equations can be only approximate. This is how the random disturbance, when added to a *nonlinear* model, prevents the method of certainty equivalence from being truly optimal (under the assumption that all other parameters in the model are known for certain). This difficulty does not occur in a linear model because the true model in the future is known exactly, whereas in the nonlinear case the linearized model for future optimization depends on future observations of y_{t-1} and x_t which are still unknown.

It is important to distinguish *two aspects* of the principle of certainty equivalence which fail in the nonlinear situation. When the model is linear and the welfare function is quadratic, the linear feedback control equations derived from using the deterministic model are also optimal in two respects for the stochastic model which has an additive disturbance. First, they are valid for the derivation of the optimal policy $\hat{x}_1$ for the first period. Recall that the computation of the optimal $\hat{x}_1$ depends on a forward-looking approach that is predicated on the optimality of future decisions. We cannot obtain an optimal first-period policy if we do not know how to behave optimally in the future. Second, the linear feedback control equations obtained by the method of certainty equivalence are indeed optimal for actual application in the future; that is to say, when the future time comes, we can actually apply these equations and achieve optimal results. We have just argued that these equations, without being optimal in the second sense, cannot be relied on to yield an optimal first-period decision, but it is important to distinguish the two aspects, one for the current policy and the other for future policies when they are actually carried out. In appraising the feedback control equations obtained by the method of certainty equivalence, we should bear in mind to which of the two uses they are to be put. Are they to be applied only for the calculation of the first-period policy or are they actually to be carried out in the future? We may choose to restrict their use to the first instance. If these equations are not truly optimal in the second respect, they can be revised in accordance with future observations when the time actually comes for implementing the decision.

Although the method of certainty equivalence in the form of linear feedback control equations is not truly optimal (in both respects just stated), it appears to be a good approximation to the optimal solution (in both respects). Its deficiency is due to the failure to locate the future values of y_{t-1} and x_t exactly for the purpose of linearizing Φ. The feedback control equations are based on the linearization of Φ around the solution path of the deterministic control problem. If this solution path is not too far from the path of y_{t-1} and x_t to be materialized or if the model is not too far from being linear, the linear approximation of Φ will be good and the linear feedback control equations that result will probably be close to being optimal, both for the derivation of the first-period policy and for actual implementation in the future. For the second use, however, there are methods to improve on these equations which are described below.

Now consider the control of the more general nonlinear stochastic model (37). It also seems reasonable to follow the certainty-equivalence approach in feedback form just recommended for the model (36). By this approach we set the random vector η_t equal to its expected value $E\eta_t$, which we assume is known, and solve the control problem for the resulting deterministic system by the method in Section 12.1. The solution can be used for setting the first-period policy, which, as we have pointed out, is not truly optimal but may well be close to being so. For the actual implementation of future policies, however, we consider three approaches based on the idea of applying certainty equivalence in the form of linear feedback control equations.

The first, designated method A, simply applies the linear feedback control equations in the form of (22), with t replacing T, to the observed data y_{t-1}. These equations were derived by assuming all future y_{t-1} to be deterministic, but they are now applied to stochastic y_{t-1}. Again, this method would be optimal if we had the actual values of x_t and y_{t-1} to perform the linear approximation of the function Φ for use in the derivation of the coefficients G_t and g_t in the feedback control equations. Because the linear expansion of Φ is not performed at the right point, the coefficients A_t, C_t, and b_t in (17) are not exactly accurate, and accordingly, the feedback control equations are not truly optimal. Depending on the accuracy of the linear approximation, these control equations may be close to being optimal.

The second method, designated as B, is described in Athans (1972). It also requires solution of the deterministic control problem by setting η_t in (37) equal to $E\eta_t$, as in method A. Athans (1972, p. 458), however, did not specifically recommend the method in Section 12.1 but referred to others described in Bryson and Ho (1969), Jacobson and Mayne (1970), and Balakrishnan and Neustadt (1964) and merely mentioned "such methods as steepest descent, conjugate directions, conjugate gradient, quasilinearization, Newton's method, etc." Whatever the method used, we assume that the deterministic control problem is solved, with the solution path for the control variables denoted by $\tilde{x}_t$ and the associated solution path for the endogenous variables denoted by $\tilde{y}_t$. Athans then recommends linearizing the stochastic model (37) around the path $(\tilde{x}_t, \tilde{y}_t)$ and applying a set of linear feedback control rules to steer the future y_t and x_t toward $\tilde{y}_t$ and $\tilde{x}_t$, respectively. (Athans also deals with the estimation of the inaccurately measured variables by the use of Kalman filtering, but we ignore this aspect of his discussion by assuming that all variables are accurately measured.)

To carry out the second step of method B we linearize the stochastic model (37) around $\tilde{x}_t$, $\tilde{y}_t$, $\tilde{y}_{t-1}$, and $E\eta_t$; z_t are assumed to be given constants. Let the scalar equation for y_{it} in (37) be

$$y_{it} = \Phi_i(y_t, y_{t-1}, x_t, z_t, \eta_t).$$ (38)

Taking the first-order expansion of (38) around $\tilde{y}_t$, $\tilde{y}_{t-1}$, $\tilde{x}_t$, and $E\eta_{it}$, where by definition

$$\tilde{y}_{it} = \Phi_i(\tilde{y}_t, \tilde{y}_{t-1}, \tilde{x}_t, z_t, E\eta_t),$$ (39)

we have

$$y_{it} = \tilde{y}_{it} + \left(\frac{\partial \Phi_i}{\partial \tilde{y}_t}\right)'(y_t - \tilde{y}_t) + \left(\frac{\partial \Phi_i}{\partial \tilde{y}_{t-1}}\right)'(y_{t-1} - \tilde{y}_{t-1})$$

$$+ \left(\frac{\partial \Phi_i}{\partial \tilde{x}_t}\right)'(x_t - \tilde{x}_t) + \left(\frac{\partial \Phi_i}{\partial E\eta_t}\right)(\eta_t - E\eta_t).$$ (40)

Using the definitions (6), (8), and

$$B'_{4t} = \left(\frac{\partial \Phi_1}{\partial E\eta_t} \cdots \frac{\partial \Phi_p}{\partial E\eta_t}\right),$$ (41)

we write the linearized system of stochastic equations as

$$(I - B_{1t})(y_t - \tilde{y}_t) = B_{2t}(y_{t-1} - \tilde{y}_{t-1}) + B_{3t}(x_t - \tilde{x}_t) + B_{4t}(\eta_t - E\eta_t). \quad (42)$$

Solving (42) for y_t, we have

$$y_t = A_t y_{t-1} + C_t x_t + b_t + v_t, \quad (43)$$

where, as in (17), the matrices A_t, C_t, and b_t are defined by (18), (19), and (20), respectively, and v_t is a random vector defined by

$$v_t = (I - B_{1t})^{-1} B_{4t}(\eta_t - E\eta_t). \quad (44)$$

Note that the matrices A_t, C_t, and b_t in (43) and (17) are identical when the method in Section 12.1 is applied to the deterministic model derived from (37) by setting $\eta_t = E\eta_t$. When the solution to the deterministic control problem is obtained, $\tilde{x}_t = x_t^0$ and $\tilde{y}_t = y_t^0$ or the solution paths equal the tentative paths used for linearizing Φ.

The next step in method B is to find a control algorithm to steer y_t toward $\tilde{y}_t$ by using the model in (43). (We can conveniently incorporate x_t as a subvector of y_t so that x_t will be steered toward $\tilde{x}_t$.) In other words, the objective is to steer $\Delta y_t = y_t - \tilde{y}_t$ (and $\Delta x_t = x_t - \tilde{x}_t$) toward 0. The model for Δy_t is, by (42) and (44),

$$\Delta y_t = A_t \Delta y_{t-1} + C_t \Delta x_t + v_t. \quad (45)$$

Given the matrices K_t for the quadratic welfare cost function $\sum_t \Delta y_t' K_t \Delta y_t$, we can obtain a set of optimal linear feedback control equations

$$\Delta x_t = G_t \Delta y_{t-1}. \quad (46)$$

These equations will then be applied to set the deviation Δx_t of x_t from the deterministic solution path $\tilde{x}_t$ in each future time period.

It is interesting to note that if the matrices K_t in the quadratic loss function used for controlling the system (45) are the same as the matrices K_t in the original quadratic loss function used to control the nonlinear stochastic model (37) the matrices G_t in (46) will be identical to the G_t in the solution of the deterministic control problem for obtaining the path $\tilde{y}_t$. To prove this theorem observe that in the deterministic control problem the model used is (17) or the deterministic part of (43), now written as

$$\tilde{y}_t = A_t \tilde{y}_{t-1} + C_t \tilde{x}_t + b_t, \quad (47)$$

with bars to indicate that the variables are deterministic and the objective is to minimize $\sum_t (\bar{y}_t - a_t)' K_t (\bar{y}_t - a_t)$. If we had set out to minimize the expectation of this quadratic function, subject to the stochastic model (43), the result from Chapter 7 would have permitted us to decompose this problem into two parts. The first is precisely the deterministic control problem. The second minimizes the expectation of $\sum_t y_t^{*'} K_t y_t^*$, subject to the model

$$y_t^* = A_t y_{t-1}^* + C_t x_t^* + v_t \tag{48}$$

obtained by subtracting (47) from (43) and defining $y_t^* = y_t - \bar{y}_t$ and $x_t^* = x_t - \bar{x}_t$. It was shown that the feedback matrix G_t is the same in the solution to either part of the problem. The matrix G_t in (46) is the same as the corresponding matrix in the solution to the second part because (45) is the same equation as (48) and the objective functions are the same. This proves the theorem.

   Under the assumption that the matrices K_t for steering the deviation Δy_t are the same as in the original welfare function, not only are the matrices G_t used in method A and in the last step of method B identical but the implementations of these two methods are the same. By method A we apply the feedback control equations $x_t = G_t y_{t-1} + g_t$ to the data y_{t-1}. According to Chapter 7, these equations are the sums of

$$\bar{x}_t = G_t \bar{y}_{t-1} + g_t \tag{49}$$

and (with $x_t^* = x_t - \bar{x}_t$ and $y_t^* = y_t - \bar{y}_t$)

$$x_t^* = G_t y_{t-1}^*; \tag{50}$$

that is, an equivalent way to apply method A is first to apply (49) to the deterministic model (47) to obtain solution paths $\bar{x}_t$ and $\bar{y}_t$ and, second to apply (50) to the deviation of y_t from the deterministic solution path. This two-step procedure is exactly what method B specifies. Of course, method B would be different if the matrices K_t in the loss function used for controlling the deviations were different. There is no compelling reason, however, to use different K_t later in method B because the diagonal elements in the original K_t matrices reflect the relative weights to be given to the squared deviations of the different variables.

   Finally, we come to method C. This is simply a policy by which the linear feedback control equation for each period t is recomputed (either by method A or B), using the observations up to the beginning of period t before this equation is actually implemented. Even if there are no unknown parameters and the random disturbance is additive, as in equation (36), method C will improve on methods A and B, however slightly, because the linear approximation of Φ will be around a point closer to the true data which were unknown before the period of decision.

12.5 CONTROL OF NONLINEAR STOCHASTIC SYSTEMS BY OPEN-LOOP POLICIES

As the last section purports to generalize Section 12.1, this section generalizes the methods in Sections 12.2 and 12.3 to the stochastic situation. These methods, as we recall, determine the values of the control variables for all periods from 1 to T simultaneously without resort to feedback control equations. We distinguish the application of these methods first for the determination of the first-period policy and second for the determination of future policies to be actually implemented.

Concerning the determination of the first-period policy alone, we can apply the method of certainty equivalence to the methods in Sections 12.2 and 12.3. This amounts to replacing the random vector η_t in (37) by its expectation $E\eta_t$ and solving the deterministic control problem that results by any of the methods stated. The solution for x_1, by whatever method, should be identical to the solution by certainty equivalence applied in feedback form as set forth in Section 12.4. In fact, the entire solution paths for x_t and y_t for all periods are identical among all methods because they all solve the same deterministic control problem.

One may do better (or perhaps only very slightly better) than certainty equivalence when applying the methods in Sections 12.2 and 12.3 to a stochastic model as far as the determination of the first-period policy is concerned. The trick is to estimate the *expectation* of the welfare cost function and then minimize this expectation with respect to $x_1, x_2, \ldots, x_T$ simultaneously. This is different from minimizing the welfare cost function itself after transforming a stochastic model to a deterministic one. Strictly speaking, the problem requires the minimization of the expectation of the welfare function, although in practice other possible deficiencies of any control solution, for example, in the specification of the welfare function, the specification of the model, the errors in estimating the model parameters, and the additivity of the welfare function, probably will make the distinction between minimizing the expectation of a nonlinear function of many variables and minimizing the nonlinear function of the expectations of these variables less important. Furthermore, minimizing the expectation of the welfare function simultaneously with respect to $x_1, x_2, \ldots, x_T$ is not the correct procedure either. Readers of Chapter 8, 10, and/or 11 on various applications of the method of dynamic programming will realize that the optimal strategy is a sequential one rather than one of simultaneous determination of the policies for all periods. These reservations aside, we may still improve on the certainty equivalence solution by minimizing the expectation of the welfare function, given a nonlinear stochastic model.

We state only the general principles involved in such a minimization, leaving the details for the interested reader to pursue. To generalize the method in Section 12.2 Fair (1974) has suggested evaluating the expectation of the objective function by stochastic simulation; for example, using the joint distribution (assumed to be known) of all random elements in the model, take a random sample of 50 multivariate observations, evaluate the objective function by using each of the 50 sets of observations, and compute the arithmetic mean of these 50 values of the objective function. This corresponds to one evaluation of the function to be minimized. Once this evaluation can be performed, the programs mentioned in Section 12.2 can be applied to minimize the function as before. Note that if a sample of 50 is used for each functional evaluation the method for the stochastic case will take at least 50 times as long as the deterministic. The estimated expected values may not be accurate because of sampling errors, and accordingly the resulting function may be more difficult to minimize.

To generalize the gradient method of Section 12.3 to the control of stochastic models, we first linearize the model and record all its random elements. The first and second moments of these random elements are assumed to exist. We then substitute this linear stochastic model in the quadratic function (31), take the expectation of the function, the expectation being dependent on the moments of course, and minimize it with respect to the vector x which consists of the control variables for all T periods. Holbrook (1974) suggests using stochastic simulation to estimate the required moments in the linearized stochastic model. The reader may refer to his work for some special assumptions concerning the stochastic elements in the linearized model. Note, however, that if the stochastic elements are only additive, as in (36), the result on certainty equivalence given in Chapter 7 implies that the first-period policy remains the same whether we minimize the expected welfare loss or the welfare loss computed on the deterministic model.

Concerning the determination of future policies to be actually applied, the methods in this section differ from those in Section 12.4. The latter, as we have pointed out, provide closed-loop policies in feedback form, so that the feedback control equations (from methods A and B, but not C, in Section 12.4) can be applied when new data on y_t become available. The feedback form has an advantage over the open-loop form precisely because it will utilize these new data. The solution for $x_1, \ldots, x_T$ which minimizes the expectation of the welfare function for all periods (from the present vantage point when all future y_t are regarded as random variables) given in this section lacks this advantage, but it does have the advantage of being derived from a minimization of the *expectation* of the welfare loss function. Which of these advantages dominates is hard to tell. Furthermore, the methods of this section can also be modified by recomputing the policy for each future period before it is actually applied, as is done in method C of the last section. In other words, apply only the solution for x_1 and leave the later decisions until the time comes.

12.6 LINEAR FEEDBACK CONTROL EQUATIONS BY RE-ESTIMATING A LINEAR MODEL

All the methods suggested in this chapter so far are based on the given nonlinear model or its linearization by Taylor expansion. One possibility, suggested by Cooper and Fisher (1974), is to obtain a linearized model for the purpose of deriving linear feedback control equations, not by differentiating the original nonlinear model but by re-estimating a new linear model, using artificial data generated by stochastic simulations of the given nonlinear model. Specifically, Cooper and Fischer (1974) summarize their approach in five steps.

First, choose tentative paths for the control variables. These could be paths with constant growth rates or solution paths to the deterministic control problem obtained by methods in Section 12.1, 12.2, or 12.3. Second, carry out stochastic simulations of the model by using a random sample for the stochastic elements in the model and assigning values for the control variables equal to the tentative paths, plus additional random disturbances. Third, with the data generated by these simulations, estimate a set of linear regressions of the target variables on the lagged target variables and current and lagged instruments. These are autoregressive-moving average equations. The set of linear equations may be much smaller than a linearized version of the entire model because only selected target variables and instruments are of interest. Fourth, using the linear regression equations from step three, obtain linear feedback control equations. Fifth, combine these linear feedback equations with the nonlinear equations in the model and perform stochastic simulations of the combined system for the purpose of studying its dynamic characteristics.

The approach of Cooper and Fischer is easy to implement. It does not require linearizing a possibly large nonlinear model. By using the data generated by stochastic simulations in the second step to estimate linear regressions in the third step, it may capture the dynamic response characteristics of the nonlinear model. Because the specification of the nonlinear model contains useful *a priori* information about the structure of the economy, data obtained by stochastic simulations of the model may form a better basis for the estimation of the linear regression function than the limited historical data themselves. If we are interested in obtaining useful feedback control equations that relate a few key variables, this approach can be recommended. Another approach would simply be to estimate a small linear model from historical data and apply it for the derivation of optimal linear feedback control equations, but this is outside the topic of nonlinear models. Cooper and Fischer have applied their approach by using the St. Louis Model of the United States economy.

12.7 SUMMARY

A number of methods have been presented in this chapter for the control of nonlinear systems, both deterministic and stochastic. For deterministic systems we have proposed the method of linearizing the nonlinear model about a tentative path, applying the familiar feedback control equations to the linear system, creating a new solution path, and iterating until the solution path converges. This method can be derived as the solution to a set of nonlinear equations obtained by the differentiation of a Lagranian expression for minimizing a twice differentiable loss function subject to the constraint of the nonlinear model. We have also mentioned the use of general computer algorithms for the minimization of the loss function with respect to the control variables, the nonlinear model being incorporated in the function. Third, we have considered particular gradient algorithms of the Newton-Raphson variety and its modification for minimizing a quadratic objective function.

The methods presented for the control of deterministic systems are only illustrative of the methods available and are far from being a comprehensive treatment of the subject. They are, however, probably sufficient for a reader who would like to solve applied problems in deterministic control. They are also suggestive of other possibilities and can serve as an introduction to an important and open field. They have included the two important categories of solution to a control problem, namely the feedback and the open-loop categories.

For the control of a stochastic system we have presented generalizations of the three methods. A convenient, and often useful, principle to follow is that of certainty equivalence. By this principle we simply replace the random elements in the model with their mathematical expectations, solve the resulting deterministic problem by whatever method available, and apply the solution to the original stochastic situation. If the solution is in a feedback form, we can take advantage of future observations when applying the certainty equivalent solution to a stochastic problem. Except for the control of a known linear system which has an additive disturbance with a quadratic welfare function, the method of certainty equivalence has been found not to be optimal in all cases discussed in this and the preceding two chapters. Nonlinearity of the model alone will prevent the method of certainty equivalence from being optimal, and we have given the reason in this chapter. Uncertainty of the parameters alone, as we have discussed in Chapters 10 and 11, will also prevent optimality by certainty equivalence. The principle is probably close to being optimal, however, in a large number of instances and is therefore extremely useful.

We have also provided ways to improve the certainty equivalence principle. We can try to minimize the expectation of a quadratic objective function as the problem stipulates rather than take expectations first before minimization. This can be accomplished by overlooking the sequential aspect of the multiperiod optimization problem under uncertainty. Thus the feedback control aspect is sacrificed in order to perform the minimization of an expectation with respect to the control variables for all T periods together. We can also insist on minimizing the expectation of welfare loss and keeping the sequential characteristic of the solution at the same time, but it would mean sacrificing the nonlinearity of the model. Approximating a nonlinear model by a linear one through stochastic simulations, the method of Section 12.6 is an example of this approach.

In evaluating a method for controlling a stochastic system, it is important to specify the application for which the method is intended. Are we mainly concerned with the first-period policy or do we also actually wish to apply the solution, in either feedback or open-loop form, for future periods? In the latter use we can always do no worse by recomputing the policy with the most recent data before its actual application.

The reader should retain two impressions from this chapter. First, control problems for nonlinear stochastic systems are not impossible to solve, although the computational burden will be increased to a significant extent by nonlinearity, as always, and the solution may not be truly optimal but only approximately so. In fact, many of the methods for linear systems and/or deterministic systems can be conveniently generalized to the nonlinear-stochastic case. Second, the subject of controlling nonlinear econometric systems is wide open for further research in terms both of improvement in methodology (including computer algorithms) and of actual applications to the study of important economic policy problems.

PROBLEMS

1. Differentiate the expression (33) with respect to Δx and show that setting the derivature to 0 will yield (34).

2. Using the notation in Section 12.3, write out the gradient and the Hessian matrix of W for use in the Newton-Raphson iteration formula (35) and compare the result with (34). *Hint.* See Holbrook (1974).

3. Simplify Problem 2 by assuming that only the vector y enters the welfare function, x being incorporated as a subvector of y. Write out the Newton-Raphson iteration formula (35) and compare the result with the corresponding revision of (34).

4. Under what circumstances and for what purposes would you prefer the method in Section 12.1 to the method in Section 12.2 (say, using a gradient algorithm) for solving a nonlinear deterministic control problem and vice versa?

5. Under what circumstances and for what purposes would you prefer the method in Section 12.1 to the method in Section 12.3 for solving a nonlinear deterministic control problem and vice versa?

6. Under what circumstances and for what purposes would you prefer method A in Section 12.4 to an adaptation of an open-loop method in Section 12.5 for solving a nonlinear stochastic control problem and vice versa?

7. Under what circumstances and for what purposes would you prefer method A in Section 12.4 to the method in Section 12.6 for solving a nonlinear stochastic control problem and vice versa?

8. What are the advantages and disadvantages of the method in Section 12.6 compared with the first method in 12.5?

9. Given that the method of certainty equivalence is optimal when applied to a stochastic control problem with a quadratic loss function and a time-dependent linear stochastic model, explain why it is not optimal when a linearized approximation of a nonlinear model is used at each time period.

10. Provide an intuitive justification of method B in Section 12.4. Is it reasonable to steer y_t toward the solution path of the deterministic control problem when in fact the model is stochastic?

11. Can you suggest another iterative method to solve (5), (7), and (9) for the unknowns y_t, x_t, and λ_t than the method described in Section 12.1?

12. Using a quadratic welfare function and the nonlinear deterministic system (2), apply the method of dynamic programming to solve a T-period control problem. Explain why you cannot carry out the solution numerically.

13. In Problem 12, if you are willing to linearize the system (2), show how you can obtain an approximate solution to the deterministic control problem by using dynamic programming. Compare your solution with that in Section 12.1.

14. Using a quadratic welfare function and the nonlinear stochastic system (36), apply the method of dynamic programming to minimize the expected value of the welfare loss for T periods. Explain why you cannot carry out the solution numerically.

15. In problem 14, if you are willing to linearize the system (36), show how you can obtain an approximate solution to the stochastic control problem with dynamic programming. Compare your solution with method A in Section 12.4.

16. The method of dynamic programming fails to provide an exact numerical solution to a stochastic control problem when the model is *nonlinear* or when the parameters of a linear model are *unknown*. Explain the crucial reason for its failure in each of these two situations. Are the reasons essentially the same?

Bibliography

Abel, A. B. (1974), "A Comparison of Three Control Algorithms as Applied to the Monetarist-Fiscalist Debate," Econometric Research Program Research Memorandum No. 166, Princeton University, Princeton, New Jersey.

Adelman, Irma (1963), "Long Cycles—A Simulation Experiment," in A. C. Hogatt and F. E. Balderston, Eds., *Symposium on Simulation Models*, South-western, Cincinnati, Ohio.

_____ (1965), "Long Cycles—Fact or Artifact?" *American Economic Review*, **55**, 457–458.

_____ , and F. L. Adelman (1959), "The Dynamic Properties of the Klein-Goldberger Model," *Econometrica*, **27**, 596–625.

Allen, R. G. D. (1959), *Mathematical Economics*, 2nd ed., Macmillan, London.

Anderson, T. W. (1958), *Introduction to Multivariate Statistical Analysis*, Wiley, New York.

_____ (1971), *The Statistical Analysis of Time Series*, Wiley, New York.

Annals of Economic and Social Measurement (1972), **1**, 4.

Annals of Economic and Social Measurement (1974), **3**, 1.

Aoki, Masanao (1967), *Optimization of Stochastic Systems*, Academic, New York.

Arzac, E. R. (1967), "The Dynamic Characteristics of Chow's Model," *Journal of Financial and Quantitative Analysis*, **2**, 383–398.

Astrom, Karl J. (1970), *Introduction to Stochastic Control Theory*, Academic, New York.

Athans, M. (1972), "The Discrete Time Linear-Quadratic-Gaussian Stochastic Control Problem," *Annals of Economic and Social Measurement*, **1**, 449–491.

_____ (1974), "The Importance of Kalman Filtering Methods for Economic Systems," *Annals of Economic and Social Measurement*, **3**, 49–64.

Bailey, Martin J. (1962), *National Income and the Price Level*; (1971), 2nd ed., McGraw-Hill, New York.

_____ (1965), "Prediction of an Autoregressive Variable Subject Both to Disturbances and to Errors of Observation," *Journal of the American Statistical Association*, **60**, 164–181.

Balakrishnan, A. V., and L. W. Neustadt, Eds. (1964), *Computing Methods in Optimization Problems*, Academic, New York.

Baumol, William J. (1970), *Economic Dynamics: An Introduction*, 3rd ed., Macmillan, New York.

Becker, Gary S., and George Stigler (1974), "Law Enforcement, Malfeasance, and Compensation of Enforcers," *The Journal of Legal Studies*, University of Chicago Law School, **3**, 1, 1–18.

Bellman, Richard (1957), *Dynamic Programming*, Princeton University Press, Princeton, New Jersey.

Blackman, R. B., and J. W. Tukey (1958), *The Measurement of Power Spectra*, Dover, New York.

Bogaard, P. J. M. van den, and H. Theil (1959), "Macrodynamic Policy-Making: An Application of Strategy and Certainty Equivalence Concepts to the Economy of the United States, 1933-1936," *Metroeconomica*, **11**, 149–167.

Bowden, Roger J. (1972), "More Stochastic Properties of the Klein-Goldberger Model," *Econometrica*, **41**, 87–98.

Box, George E. P., and Gwilym M. Jenkins (1970), *Time Series Analysis Forecasting and Control*, Holden-Day, San Francisco.

Brainard, W. C. (1967), "Uncertainty and the Effectiveness of Policy," *American Economic Review*, **62** (2), 411–425.

Brockett, Roger W. (1970), *Finite Dimensional Linear Systems*, Wiley, New York.

Bronfenbrenner, M. (1961), "Statistical Test of Rival Monetary Rules," *Journal of Political Economy*, **69**, 1–14.

Brown, E. C. (1955), "The Static Theory of Automatic Fiscal Stabilization," *Journal of Political Economy*, **63**, 427–440.

Bryson, A. E., and Y. C. Ho (1969), *Applied Optimal Control*, Blaisdell, Waltham, Massachusetts.

Cagan, Philip (1956), "The Monetary Dynamics of Hyper-inflation," in Milton Friedman, Ed., *Studies in the Quantity Theory of Money*, University of Chicago Press, Chicago.

Canon, M. D., C. D. Cullum, Jr., and E. Polak (1970), *Theory of Optimal Control and Mathematical Programming*, McGraw-Hill, New York.

Chang, S. S. L. (1961), *Synthesis of Optimal Control Systems*, McGraw-Hill, New York.

Chitre, Vikas (1972), "A Dynamic Programming Model of Demand for Money with Planned Total Expenditure," *International Economic Review*, **13** (2), 303–323.

Chow, Gregory C. (1957), *Demand for Automobiles in the United States: A Study in Consumer Durables*, North-Holland, Amsterdam.

――― (1960), "Statistical Demand Functions for Automobiles and Their Use for Forecasting," *The Demand for Durable Goods*, A. C. Harberger, Ed., University of Chicago Press, Chicago, pp. 147–78.

――― (1967), "Multiplier, Accelerator, and Liquidity Preference in the Determination of National Income in the United States," *The Review of Economics and Statistics*, **49**, 1, 1–15.

――― (1968), "The Acceleration Principle and the Nature of Business Cycles," *Quarterly Journal of Economics*, **82** (3), 403–418.

――― (1970), "Optimal Stochastic Control of Linear Economic Systems," *Journal of Money, Credit and Banking*, **2**, 291–302.

――― (February 1972), "Optimal Control of Linear Econometric Systems with Finite Time Horizon," *International Economic Review*, **13** (1), 16–25.

――― (October 1972), "How Much Could be Gained by Optimal Stochastic Control Policies?" *Annals of Economic and Social Measurement*, **1** (4), 391–406.

――― (February 1973), "On the Computation of Full-Information Maximum Likelihood Estimates for Nonlinear Equations Systems," *The Review of Economics and Statistics*, **55** (7), 104–109.

_____ (October 1973), "Effect of Uncertainty on Optimal Control Policies," *International Economic Review*, **14** (13), 632–645.

_____ (December 1973), "Problems of Economic Policy from the Viewpoint of Optimal Control," *American Economic Review*, **63** (5), 826–837.

_____ (1973d), "A Solution to Optimal Control of Linear Systems with Unknown Parameters," Econometric Research Program Research Memorandum no. 157, Princeton University, Princeton, New Jersey.

_____, and R. C. Fair (1973), "Maximum Likelihood Estimation of Linear Equations Systems with Auto-Regressive Residuals," *Annals of Economic and Social Measurement*, **2** (1), 17–28.

_____, and R. E. Levitan (June 1969), "Spectral Properties of Nonstationary Systems of Linear Stochastic Difference Equations," *Journal of the American Statistical Association*, **64**, 581–590.

_____ (August 1969), "Nature of Business Cycles Implicit in a Linear Economic Model," *Quarterly Journal of Economics*, **83** (3), 504–517.

Christ, Carl (1973), "The 1973 Report of the President's Council of Economic Advisors: A Review," *American Economic Review*, **63** (4), 515–526.

Clark, J. M. (1917), "Business Acceleration and the Law of Demand: A Technical Factor in Economic Cycles," *Journal of Political Economy*, **25**, 217–235.

Cochran, W. C. (1970), "Some Effects of Errors of Measurement on Multiple Correlation," *Journal of the American Statistical Association*, **65** (329), 22–34.

Cochrane, D., and G. H. Orcutt (1947), "Application of Least Squares Regression to Relationships Containing Autocorrelated Error Terms," *Journal of the American Statistical Association*, **44**, 32–61.

Cooper, J. P., and S. Fischer (1974), "A Method for Stochastic Control of Non-linear Econometric Models and an Application," *Annals of Economic and Social Measurement*, **3**, 205–206.

Cox, D. R., and H. D. Miller (1965), *The Theory of Stochastic Processes*, Wiley, New York.

De Leeuw, Frank (1964), "Financial Markets in Business Cycles: A Simulation Study," *American Economic Review Papers and Proceedings*, **54**, 309–323.

Dhrymes, P. J. (1970), *Econometrics*, Harper and Row, New York.

_____ (1973), "Restricted and Unrestricted Reduced Forms: Asymptotic Distribution and Relative Efficiency," *Econometrica*, **41** (1), 119–134.

Duesenberry, James S., Otto Eckstein, and Gary Fromm (1960), "A Simulation of United States Economy in Recession," *Econometrica*, **28**, 749–809.

Eppen, Gary D., and Eugene F. Fama (1969), "Cash Balance and Simple Portfolio Problems with Proportional Costs," *International Economic Review*, **10** (2), 119–133.

Fair, R. C. (1974), "On the Solution of Optimal Control Problems as Maximization Problems," *Annals of Economic and Social Measurement*, **3**, 135–154.

Fama, Eugene E. (1970), "Multiperiod Consumption-Investment Decisions," *American Economic Review*, **60** (1), 163–174.

Fand, D. A. (1966), "A Time-Series Analysis of the 'Bills Only' Theory of Interest Rates," *The Review of Economics and Statistics*, **48**, 364.

Fishman, G. S. (1969), *Spectral Methods in Econometrics*, Harvard University Press, Cambridge, Massachusetts.

Fox, Karl, J. K. Sengupta, and E. Thorkecke (1966), *The Theory of Quantitative Economic Policy*, North-Holland, Amsterdam.

Friedman, Benjamin M. (1973), *Methods in Optimization for Economic Stabilization Policy*, North-Holland, Amsterdam.

Friedman, Milton (1957), *A Theory of the Consumption Function*, Princeton University Press, Princeton, New Jersey.

Frisch, Ragnar (1933), "Propagation Problems and Impulse Problems in Dynamic Economics," *Economic Essays in Honour of Gustav Cassel*, Allen and Unwin, London, pp. 197 and 202–203.

Garbade, K. (1974a), "Discretion in the Choice of Macroeconomic Policies," Working Papers Series 74–42, New York University Graduate School of Business Administration, New York.

_____ (1974b), "Optimal Control Policies for a Nonlinear Structural Model," Econometric Research Program Research Memorandum No. 170, Princeton University, Princeton, New Jersey.

Goldberg, Samuel (1958), *Introduction to Difference Equations*, Wiley, New York.

Goldberger, A. C. (1959), *Impact Multipliers and Dynamic Properties of the Klein-Goldberger Model*, North-Holland, Amsterdam.

_____, A. L. Nagar, and H. S. Odeh (1961), "The Covariance Matrices of Reduced-Form Coefficients and of Forecasts for a Structural Econometric Model," *Econometrica*, **29** (4), 556–573.

Goldfeld, S. M., R. E. Quandt, and H. F. Trotter (1966), "Maximization by Quadratic Hill-Climbing," *Econometrica*, **34** (3), 541–551.

Granger, C. W. J. (1966), "The Typical Spectral Shape of an Economic Variable," *Econometrica*, **34**, 150.

_____ (1967), "New Techniques for Analyzing Economic Time Series and Their Place in Econometrics," in M. Shubik, ed., *Essays in Mathematical Economics in Honor of Oskar Morgenstern*, Princeton University Press, Princeton, New Jersey.

_____, and M. Hatanaka (1964), *Spectral Analysis of Economic Time Series*, Princeton University Press, Princeton, New Jersey.

_____, and Oskar Morgenstern (1970), *Predictability of Stock Market Prices*, Heath, Lexington, Massachusetts.

Hakansson, Nils H. (1970), "Optimal Investment and Consumption Strategies Under Risk for a Class of Utility Functions," *Econometrica*, **38**, 587–607.

Hammersley, J. M., and D. C. Handscomb (1964), *Monte Carlo Methods*, Methuen, London.

_____, and K. W. Morton (1956), "A New Monte Carlo Technique: Antithetic Variates," *Proceedings of the Cambridge Philosophical Society*, **52**, 449–475.

Hannan, E. J. (1960), *Time Series Analysis*, Methuen, London.

Hatanaka, M., and M. Suzuki (1967), "A Theory of the Pseudospectrum and Its Application to Nonstationary Dynamic Econometric Models," in M. Shubik, ed., *Essays in Mathematical Economics in Honor of Oskar Morgenstern*, Princeton University Press, Princeton, New Jersey, pp. 443–466.

Holbrook, R. S. (1972), "Optimal Economic Policy and the Problems of Instrument Instability," *The American Economic Review*, **62** (1), 57–65.

_____ (1974), "A Practical Method for Controlling a Large Non-linear Stochastic System," *Annals of Economic and Social Measurement*, **3**, 155–175.

Holt, Charles C., Franco Modigliani, John F. Muth, and Herbert A. Simon (1963), *Planning Production, Inventories and Work Force*, Prentice-Hall, Englewood Cliffs, New Jersey.

Howrey, E. Philip (1967), "Dynamic Properties of Stochastic Linear Econometric Models," Princeton University Econometric Research Program Research Memorandum No. 87.

_____ (1968), "A Spectrum Analysis of the Long-Swing Hypothesis," *International Economic Review*, **9** (2), 228–252.

_____ (1971), "Stochastic Properties of the Klein-Goldberger Model," *Econometrica*, **39**, 73–88.

_____, and L. R. Klein (1972), "Dynamic Properties of Nonlinear Econometric Models," *International Economic Review*, **13**, 599–618.

Jacobson, D. H., and D. Q. Mayne (1970), *Differential Dynamic Programming*, Elsevier, New York.

Johnston, J. (1955), "Econometric Models and the Average Duration of Business Cycles," The Manchester School of Economic and Social Studies, **23**, 3.

Kalman, R. E. (1960), "A New Approach to Linear Filtering and Prediction Problems," *Journal of Basic Engineering Trans.*, ASME, **82D**, 33–45.

Katz, S. (1962), "A Discrete Version of Pontryagin's Maximum Principle," *Journal of Electronics and Control*, **13**, 179–184.

Kaufman, G. M. (1967), "Some Bayesian Moment Formulae," Center for Operations Research and Econometrics Discussion Paper No. 6710, Catholic Univeristy of Louvain.

Kendall, Maurice G. (1945), *Contributions to the Study of Oscillatory Time Series*, Occasional Papers, **9**, Cambridge University Press, Cambridge.

_____, and Alan Stuart (1966), *The Advanced Theory of Statistics*, Vol. 3, Griffin, London, and Hafner, New York.

Kenkel, James L. (1974), *Dynamic Linear Economic Models*, Gordon and Breach, London.

Keynes, John M. (1936), *A General Theory of Employment Interest and Money*, Harcourt, Brace, New York.

Klein, L. R., and A. S. Goldberger (1955), *An Econometric Model of the United States, 1929-1952*, North-Holland, Amsterdam.

Kmenta, J., and P. E. Smith (1973), "Autonomous Expenditures Versus Money Supply: An Application of Dynamic Multipliers," *Review of Economics and Statistics*, **55** (3), 299–307.

Kuhn, H. W., and A. W. Tucker (1951), "Nonlinear Programming," in *Proceedings of the Second Berkeley Symposium on Mathematical Statistics and Probability*, J. Neyman, Ed., University of California Press, Berkeley.

Kushner, Harold J. (1971), *Introduction to Stochastic Control*, Holt, Rinehart and Winston, New York.

Liu, T. C. (1963), "An Exploratory Quarterly Econometric Model of Effective Demand in the Postwar U. S. Economy," *Econometrica*, **31**, 301–348.

MacRae, E. C. (1972), "Linear Decision with Experimentation," *Annals of Economic and Social Measurement*, **1**, 437–448.

Mariano, R. S., and S. Schleicher (1972), "On the Use of Kalman Filters in Economic Forecasting," University of Pennsylvania, Department of Economics Discussion Paper 247, Philadelphia.

Modigliani, Franco (1964), "Some Empirical Tests of Monetary Management and of Rules versus Discretion," *Journal of Political Economy*, **72**, 211–245.

Moore, G. H., and J. Shiskin (1967), *Indicators of Business Expansions and Contractions*, National Bureau of Economic Research, Columbia University Press, New York.

Morishima, Michio, and M. Saito (1964), "A Dynamic Analysis of the American Economy, 1902-1952," *International Economic Review*, **5**, 125-164.

Mossin, J. (1968), "Optimal Multiperiod Portfolio Policies," *Journal of Business*, **41**, 215-229.

Muth, John F. (1960), "Optimal Properties of Exponentially Weighted Forecasts of Time Series with Permanent and Transitory Components," *Journal of American Statistical Association*, **55**, 299-306.

_____ (1961), "Rational Expectations and the Theory of Price Movements," *Econometrica*, **29**, 315-335.

Naylor, Thomas H., Kenneth Wertz, and Thomas H. Wonnacott (1969), "Spectral Analysis of Data Generated by Simulation Experiments with Econometric Models," *Econometrica*, **37**, 333-352.

Nerlove, M. (1964), "Spectral Analysis of Seasonal Adjustment Procedures," *Econometrica*, **32**, 241-286.

_____ (1972), "Lags in Economic Behavior," *Econometrica*, **40** (2), 221-252.

Norman, Alfred (1974), "On the Relationship Between Linear Feedback Control and First Period Certainty Equivalence," *International Economic Review*, **15** (1), 209-215.

Pagan, A. (1973), "Optimal Control of Econometric Models with Autocorrelated Disturbance Terms," Econometric Research Program Research Memorandum No. 156, Princeton University, Princeton, New Jersey.

Paryani, K. (1971), *Optimal Control of Linear Discrete Macroeconomic Systems*, University Microfilms, Ann Arbor, Michigan.

Parzen, E. (1961), "Mathematical Considerations in the Estimation of Spectra," *Technometrics*, **3**, 167-190.

Phelps, Edmund (1962), "The Accumulation of Risky Capital: A Sequential Utility Analysis," *Econometrica*, **30** (4), 729-743.

Phillips, A. W. (1954), "Stabilization Policy in a Closed Economy," *The Economic Journal*, **64**, 290-323.

_____ (1957), "Stabilization Policy and the Time-Form of Lagged Responses," *The Economic Journal*, **67**, 265-277.

_____ (1958), "La cybernetique et le contrôle des systems économique," *Cahiers de l'Institut de Science Economique Appliquée*, Série N, No. 2, 41-50.

Pindyck, Robert S. (1973), *Optimal Planning for Economic Stabilization*, North-Holland, Amsterdam.

Pontryagin, L. S., et al. (1962), *The Mathematical Theory of Optimal Processes*, Wiley, New York.

Powell, M. J. D. (1964), "An Efficient Method for Finding the Minimum of Several Variables without Calculating Derivatives," *Computer Journal*, **7**, 155-162.

Prescott, E. C. (1972), "The Multiperiod Control Problem Under Uncertainty," *Econometrica*, **40**, 1043-1058.

Quenouille, M. H. (1947), *The Analysis of Multiple Time-Series*, Charles Griffin, London.

Rausser, G. C., and J. W. Freebairn (1974), "Approximate Adaptive Control Solutions to U. S. Beef Trade Policy," *Annals of Economic and Social Measurement*, **3**, 177-203.

Rothschild, Michael (1974), "Searching for the Lowest Price when the Distribution of Prices is Unknown: A Summary," *Annals of Economic and Social Measurement*, **3**, 293-294.

Samuelson, Paul A. (1939), "Interactions between the Multiplier Analysis and the Principle of Acceleration," *The Review of Economic Statistics*, **21**, 75-78.

_____ (1948), *Foundations of Economic Analysis*, Harvard University Press, Cambridge, Massachusetts.

_____ (1969), "Lifetime Portfolio Selection by Dynamic Stochastic Programming," *The Review of Economics and Statistics*, **51** (3), 239–246.

Simon, Herbert A. (1956), "Dynamic Programming Under Uncertainty with a Quadratic Criterion Function," *Econometrica*, **24** (1), 74–81.

Taylor, Lance (1970), "The Existence of Optimal Distributed Lags," *The Review of Economic Studies*, **37**, 95–106.

Theil, Henri (1958), *Economic Forecasts and Policy*, North-Holland, Amsterdam.

_____ (1964), *Optimal Decision Rules for Government and Industry*, North-Holland, Amsterdam.

Tinbergen, Jan (1952), *On the Theory of Economic Policy*, North-Holland, Amsterdam.

_____ (1956), *Economic Policy: Principles and Design*, North-Holland, Amsterdam.

Tintner, Gerhard, and J. K. Sengupta (1972), *Stochastic Economics*, Academic, New York.

Tse, E. (1974), "Adaptive Dual Control Methods," *Annals of Economic and Social Measurement*, **3**, 65–83.

Turvey, R. (1965), "On the Demand for Money," *Econometrica*, **33**, 459–460.

Vishwakarma, K. P. (1974), *Macro-economic Regulation*, Rotterdam University Press, Rotterdam.

_____, P. M. C. de Boer, and F. A. Palm (1970), "Optimal Prediction of Inter-Industry Demand," Econometric Institute Report 7022, Netherlands School of Economics, Rotterdam.

Whittle, Peter (1963), *Prediction and Regulation by Linear Least Square Methods*, Van Nostrand, New York.

Zellner, A. (1971), *An Introduction to Bayesian Inference in Econometrics*, Wiley, New York.

Index